DESIGN AUTOMATION METHODS AND TOOLS
FOR MICROFLUIDICS-BASED BIOCHIPS

Design Automation Methods and Tools for Microfluidics-Based Biochips

Edited by

KRISHNENDU CHAKRABARTY
Duke University, Durham, NC, U.S.A.

and

JUN ZENG
Coventor Inc., Cambridge, MA, U.S.A.

Springer

A C.I.P. Catalogue record for this book is available from the Library of Congress.

ISBN-10 1-4020-5122-0 (HB)
ISBN-13 978-1-4020-5122-7 (HB)
ISBN-10 1-4020-5123-9 (e-book)
ISBN-13 978-1-4020-5123-4 (e-book)

Published by Springer,
P.O. Box 17, 3300 AA Dordrecht, The Netherlands.

www.springer.com

Printed on acid-free paper

TABLE OF CONTENTS

PREFACE

Microfluidics-based biochips, also called lab-on-a-chip, are becoming increasingly popular as a technology platform for the detection, analysis and manipulation of biochemical samples for genomics, proteomics, clinical diagnostics, environmental monitoring, and bio-defense. Biochips automate highly repetitive laboratory tasks by replacing cumbersome equipment with miniaturized and integrated systems, and they enable the handling of small amounts, e.g., nanoliters, of fluids. Thus they are able to provide ultra-sensitive detection at much faster speed and significantly lower costs per assay than traditional methods.

As the use of microfluidics-based biochips increases, their complexity is expected to become significant due to the need for multiple and concurrent assays on the chip, as well as more sophisticated control mechanisms for resource management. Time-to-market and fault tolerance are also expected to emerge as design considerations. As a result, current full-custom design techniques will not scale well for larger designs. There is a need to deliver the same level of CAD support to the biochip designer that the semiconductor industry now takes for granted. These CAD tools will allow designers to harness the new technology that is rapidly emerging for integrated microfluidics. The 2003 International Technology Roadmap for Semiconductors (ITRS) clearly identifies the integration of electrochemical and electro-biological techniques as one of the system-level design challenges that will be faced beyond 2009, when feature sizes shrink below 50 nm. Efforts are underway in the CAD community to identify synergies between biochips and microelectronics CAD. The 2005 *Design, Automation, and Test in Europe (DATE) Conference* included a well-attended "Biochips Day" event. A special session on BioMEMS was organized at the 2004 *IEEE/ACM Design Automation Conference.* The 2005 *IEEE/ACM/IFIP International Conference on Hardware - Software Co-design and System Synthesis* (*CODES-ISSS*) included a special session on biochips and bioinformatics.

This book is based on the biochips special issue of *IEEE Transactions on Computer-Aided Design of Integrated Circuits and Systems*, published in February 2006. It is devoted to several aspects of design automation for microfluidics-based biochips. The chapters in this book cover technology issues and design automation challenges, modeling and simulation methods, synthesis, layout, and system control, and computer-aided data analysis.

In the first chapter in this book, Su at al. present an overview of the underlying biochip technologies and relevant design automation issues. The next eight chapters are focused on modeling and simulation methods for microfluidics. Zeng presents electrohydrodynamic simulations for droplets, and analyzes the operating mechanisms underlying electrowetting-on-dielectric and dielectrophoresis. Lienemann et al. discuss the simulation of microfluidic arrays based on electrowetting-on-dielectric. Next, Wang et al. describe a three-dimensional fluid analysis program called FastStokes, which can rapidly computes drag forces on complicated structures by solving an integral formulation of the Stokes equation. In the next paper, Wang et al. present composable behavioral models and the schematic-Based simulation of biochips based on electrokinetics. Altman et al. present a multi-scale fast boundary element algorithm called FFTSVD, and its application to MEMS and microfluidics simulation. Vasilyev et al. present a model-order-reduction technique based on a modified trajectory piecewise-linear algorithm, and its application to automatic macromodel extraction for microfluidic devices. Finally, Bedekar et al. present models for the system-level simulation of fluid flow, electric field and analyte dispersion in microfluidic devices. Compact models used to compute the pressure-driven and electroosmotic flow rates are based on the integral formulation of the mass, momentum and current conservation equations. Magargle at al describe how we can model the injector device in microfluidic systems using artificial neural networks that are trained with finite element simulations of the underlying mass transport PDEs.

The next four chapters describe techniques for the synthesis and layout of microfluidic biochips, as well system-level droplet control issues for droplet-based biochips. Kahng et al. leverage CAD techniques for electronic design for probe selection, probe placement, and embedding in DNA arrays. Next, Pfeiffer et al. describe a physical design approach for multiplexed capillary electrophoresis (CE) separation microchips. Böhringer describes algorithms to generate efficient sequences of control signals for moving droplets on a microfluidic array. Griffith et al. describe an alternative approach for solving a similar problem.

Biochips enable high-throughput biological data acquisition. The final chapter in this book articulates the need for computer-aided analysis tools to process colossal amounts of information collected by biochips, and it present a pattern-mining algorithm and its example application to large-scale biochip data.

We thank Mark De Jongh at Springer for encouraging us to proceed with this book. We thank all the chapter contributors for their submissions and interest in this book. We hope that this book will generate more interest in this emerging technology area and serve as a bridge between the CAD, MEMS, and biochemistry communities.

Krishnendu Chakrabarty and Jun Zeng
April 6, 2006

Chapter 1

MICROFLUIDICS-BASED BIOCHIPS: TECHNOLOGY ISSUES, IMPLEMENTATION PLATFORMS, AND DESIGN AUTOMATION CHALLENGES*

Fei Su, Krishnendu Chakrabarty and Richard B. Fair[1]

Department of Electrical & Computer Engineering , Duke University, Durham, NC 27708, E-mail: {fs, krish, rfair}@ee.duke.edu

Abstract: Microfluidics-based biochips are soon expected to revolutionize clinical diagnosis, DNA sequencing, and other laboratory procedures involving molecular biology. In contrast to continuous-flow systems that rely on permanently-etched microchannels, micropumps, and microvalves, digital microfluidics offers a scalable system architecture and dynamic reconfigurability; groups of unit cells in a microfluidics array can be reconfigured to change their functionality during the concurrent execution of a set of bioassays. As more bioassays are executed concurrently on a biochip, system integration and design complexity are expected to increase dramatically. We present an overview of an integrated system-level design methodology that attempts to address key issues in the synthesis, testing and reconfiguration of digital microfluidics-based biochips. Different actuation mechanisms for microfluidics-based biochips, and associated design automation trends and challenges are also discussed. The proposed top-down design automation approach is expected to relieve biochip users from the burden of manual optimization of bioassays, time-consuming hardware design, and costly testing and maintenance procedures, and it will facilitate the integration of fluidic components with microelectronic component in next-generation SOCs.

Keywords: Biochips, design automation, microfluidics, reconfiguration, synthesis, testing.

* This research was supported by the National Science Foundation under grant number IIS-0312352.

K. Chakrabarty and J. Zeng (eds.),
Design Automation Methods and Tools for Microfluidics-Based Biochips, 1–29.

1. INTRODUCTION

Microfluidics-based biochips for biochemical analysis are receiving much attention nowadays [1, 2, 3, 4]. These composite microsystems, also known as lab-on-a-chip or bio-MEMS, offer a number of advantages over conventional laboratory procedures. They automate highly repetitive laboratory tasks by replacing cumbersome equipment with miniaturized and integrated systems, and they enable the handling of small amounts, e.g., micro- and nano-liters, of fluids. Thus they are able to provide ultra-sensitive detection at significantly lower costs per assay than traditional methods, and in a significantly smaller amount of laboratory space.

Advances in microfluidics technology offer exciting possibilities in the realm of enzymatic analysis (e.g., glucose and lactate assays), DNA analysis (e.g., PCR and nucleic acid sequence analysis), proteomic analysis involving proteins and peptides, immuno-assays, and toxicity monitoring. An emerging application area for microfluidics-based biochips is clinical diagnostics, especially immediate point-of-care diagnosis of diseases [5, 6]. Microfluidics can also be used for countering bio-terrorism threats [7, 8]. Microfluidics-based devices, capable of continuous sampling and real-time testing of air/water samples for biochemical toxins and other dangerous pathogens, can serve as an always-on "bio-smoke alarm" for early warning.

The first generation of microfluidic biochips contained permanently-etched micropumps, microvalves, and microchannels, and their operation was based on the principle of continuous fluid flow [3, 4]. A promising alternative is to manipulate liquids as discrete droplets [9, 10]. Following the analogy of microelectronics, this novel approach is referred to as "digital microfluidics". In contrast to continuous-flow biochips, digital microfluidics-based biochips, which we also refer to as second-generation biochips, offer scalable system architecture based on a two-dimensional microfluidic array of identical basic unit cells. Moreover, because each droplet can be controlled independently, these "digital" systems also have dynamic reconfigurability, whereby groups of unit cells in a microfluidic array can be reconfigured to change their functionality during the concurrent execution of a set of bioassays. The advantages of scalability and reconfigurability make digital microfluidic biochips a promising platform for massively parallel DNA analysis, automated drug discovery, and real-time biomolecular detection.

As the use of digital microfluidics-based biochips increases, their complexity is expected to become significant due to the need for multiple and concurrent assays on the chip, as well as more sophisticated control for resource management. Time-to-market and fault tolerance are also expected to emerge as design considerations. As a result, current full-custom design techniques will not scale well for larger designs. There is a pressing need to deliver the same level of computer-aided design (CAD) support to the biochip designer that the semiconductor industry now takes for granted. Moreover, it is expected that these microfluidic biochips will be integrated with microelectronic components in next-generation system-on-chip (SOC) designs. The 2003 International Technology Roadmap for Semiconductors (ITRS) clearly identifies the integration of electrochemical and electro-biological techniques as one of the system-level design challenges that will be faced beyond 2009, when feature sizes shrink below 50 nm [11].

As digital microfluidics-based biochips become widespread in safety-critical biochemical applications, the reliability of these systems will emerge as a critical performance parameter. These systems need to be tested adequately not only after fabrication, but also continuously during in-field operation. For instance, for detectors monitoring for dangerous pathogens in critical locations such as airports, field testing is critical to ensure low false-positive and false-negative detection rates. In such cases, concurrent testing, which allows testing and normal bioassays to run simultaneously on a microfluidic system, can play an important role. It consequently facilitates built-in self-test (BIST) of digital microfluidic biochips and makes them less dependent on costly manual maintenance on a regular basis. Therefore, there exists a need for efficient testing methodologies for these microsystems. Due to the underlying mixed technology and multiple energy domains, the microfluidic biochip exhibits unique failure mechanisms and defects. In fact, the ITRS 2003 document recognizes the need for new test methods for disruptive device technologies that underly microelectromechanical systems and sensors, and highlights it as one of the five difficult test challenges beyond 2009 [11].

The reconfigurability inherent in digital microfluidic biochips can be utilized to achieve longer system lifetime through on-line reconfiguration to avoid operational faults. It can also be used to increase production yield through production-time reconfiguration to bypass manufacturing faults. System reliability motivates the need for on-line reconfiguration techniques to tolerate faults during field operation. Reconfiguration is also useful for yield enhancement because it can be used to tolerate manufacturing faults. In this scenario, we assume that a microfluidic biochip has been fabricated for a set of bioassays, but some defective unit cells are identified prior to its deployment. The configuration of the microfluidic array must therefore be changed in such a way that the functionality of the bioassays is not compromised.

In this paper, we present an overview of an integrated methodology that addresses key issues in the synthesis, testing and reconfiguration of digital microfluidic biochips. The goal here is to provide top-down system-level design automation tools to biochip users, which can relieve them from the burden of manual optimization of assays, time-consuming hardware design, and costly testing and maintenance procedures. Users will be able to describe bioassays at a sufficiently high level of abstraction; synthesis tools will then map the behavioral description to a microfluidic biochip and generate an optimized schedule of bioassay operations, the binding of assay operations to resources, and a layout of the microfluidic biochip. For fabricated microfluidic biochips, cost-effective testing techniques will be available to detect faulty unit cells after the manufacture and during field operation. On-line and off-line reconfiguration techniques, incorporated in these design automation tools, will then be used to easily bypass faulty unit cells once they are detected, and remap bioassays operations to other fault-free resources, thereby supporting defect/fault tolerance. Thus the biochip user can concentrate on the development of the nano- and micro-scale bioassays, leaving implementation details to the design automation tools. Therefore, these tools will reduce human effort and enable high-volume production.

The organization of the remainder of the paper is as follows. Section 2 reviews biochip and microfluidics technology. Different actuation mechanisms for microfluidics-based biochips are discussed. We also present an overview of digital microfluidic biochips based on electrowetting. Next, Section 3 discusses design trends and challenges for digital microfluidics-based biochips. After reviewing today's design techniques, we propose a top-down design methodology for digital microfluidic biochips. This methodology encompasses synthesis, testing and reconfiguration. Challenges in the proposed system-level design method are also identified and discussed. Finally, conclusions are drawn in Section 4.

2. BIOCHIP AND MICROFLUIDICS TECHNOLOGY

2.1 Biochips

Early biochips were based on the concept of a DNA microarray, which is a piece of glass, plastic or silicon substrate on which pieces of DNA have been affixed in a microscopic array. Scientists use such chips to screen a biological sample simultaneously for the presence of many genetic sequences at once. The affixed DNA segments are known as *probes.* Thousands of identical probe molecules are affixed at each point in the array to make the chips effective detectors. The flowchart of DNA microarray production and operation is shown in Figure 1-1. Note that sample preparation need to be carried out off chip. There are a number of commercial microarrays available in the market place such as GeneChip® DNAarray from Affymetrix, DNA microarray from Infineon AG, or NanoChip® microarray from Nanogen [12, 13, 14]. Similar to a DNA microarray, a protein array is a miniature array where a multitude of different capture agents, most frequently monoclonal antibodies, are deposited on a chip surface (glass or silicon); they are used to determine the presence and/or amount of proteins in biological samples, e.g., blood. A drawback of DNA and protein arrays is that they are neither reconfigurable nor scalable after manufacture. Moreover, they lack the ability to carry out sample preparation, which is critical to biochemical applications.

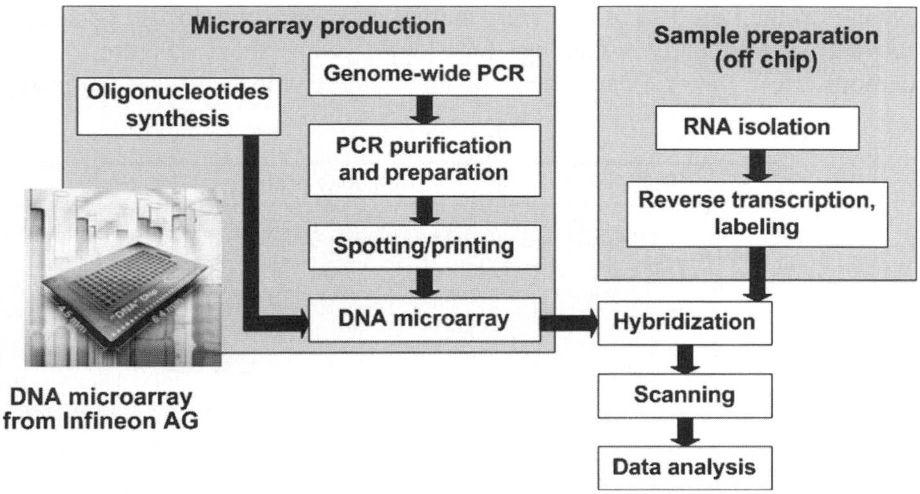

Figure 1-1. Steps in the production and operation of a DNA microarray.

The basic idea of microfluidic biochips is to integrate all necessary functions for biochemical analysis onto one chip using microfluidics technology. These micro-total-analysis-systems (μTAS) are more versatile and complex than microarrays. Integrated functions include microfluidic assay operations and detection, as well as sample pre-treatment and preparation. So far there are two different generations of microfluidic biochips, namely continuous-flow biochips and droplet-based microfluidic biochips.

2.2 Microfluidics

2.2.1 Continuous-flow microfluidics

The first generation microfluidic technologies are based on the manipulation of continuous liquid flow through microfabricated channels. Actuation of liquid flow is implemented either by external pressure sources, integrated mechanical micropumps, or by electrokinetic mechanisms [3, 4]. For example, electro-osmosis is a commonly-used electrokinetic method, which refers to the motion of an ionic fluid solution by means of an electrical field. As shown in Figure 1-2(a), a double layer of ions, consisting of a compact immobile layer and a mobile diffuse layer, is formed in the liquid sandwiched between two glass plates [15]. If an electric field is applied parallel to the liquid-solid interface, mobile charges in the diffuse layers are moved, consequently dragging the liquid with them. Figure 1-2(b) demonstrates the forward and reverse liquid flow in a fabricated

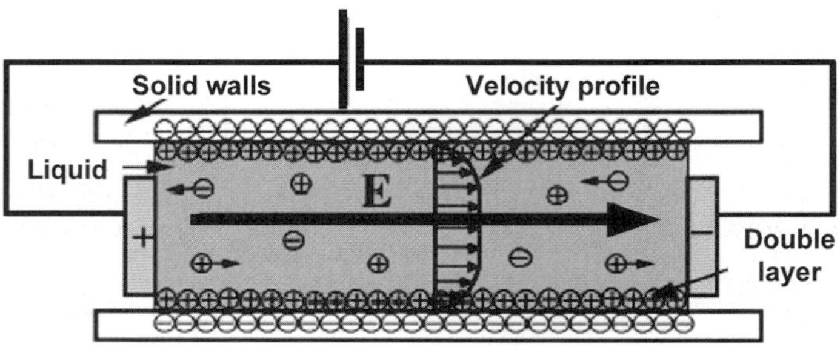

(a)

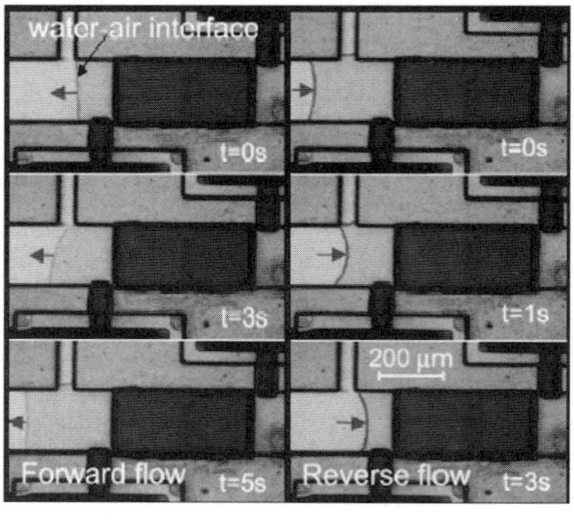

(b)

Figure 1-2. (a) Depiction of electro-osmotic flow; (b) forward and reverse fluid flow with DC voltages applied and polarities reversed respectively [15].

microchannel when forward and reversed DC voltages are applied, respectively; this continuous-flow microfluidic system based on electro-osmosis was developed at the University of Michigan [15].

Continuous-flow systems are adequate for many well-defined and simple biochemical applications, but they are unsuitable for more complex tasks requiring a high degree of flexibility or complicated fluid manipulations [3, 4]. These closed-channel systems are inherently difficult to integrate and scale because the parameters that govern the flow field (e.g., pressure, fluid resistance, and electric field) vary along the flow path making the fluid flow at any one location dependent on the properties of the entire system. Moreover, unavoidable shear flow and diffusion in microchannels makes it difficult to eliminate intersample contamination and dead volumes. Permanently-etched microstructures also lead to limited reconfigurability and poor fault tolerance capability. Therefore, the fabrication of complex yet reliable continuous-flow biochips remains a major technical challenge.

2.2.2 Droplet-based microfluidics

Alternatives to the above closed-channel continuous-flow systems include novel open structures, where the liquid is divided into discrete, independently controllable droplets, and these droplets can be manipulated to move on a substrate [9, 10, 22]. By using discrete unit-volume droplets, a microfluidic function can be reduced to a set of repeated basic operations, i.e., moving one unit of fluid over one unit of instance. This "digitization" method facilitates the use of a hierarchical and cell-based approach for microfluidic biochip design. In this scenario, we envisage that a large-scale integrated digital microfluidic biochip can be constructed out of repeated instances of well-characterized unit cells in the same way that complex VLSI circuits may be built upon well-characterized transistors. Moreover, the constituent microfluidic unit cells, referred to as *microfluidic modules*, can be reorganized at different levels of hierarchy to support biochemical applications of various scales. Defect/fault tolerance is also easily incorporated in the design due to the inherent dynamic reconfigurability. Therefore, in contrast to continuous fluid flow, digital microfluidics offer a flexible and scalable system architecture as well as high defect/tolerance capability.

A number of methods for manipulating microfluidic droplets have been proposed in the literature [16, 17, 18, 20, 21, 22]. These techniques can be classified as chemical, thermal, acoustical and electrical methods. For example, Gallardo *et al.* proposed an electrochemical method, whereby they used a voltage-controlled, reversible electrochemical reaction that creates or consumes redox-active surfactants (i.e., surface-active molecules). This reaction generates a surface tension gradient along a channel [16]. The surface tension gradient is capable of driving liquid droplets through a simple fluidic network; an example is shown in Figure 1-3 [16]. Time-lapse images in Figure 1-3(a)-3(c) demonstrate the movement of liquid crystal (LC) droplets based on the electrochemical method. As shown in Figure 1-3(d), the velocity of fluid motion is a function of the applied potential; moderate velocities of 2.5 mm/s were obtained at low voltages (< 1 V). Figure 3(e) also illustrates the image of the transportation of sulfur microparticles across the surface of an aqueous solution. However, since the electrochemical gradient must be established along the entire length of the channel, this technique, like electrokinetic methods used in continuous-flow systems, does not provide a convenient way to independently control multiple droplets.

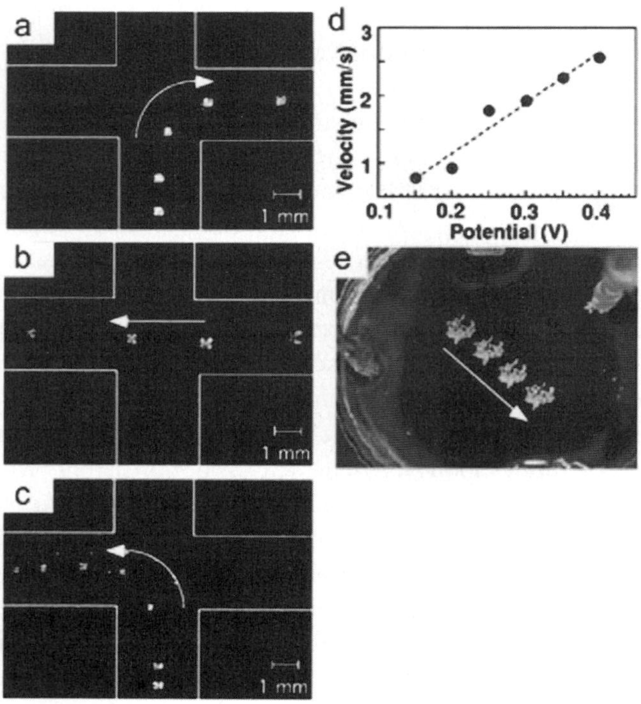

Figure 1-3. (a)-(c) Movement of LC droplets through a simple fluidic network; (d) movement velocity as a function of the applied potential; (e) movement of sulfur microparticles [16].

In another electrochemical method, Ichimura *et al.* used a photoresponsive surface to generate surface-energy gradients to drive droplets [17]. Photographs of light-driven motion of an olive oil droplet on a silica plate, which is modified with macrocyclic amphiphile tethering photochromic azobenzene units, are shown in Figure 1-4 [17]. However, the reported droplet movement velocities of 50 μm/s are very slow and many liquids including water cannot be transported by this technique due to contact angle hysteresis.

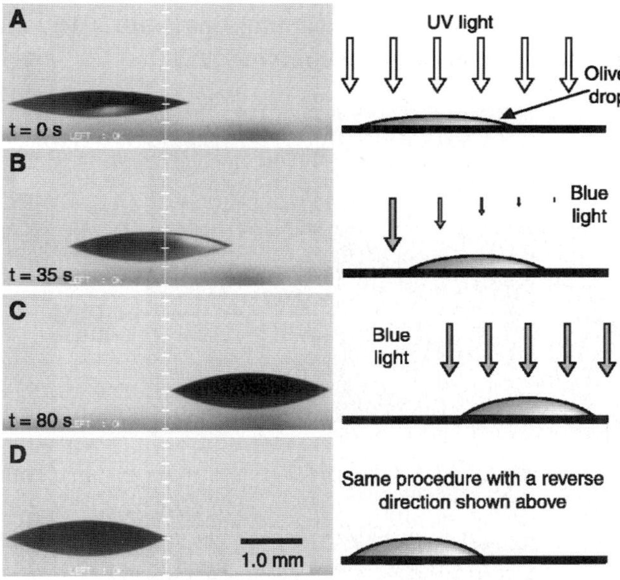

Figure 1-4. Lateral photographs of light-driven motion of an olive oil droplet on a silica plate by asymmetrical irradiation with 436-nm light perpendicular to the surface [17].

Another type of effect, namely thermocapillarity, exploits the temperature dependence of surface tension to drive droplet motion [18]. Thermocapillarity-based systems incorporate multiple independently controllable micromachined heaters into a substrate to control multiple droplets. However, the design and analysis of these systems is complex due to the critical requirement of complete and complicated heat-transfer analysis. Moreover, to achieve a modest velocity (e.g., 20 mm/s), a relatively high temperature gradient (e.g., a differential of 20-40 °C) is needed. Unfortunately, such large temperature variations are unacceptable for many biochemical applications where temperature control to within 1 °C range is desired [19].

Surface acoustic waves (SAW) can be used to propagate across the piezoelectric substrate just like earthquakes do, driving droplets to move on the chip surface, as shown in Figure 1-5 [20]. Given the right frequency of the signal, a mechanical wave is launched across the chip; the forces within this "nano-earthquake" are sufficient to actuate the droplet on the surface. SAW-based technology can also be used to perform droplet mixing. At low power levels, the SAW is converted into an internal streaming in the droplet. In contrast to the diffusion, streaming induces a very efficient mixing and stirring within the droplet. Furthermore, if different frequency is applied during this process, different streaming patterns are induced and superimposed, leading to quasi-chaotic mixing [20].

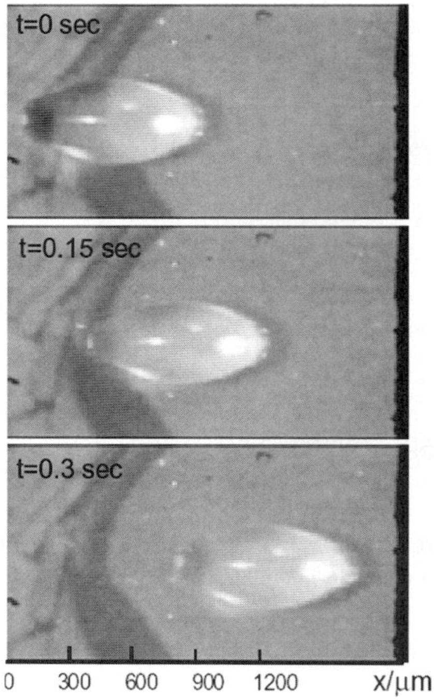

Figure 1-5. Photos of droplet motion caused by SAW forces [20].

In addition to above chemical and thermal methods, electrical methods to actuate droplets have received considerable attention in recent years [9, 10, 21, 22, 23]. Dielectrophoresis (DEP) and electrowetting-on-dielectric (EWOD) are the two most common electrical methods. DEP relies on the application of high-frequency AC voltages [22, 23], while EWOD is based on DC (or low-frequency AC) voltages [9, 10]. Both these techniques take advantage of electrohydrodynamic forces, and they can provide high droplet speeds with relatively simple geometries. Liquid DEP actuation is defined as the attraction of polarizable liquid masses into the regions of higher electric field intensity, as shown in Figure 1-6 [24]. DEP-based microfluidics relies on coplanar electrodes patterned on a substrate, coated with a thin dielectric layer, and energized with AC voltage (200-300 V-rms at 50-200 kHz). Rapid dispensing of large numbers of picoliter-volume droplets and a voltage-controlled array mixer have been demonstrated using DEP [22]. Images of multiple droplet movement on an 8×8 two-dimensional electrode array driven by DEP forces are shown in Figure 1-7 [23]; this DEP-driven microfluidic array was developed at the University of Texas M. D. Anderson Cancer Center. However, excessive Joule heating may be a problem for DEP actuation, even though it can be reduced by using materials of higher

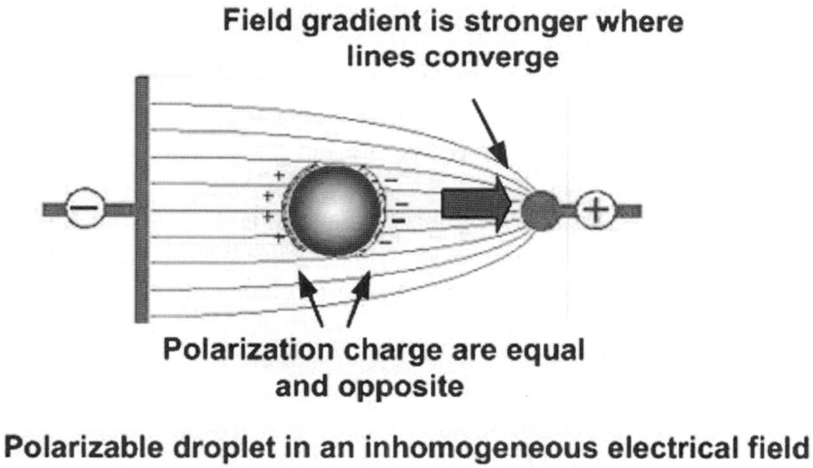

Figure 1-6. Liquid DEP actuation [24].

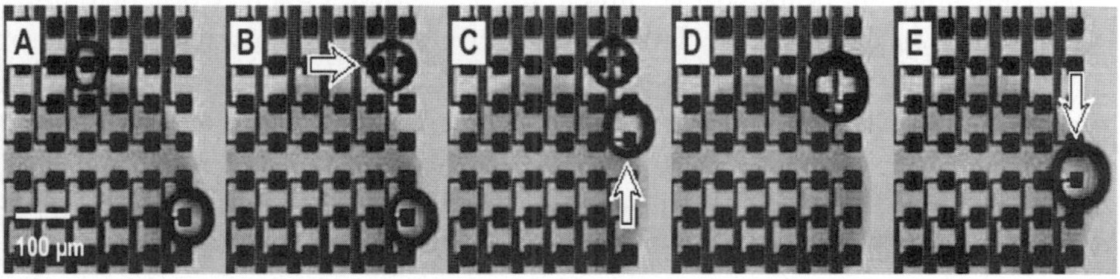

Figure 1-7. Droplets are driven by DEP forces on the surface of a two-dimensional array [23].

thermal conductivity or by reducing structure size [22, 25]. EWOD uses DC (or low- frequency AC) electric fields to directly control the interfacial energy between a solid and liquid phase. In contrast to DEP actuation, Joule heating is virtually eliminated in EWOD because the dielectric layer covering the electrodes blocks DC electric current. As a consequence, aqueous solutions with salt concentration as high as 0.15 M can be actuated without heating [25]. The EWOD technique for digital microfluidic biochips forms the basis of the work reported in this paper; we describe it in more detail in the next section.

2.2.3 Digital microfluidics-based biochips

The digital microfluidic biochips discussed in this paper are based on the manipulation of nanoliter droplets using the principle of electrowetting. Electrowetting-on-dielectric (EWOD) refers to the modulation of the interfacial tension between a conductive fluid and a solid electrode coated with a dielectric layer by applying an electric field between them. An imbalance of surface tension is created if an electric field is applied to only one side of the droplet; this interfacial tension gradient forces the droplet to move.

The basic unit cell of a EWOD-based digital microfluidic biochip consists of two parallel glass plates, as shown in Figure 1-8(a). The bottom plate contains a patterned array of individually controllable electrodes, and the top plate is coated with a continuous ground electrode. All electrodes are formed by indium tin oxide (ITO). A dielectric insulator, e.g., parylene C, coated with a hydrophobic film of Teflon AF, is added to the plates to decrease the wettability of the surface and to add capacitance between the droplet and the control electrode. The detailed fabrication process is described in [26]. The droplet containing biochemical samples and the filler medium, such as the silicone oil, are sandwiched between the plates; the droplets travel inside the filler medium. In order to move a droplet, a control voltage is applied to an electrode adjacent to the droplet and at the same time the electrode just under the droplet is deactivated. The EWOD effect causes an accumulation of charge in the droplet/insulator interface, resulting in a surface tension gradient across the gap between the adjacent electrodes, which consequently causes the transportation of the droplet. By varying the electrical potential along a linear array of electrodes, electrowetting can be used to move nanoliter volume liquid droplets along this line of electrodes [26]. The velocity of the droplet can be controlled by adjusting the control voltage (0 ~ 90 V), and droplets can be moved at speeds of up to 20 cm/s [27]. Droplets can also be transported, in user-defined patterns and under clocked-voltage control, over a two-dimensional array of electrodes shown in Figure 1-8(b) without the need for micropumps and microvalves. In the remainder of this paper, EWOD-based digital microfluidic biochips are simply referred to as "digital microfluidic biochips".

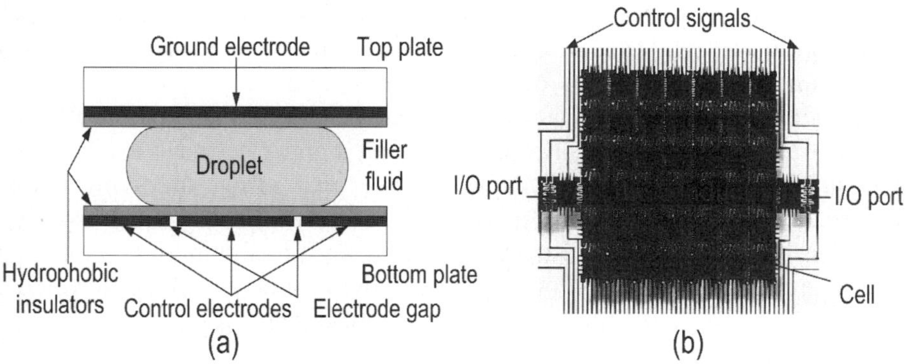

Figure 1-8. (a) Basic unit cell used in an EWOD-based digital microfluidic biochip; (b) a two-dimensional array for digital microfluidics.

The *in-vitro* measurement of glucose and other metabolites, such as lactate, glutamate and pyruvate, is of great importance in clinical diagnosis o f metabolic disorders. A colorimetric enzyme-kinetic glucose assay has been recently demonstrated in lab experiments on a digital microfluidic biochip [6, 28, 29]. This biochip uses a digital microfluidic array, which moves and mixes droplets containing biochemical samples and reagents, and an integrated optical detection system consisting of a LED and a photodiode; see Figure 1-9 [6, 28, 29].

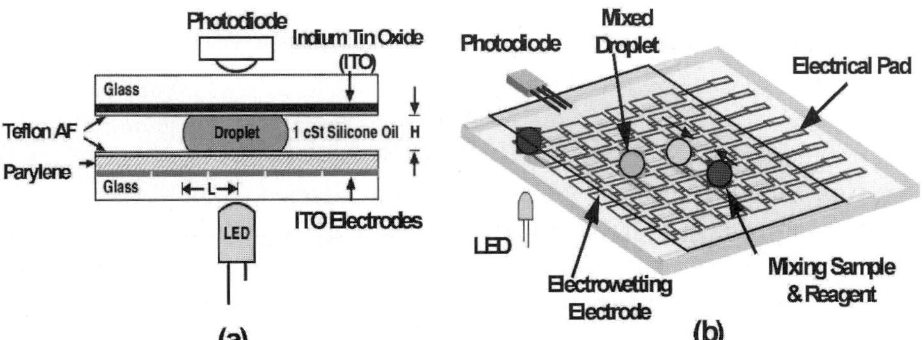

Figure 1-9. Schematic of a digital microfluidic biochip used for colorimetric assays: (a) basic unit cell; (b) Top view of microfluidic array.

In addition to glucose assays, the detection of other metabolites such as lactate, glutamate and pyruvate in a digital microfluidics-based biochip has also been demonstrated recently [6, 28, 29]. Furthermore, these assays as well as the glucose assay can be integrated to form a set of multiplexed bioassays that are performed concurrently on a microfluidic platform. Figure 1-10 illustrates a fabricated microfluidic system used for multiplexed bioassays [6]. For example, Sample 1 contains glucose and Reagent1 contains glucose oxidase and other chemicals. Similarly, Sample 2 contains lactate and Reagent2 consists of lactate oxidase and other chemicals. In this way, both glucose assay and lactate assay can be carried out concurrently. To demonstrate multiplexed assays, only unit cells and electrodes used for the bioassay have been fabricated. Note however that assays involving whole blood cells have not yet been successfully demonstrated by electrowetting [30]. Despite these limitations, advances in design automation tools will allow the design and fabrication of generic microfluidic platforms to which a set of assays can be mapped for optimized throughput, resource utilization, and fault tolerance.

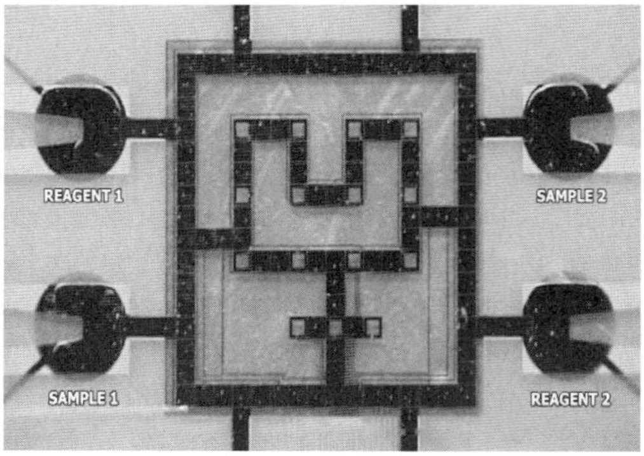

Figure 1-10. Fabricated microfluidic array used for multiplexed bioassays [6].

3. DESIGN TRENDS AND CHALLENGES

3.1 Typical Design Methodology: Bottom-Up

MEMS design is a relatively young field compared to integrated circuit design. Since the concept of special CAD systems for MEMS was first proposed at Transducer'87 [31], several research groups have reported significant progress in this area, and a number of commercial MEMS CAD tools are now available [32, 33]. Many of these tools are focused solely on the modeling of thermal and electro/mechanical properties. Recently, synthesis tools for MEMS have also been developed [34]. However, because of the differences in actuation methods between MEMS and microfluidics, they cannot be directly used for the design of microfluidic biochips. While MEMS design tools have reached a certain level of maturity, CAD tools for biochips are still in their infancy. Some design automation techniques have been proposed for DNA probe arrays [35]; however, as indicated in Section 2.2, microfluidics-based biochips are more versatile and complex than DNA arrays.

Current design methodologies for microfluidics-based biochips are typically full-custom and bottom-up in nature. Since much microfluidics work to date has been focused on device development, most design automation research for microfluidic biochips has been limited to device-level physical modeling of components [36, 37, 38]. For example, a combined circuit/device model for the analysis of microfluidic devices that incorporate fluidic transport, chemical reaction, reagent mixing and separation is presented in [36]. In the proposed circuit/device model, the continuous fluidic network is represented by a circuit model and the functional units of the microfluidic system are represented by appropriate device models. In addition, there are also some available commercial computational fluid dynamics (CFD) tools, such as CFD-ACE+ from CFD Research Corporation and FlumeCAD from Coventor, Inc. that support 3D simulation of microfluidic transport. Recently, physical modeling for digital microfluidics-based biochip has begun to receive much attention [37, 38]. For example, a unified framework of droplet electrohydrodynamics (EHD) to analyze the two major operating principles of droplet-based microfluidics, i.e., dielectrophoresis (DEP) and electrowetting-on-dielectric (EWOD), is presented in [38]. The numerical simulations based on droplet EHD are validated against analytical and experimental results, and they are then used to illustrate the operation of digital microfluidics-based devices.

Once the devices are optimized using detailed physical simulation, they can be used to assemble a complete microfluidics-based biochip. Therefore, a bottom-up development approach is rather natural, which involves the development of each block from the device to the system level. Microfluidic devices (e.g., electrodes and glass plates) are combined to form microfluidic modules (e.g., mixers or storage units), which are then combined to obtain the complete system (e.g., microfluidics-based glucose detectors). Since the system behavior can only be verified at this stage, costly and time-consuming redesign effort is required at the circuit level if the system does not satisfy design constraints.

Although these full-custom and bottom-up methodologies have been employed successfully in the past, they are clearly inadequate for the design of complex microfluidics-based biochips. As developments in microfluidics continue, it is likely that future microfluidics-based biochips will contain more than hundreds or thousands of basic components. Thus, an efficient design methodology and framework are required. While top-down system-level design tools are now commonplace in IC design, few such efforts have been reported for digital microfluidics-based biochips. A recent release of CoventorWare from Coventor, Inc. includes microfluidic behavioral models to allow top-down system-level design [39]. However, this CAD tool is only able to deal with continuous flow systems, and it is therefore inadequate for the design of digital microfluidic biochips.

3.2 Proposed Design Methodology: Top-Down

3.2.1 Overview

Motivated by the analogy between digital microfluidics-based biochips and digital integrated circuits, we aim to leverage advances in classical integrated circuit CAD techniques to address the design challenges associated with large-scale biochemical applications. The proposed system-level top-down design methodology is not only used to reduce biochip design complexity and time-to-market with the aid of design automation tools, but it can also be extended to enhance yield and system reliability.

The framework of this design methodology is illustrated in Figure 1-11. First the biochip users (e.g., biochemists) provide the protocol for nano- and micro-scale bioassays. We anticipate that advances in micro-scale chemistry will lead to such well-defined protocols. A sequencing graph $G(V, E)$ can directly be applied to describe this assay protocol, where vertex set $V = \{v_i: i = 0, 1,\ldots, k\}$ in one-to-one correspondence with the set of assay operations and edge set $E = \{(v_i, v_j): i, j = 0, 1,\ldots, k\}$ represents dependencies between assay operations. We can also use a high-level description language such as SystemC to model the protocol, and then derive a sequencing graph model from it. Moreover, this model can be used to perform behavioral-level simulation to verify the assay functionality at the high level [2]. Next, a synthesis tool is used to generate detailed implementations of digital microfluidic biochips from the sequencing graph model. A microfluidic module library is also provided as an input of the synthesis procedure. This module library, analogous to a standard/custom cell library used in cell-based VLSI design, includes different microfluidic functional modules, such as mixers and storage units. Each module is characterized by its function (mixing, storing, detection, etc.) and parameters such as width, length and operation duration. The microfluidic modules can be characterized through experiments, and their parameters can be stored for use by CAD tools that support large-scale biochip design. In addition, some design specifications are also given *a priori*, e.g., an upper limit on the completion time, an upper limit on the size of microfluidic array, and the set of non-reconfigurable resources such as on-chip reservoirs/dispensing ports and integrated optical detectors.

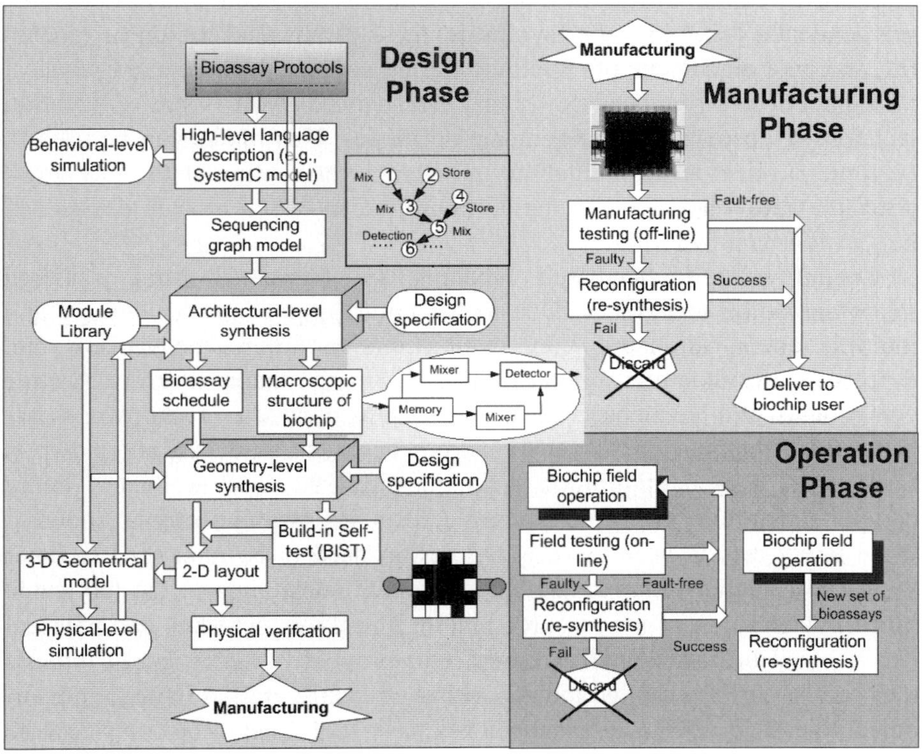

Figure 1-11. Overview of top-down design methodology.

The proposed synthesis tool performs both architectural-level synthesis (e.g., scheduling and resource binding) and geometry-level synthesis (e.g., module placement and routing); its details will be discussed in the next section [40, 41]. The output of the synthesis tools includes the mapping of assay operation to on-chip resources, a schedule for the assay operations, and a 2-D biochip physical design (e.g., the placement of the modules). The synthesis procedure attempts to find a desirable design point that satisfies the input specifications and also optimizes some figures of merit, such as performance and area. Moreover, since digital microfluidics-based biochips need to be tested adequately not only after fabrication, but also continuously during in-field operation, self-testing plays an important role in yield enhancement and reliability. Thus design-for-test (DFT) is also incorporated in the proposed synthesis procedure, whereby a test plan and a set of test hardware (e.g., test droplet sources/sinks and capacitive detection circuits) associated with the synthesized assay operation and biochip physical design are generated [42, 43]. After synthesis, the 2-D physical design of biochip (i.e., module placement and routing) can be coupled with detailed physical information from a module library (associated with some fabrication technology) to obtain a 3-D geometrical model. This model can be used to perform physical-level simulation and design verification at the low level. After physical verification, a digital microfluidics-based biochip design can be sent for manufacturing.

Digital microfluidics-based biochips are fabricated using standard microfabrication techniques. Due to the underlying mixed technology and multiple energy domains, they exhibit unique failure mechanisms and defects. A manufactured microfluidic array may contain several defective components. We have observed defects such as dielectric breakdown, shorts between adjacent electrodes, and electrode degradation; details are shown in Section 3.2.3. Reconfiguration techniques can be used to bypass faulty components to tolerate manufacturing defects. Bioassay operations bound to these faulty resources in the original design need to be remapped to other fault-free resources. Due to the strict resource constraints in the fabricated biochip, alterations in the resource binding operation, schedule and physical design must be carried out carefully. Our proposed system-level synthesis tool can be easily modified to deal with the reconfiguration issue to support defect tolerance. Using the enhanced synthesis tool, a set of bioassays can be easily mapped to a biochip with a few defective unit cells. Thus we do not need to discard the defective biochip, thereby leading to higher yield.

As digital microfluidics-based biochips are widely deployed in safety-critical applications, the field testing is also required to ensure the high reliability of biochips. Once the testing procedure determines the faulty status of biochips, the operation of the normal bioassay is stopped. Then reconfiguration techniques are applied to tolerate operational faults; the biochip is redesigned with the help of the proposed system-level design automation tools. In addition, the similar reconfiguration and design automation techniques can also be applied to remap a new set of bioassays to a fabricated microfluidic biochip, thereby increasing resource utilization and reducing the manufacturing cost.

Compared to the full custom and bottom-up design methods, this top-down system-level design methodology not only reduces the design cycle time and time-consuming redesign efforts, but it can also deal with design-for-test (DFT) and design-for-reliability (DFR) issues efficiently. Some important details of this system-level design methodology are discussed below.

3.2.2 Synthesis techniques

As more bioassays are executed concurrently on a digital microfluidics-based biochip, system integration and design complexity are expected to increase steadily. Thus system-level design automation tools, e.g., synthesis tools, are needed to handle design complexity. Synthesis research for digital microfluidic biochips can benefit from classical CAD techniques, which is a well-studied problem and advances in synthesis techniques for integrated circuits continue even today [44, 45].

We envisage that the synthesis of a digital microfluidic biochip can be divided into two major phases, referred to as architectural-level synthesis (i.e., high-level synthesis) and geometry-level synthesis (i.e., physical design) [40, 41]. A behavioral model for a biochemical assay is first obtained from the protocol for that assay. Next, architectural-level synthesis is used to generate a macroscopic structure of the biochip; this structure is analogous to a structural RTL model in electronic CAD. This macroscopic model provides an assignment of assay functions to biochip resources, as well as a mapping of assay functions to time-steps, based in part on the dependencies between them. Finally, geometry-level synthesis creates a physical representation at the geometrical level, i.e., the final layout of the biochip consisting of the configuration of the microfluidic array, locations of reservoirs and dispensing ports, and other geometric details.

The goal of a synthesis procedure is to select a design that minimizes a certain cost function under resource constraints. For example, architectural-level synthesis for microfluidic biochips can be viewed as the problem of scheduling assay functions and binding them to a given number of resources so as to maximize parallelism, thereby decreasing response time. On the other hand, geometry-level synthesis addresses the placement of resources and the routing of droplets to satisfy objectives such as area or throughput. Defect/fault tolerance can also be included as a critical objective in the proposed synthesis method.

In architectural-level synthesis, both resource binding problem and scheduling problem are addressed to generate a structural view of biochip design. As in the case of high-level synthesis for integrated circuits, resource binding in the biochip synthesis flow refers to the mapping from bioassay operations to available functional resources. Note that there may be several types of resources for any given bioassay operation. For example, a 2×2-array mixer, a 2×3-array mixer and a 2×4-array mixer can be used for a droplet mixing operation. In such cases, a resource selection procedure must be used. On the other hand, due to the resource constraints, a resource binding may associate one functional resource with several assay operations; this necessitates resource sharing. Once resource binding is carried out, the time duration for each bioassay operation can be easily determined. Scheduling determines the start times and stop times of all assay operations, subject to the precedence constraints imposed by the sequencing graph. In a valid schedule, assay operations that share a microfluidic module cannot execute concurrently. We have developed an optimal strategy based on integer linear programming for scheduling assay operations under resource constraints [40]. Since the scheduling problem is NP-complete, we have also developed two heuristic techniques that scale well for large problem instances. While the heuristic based on *list scheduling* is computationally more efficient, the second heuristic based on *genetic algorithms* yields lower completion times for bioassays. In addition, the heuristic based on *genetic algorithms* is also able to handle resource binding. Experiments show that the results obtained from the heuristics are close to provable lower bound for a bioassay of large size [40].

A key problem in the geometry-level synthesis of biochips is the placement of microfluidic modules such as different types of mixers and storage units. Based on the results obtained from architectural-level synthesis (i.e., a schedule of bioassay operation, a set of microfluidic modules, and the binding of bioassay operations to modules), placement determines the locations of each module on the microfluidic array in order to optimize some design metrics. Since digital microfluidics-based biochips enable dynamic reconfiguration of the microfluidic array during run-time, they allow the placement of different modules on the same location during different time intervals. Thus, the placement of modules on the microfluidic array can be modeled as a 3-D packing problem. Each microfluidic module is represented by a 3-D box, the base of which denotes the rectangular area of the module and the height denotes the time-span of its operation. The microfluidic biochip placement can now be viewed as the problem of packing these boxes to minimize the total base area, while avoiding overlaps. Since the placement problem is known to be NP-complete [44], a simulated annealing-based heuristic approach has been developed to solve the problem in a computationally efficient manner [41]. Solutions for the placement problem can provide the designer with guidelines on the size of the array to be manufactured. If module placement is carried out for a fabricated array, area minimization frees up more unit cells for sample collection and preparation.

3.2.3 Testing techniques and design-for-test (DFT)

Over the past decade, the focus in testing research has broadened from logic and memory test to include the testing of analog and mixed-signal circuits. Compared to relatively mature IC testing field, MEMS testing is still in its infancy. Recently, fault modeling and fault simulation in surface micromachined MEMS has received attention [46, 47]. However, test techniques for MEMS cannot be directly applied to microfluidics-based biochips, since the techniques and tools currently in use for MEMS testing do not handle fluids. Recently, fault modeling, fault simulation, and a DFT methodology for continuous-flow microfluidic systems have been proposed [48, 49, 50]. Although advances in test technology are now required to facilitate the continued growth of composite microfluidic systems based on droplet flow, very limited work on the testing for such "digital" microfluidic biochips has been reported to date.

We can classify the faults in these systems as being either catastrophic or parametric, along the line of fault classification for analog circuits [51]. Catastrophic (hard) faults lead to a complete malfunction of the system, while parametric (soft) faults cause a deviation in the system performance. A parametric fault is detectable only if this deviation exceeds the tolerance in system performance. Due to their underlying mixed technology and multiple energy domains, digital microfluidics-based biochips exhibit failure mechanisms and defects that are significantly different from the failure modes in integrated circuits.

Catastrophic faults in digital microfluidics-based biochips may be caused by the following physical defects:

Dielectric breakdown: The breakdown of the dielectric at high voltage levels creates a short between the droplet and the electrode. When this happens, the droplet undergoes electrolysis, thereby preventing further transportation.

Short between the adjacent electrodes: If a short occurs between two adjacent electrodes, the two electrodes shorted effectively form one longer electrode. When a droplet resides on this electrode, it is no longer large enough to overlap the gap between adjacent electrodes. As a result, the actuation of the droplet can no longer be achieved.

Degradation of the insulator: This degradation effect is unpredictable and may become apparent gradually during the operation of the microfluidic system. Figure 1-12 illustrates the electrode degradation due to insulator degradation defect [26]. A consequence of insulator degradation is that droplets often fragment and their motion is prevented because of the unwanted variation of surface tension forces along their flow path.

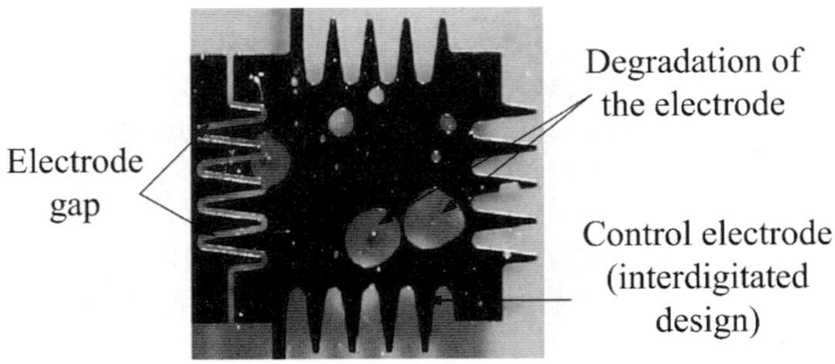

Figure 1-12. Top view of a faulty unit cell: electrode degradation.

Open in the metal connection between the electrode and the control source: This defect results in a failure in activating the electrode for transport.

Physical defects that cause parametric faults include the following:

Geometrical parameter deviation: The deviation in insulator thickness, electrode length and height between parallel plates may exceed their tolerance value.

Change in viscosity of droplet and filler medium. These deviations can occur during the operation due to an unexpected biochemical reaction, or changes in operational environment, e.g., temperature variation.

Faults in microfluidics-based biochips can also be classified based on the time at which they appear. Therefore, system failure or degraded performance can either be caused by manufacturing defects or they might be due to parametric variations. Testing of manufacturing defects, such as a short between the adjacent electrodes or a deviation in the value of the geometrical parameters, should be performed immediately after production. However, operational faults, such as degradation of the insulator or change in fluid viscosity, can occur throughout the lifetime of the system. Therefore, concurrent testing during system operation is necessary for such faults.

We have proposed a unified test methodology for digital microfluidic biochips, whereby faults can be detected by controlling and tracking droplet motion electrostatically [52, 53]. Based on this unified detection mechanism, we can dispense the test stimuli droplet containing the normal conductive fluid (e.g., KCL solution) into the microfluidic system-under-test from the droplet source. These droplets are guided through the unit cells following the test plan towards the droplet sink, which is connected to an integrated capacitive detection circuit. Most catastrophic faults result in a complete cessation of droplet transportation [52, 53]. Thus, for the faulty system, the test stimuli droplet is stuck during its motion. On the other hand, for the fault-free system, all the test stimuli droplets can be observed at the droplet sink by the capacitive detection circuit. Therefore, we can easily determine the fault-free or faulty status of the droplet-based microfluidic system by simply observing the arrival of test stimuli droplets at some selected ports of the system.

An efficient test plan not only ensures that the testing operation does not conflict with the normal biomedical assay, but it also guides test stimuli droplets to cover all the unit cells available for testing. This test plan can be optimized to minimize the total testing time cost for a given test hardware overhead, which refers here to the number of droplet sources and droplet sinks. We can formulate the test planning problem in terms of the graph partitioning and the Hamiltonian path problems from graph theory [42]. Since this optimization problem can be proven to be NP-complete, we also develop heuristic approaches to solve the test planning problem [42]. Experimental results indicate that for large array sizes, heuristic methods yield solutions close to provable lower bounds while ensuring scalability and low computation cost.

The proposed testing methodology can be used for field-testing of digital microfluidics-based systems; as a result, it increases the system reliability during everyday operation [43]. With negligible hardware overhead, this method also offers an opportunity to implement BIST for microfluidic systems and therefore eliminate the need for costly, bulky, and expensive external test equipment. Furthermore, after detection, droplet flow paths for biomedical assays can be reconfigured dynamically such that faulty unit cells are bypassed without interrupting the normal operation. Thus, this approach increases fault-tolerance and system lifetime when such systems are deployed for safety-critical applications.

3.2.4 Reconfiguration techniques and design-for-reliability (DFR)

As in the case of integrated circuits, increase in the density and area of microfluidic biochips may reduce yield, especially for smaller feature sizes. It will take time to ramp up the yield based on an understanding of defects in such biochips. Therefore, defect tolerance for digital microfluidic biochips is especially important for the emerging marketplace. Moreover, some manufacturing defects are expected to be latent and they may manifest themselves during field operation of the biochips. Since many microfluidic biochips are intended for safety-critical applications, system dependability is an essential performance parameter. Thus fault tolerance techniques will play a critical role in field applications, especially in harsh operational environments. Efficient reconfiguration techniques are motivated by the need for defect/fault tolerance.

A digital microfluidics-based biochip can be viewed as a dynamically reconfigurable system. If a unit cell becomes faulty during the operation of the biochip, and the fault is detected using the proposed testing technique, the microfluidic module containing this unit cell can easily be relocated to another part of the microfluidic array by changing the control voltages applied to the corresponding electrodes. Fault-free unused unit cells in the array are utilized to accommodate the faulty module. Hence, the configuration of the microfluidic array, i.e., the placement of the microfluidic modules, influences the fault tolerance capability of the biochip. Thus we introduce a simple measure, referred to as the fault tolerance index, to evaluate the fault tolerance capability of the microfluidic biochip; this measure is incorporated into the placement procedure. This design-for-reliability (DFR) procedure leads to small biochip area due to efficient utilization of dynamic reconfigurability, as well as high fault tolerance due to the efficient use of spare unit cells.

Defect/fault tolerance can also be achieved by including redundant elements in the system; these elements can be used to replace faulty elements through reconfiguration techniques. Another method is based on graceful degradation, in which all elements in the system are treated in a uniform manner, and no element is designated as a spare. In the presence of defects, a subsystem with no faulty element is first determined from the faulty system. This subsystem provides the desired functionality, but with a gracefully-degraded level of performance (e.g., longer execution times). Due to the dynamic reconfigurability of digital microfluidics-based biochips, the microfluidic components (e.g., mixers) used during the bioassay can be viewed as reconfigurable virtual devices. For example, a 2×4 array mixer (implemented using a rectangular array of control electrodes — two in the X-direction and four in Y-direction) can easily be reconfigured to a 2×3 array mixer or a 2×2 array mixer. This feature facilitates the use of graceful degradation to achieve defect tolerance in digital biochips. Since a high-level scheme is required to efficiently reconfigure and reallocate the assay operations, our proposed system-level design automation tools can be utilized to support defect/fault tolerance, thereby leading to a high system reliability.

3.3 Challenges

A number of open problems remain to be tackled in the development of the proposed top-down system-level design methodology. First, we note that, following the geometry-level synthesis, the automatically-generated layout of digital microfluidics-based biochips need to be coupled with more detailed geometrical data for 3-D physical simulation. Although this detailed simulation-based approach can be used for physical verification, it is time-consuming and highly dependent on the accuracy of the geometrical model. We can speed up and automate the physical verification procedure for biochip designs by leveraging classical integrated circuit verification techniques (e.g., design rule checking). As in circuit design, the layered microfabrication process information can be encapsulated in a layout design rule file. The synthesized layout of microfluidic biochip is verified to satisfy an abstraction of geometric design constraints, which consequently ensures robust manufacturing. However, the design rules that need to be checked in the microfluidics-based biochips are significantly different from those in circuit area. They are also unlike classical MEMS due to the fluidic domain [54]. The determination of accurate and efficient design rules for physical verification of digital microfluidics-based biochips remains an open problem.

Effective testing of biochips also needs to be investigated. Some physical failure mechanisms are not yet well-understood. For example, due to the unknown thermal effects on microfluidic assay operation, the defects associated with power supply or environmental temperature variation are hard to detect. Efficient fault models and test stimuli generation techniques are required for the testing of biochips. Moreover, while catastrophic faults have the highest priority for detection as they result in complete malfunction, parametric faults are much harder to detect and may result in malfunction depending on the application domain and specification. As a result, design-for-test (DFT) techniques to handle parametric faults are more complicated than those for the detection of catastrophic faults.

Coupling of energy domains also affect the synthesis and performance optimization of biochips. Due to coupling between different energy domains (e.g., electrical, fluidic and thermal domains) [2], multiple-objective optimization problems must be solved during synthesis. For example, we should not only aim to minimize the assay operation time, but we should also keep the power consumption low to avoid fluid overheating. Such optimization problems that span several energy domains appear to be extremely difficult. Efficient solutions to such optimization problem are nevertheless essential to ensure the quality of biochips designed using automated synthesis techniques.

4. CONCLUSION

We have presented a new system-level design automation methodology for droplet-based microfluidic biochips. Technology issues underlying biochips and microfluidics have first been reviewed. We focused here on a new implementation platform for digital microfluidic biochips based on electrowetting-on-dielectric (EWOD). The level of system integration and the complexity of digital microfluidics-based biochips are expected to increase in the near future due to the growing need for multiple and concurrent bioassays on a chip. To address the associated design challenges, we have proposed a top-down design methodology for digital microfluidic biochips. In this proposed method, synthesis tools are used to map the behavioral description of bioassays to a microfluidic biochip and generate an optimized schedule of bioassay operations, the binding of assay operations to resources, and a layout of the microfluidic biochip. Compared to the current full custom and bottom-up design methods, this top-down system-level design methodology not only reduces the design cycle time and time-consuming redesign efforts, but it can also deal with design-for-test (DFT) and design-for-reliability (DFR) issues efficiently. For fabricated microfluidic biochips, cost-effective testing techniques have been proposed to detect faulty unit cells after the manufacture and during field operation. Dynamic reconfiguration techniques, incorporated in these design automation tools, are also used to easily bypass faulty unit cells once they are detected, and remap bioassays operations to other fault-free resources, thereby supporting defect/fault tolerance. This work is expected to reduce human efforts and enable high-volume productions and applications of microfluidics-based biochips, thereby paving the way for the integration of biochip components in the next generation of system-on-chip designs, as envisaged by the 2003 ITRS document.

REFERENCES

1. M. A. Burns, B. N. Johnson, S. N. Brahmasandra, K. Handique, J. R. Webster, M. Krishnan, T. S. Sammarco, P. M. Man, D. Jones, D. Heldsinger, C. H. Mastrangelo and D. T. Burke, "An integrated nanoliter DNA analysis device", *Science*, vol. 282, pp. 484-487, 1998.
2. T. Zhang, K. Chakrabarty and R. B. Fair, *Microelectrofluidic Systems: Modeling and Simulation*, CRC Press, Boca Raton, FL, 2002.
3. T. Thorsen, S. Maerkl and S. Quake, "Microfluidic large-scale integration", *Science*, vol. 298, pp. 580-584, 2002.
4. E. Verpoorte and N. F. De Rooij, "Microfluidics meets MEMS", *Proceedings of the IEEE*, vol. 91, pp. 930-953, 2003.
5. T. H. Schulte, R. L. Bardell and B. H. Weigl "Microfluidic technologies in clinical diagnostics", *Clinica Chimica Acta*, vol. 321, pp. 1-10, 2002.
6. V. Srinivasan, V. K. Pamula, and R. B. Fair, "An integrated digital microfluidic lab-on-a-chip for clinical diagnostics on human physiological fluids," *Lab on a Chip*, pp. 310-315, 2004.
7. H. F. Hull, R. Danila and K. Ehresmann, "Smallpox and bioterrorism: Public-health responses", *Journal of Laboratory and Clinical Medicine*, vol. 142, pp. 221-228, 2003.
8. S. Venkatesh and Z. A. Memish, "Bioterrorism: a new challenge for public health", *International Journal of Antimicrobial Agents*, vol. 21, pp. 200-206, 2003.
9. M. G. Pollack, R. B. Fair and A. D. Shenderov, "Electrowetting-based actuation of liquid droplets for microfluidic applications", *Applied Physics Letters*, vol. 77, pp. 1725-1726, 2000.
10. S. K. Cho, S. K. Fan, H. Moon, and C. J Kim, "Toward digital microfluidic circuits: creating, transporting, cutting and merging liquid droplets by electrowetting-based actuation", *Proc. IEEE MEMS Conf.*, pp. 32-52. 2002.
11. International Technology Roadmap for Semiconductors (ITRS), http://public.itrs.-net/Files/2003ITRS/Home2003.htm.
12. Affymetrix GeneChip®, http://www.affymetrix.com
13. Infineon Electronic DNA Chip, http://www.infineon.com
14. Nanogen NanoChip®, http://www.nanogen.com
15. S. Mutlu, F. Svec, C. H. Mastrangelo, J. M. J. Frechet and Y. B. Gianchandani, "Enhanced electro-osmosis pumping with liquid bridge and field effect flow rectification", *Proc. IEEE MEMS Conf.*, pp. 850-853, 2004.
16. B. S. Gallardo, V. K. Gupta, F. D. Eagerton, L. I. Jong, V. S. Craig, R. R. Shah, and N. L. Abbott, "Electrochemical principles for active control of liquids on submillimeter scales," *Science*, vol. 283, pp. 57-60, 1999.
17. K. Ichimura, S. Oh, and M. Nakagawa, "Light-driven motion ofliquids on a photoresponsive surface," *Science*, vol. 288, pp. 1624-1626, 2000.
18. T. S. Sammarco and M. A. Burns, "Thermocapillary pumping of discrete droplets in microfabricated analysis devices," *AI Che J.*, vol. 45, 350-366, 1999.
19. G. N. Somero, "Proteins and temperature", *Annual Review of Physiology*, vol. 57, pp.43-68, 1995.
20. Wixforth and J. Scriba, "Nanopumps for programmable biochips", http://www.advalytix.de
21. M. Washizu, "Electrostatic actuation of liquid droplets for microreactor applications," *IEEE Trans. Ind. Appl.*, vol. 34, pp. 732-737, 1998.

22. T. B. Jones, M. Gunji, M. Washizu and M. J. Feldman, "Dielectrophoretic liquid actuation and nanodroplet formation," *J. Appl. Phys.,* vol. 89, pp. 1441-1448, 2001.
23. J. Vykoukal et al., "A programmable dielectric fluid processor for droplet-based chemistry", *Micro Total Analysis Systems 2001*, 72-74, 2001.
24. A DEP Primer, http://www.dielectrophoresis.org
25. T. B. Jones, K. L. Wang, and D. J. Yao, "Frequency-dependent electromechanics of aqueous liquids: electrowetting and dielectrophoresis," *Langmuir*, vol. 20, pp. 2813-2818, 2004.
26. M. Pollack, *Electrowetting-Based Microactuation of Droplets for Digital Microfluidics*, PhD thesis, Duke University. 2001.
27. M. G. Pollack, A. D. Shenderov and R. B. Fair, "Electrowetting-based actuation of droplets for integrated microfluidics", *Lab on a Chip*, vol. 2, pp. 96-101, 2002.
28. V. Srinivasan, V. K. Pamula, M. G. Pollack, R. B. Fair, "A digital microfluidic biosensor for multianalyte detection", *Proc. IEEE MEMS Conf.*, pp. 327-330, 2003.
29. V. Srinivasan, V. K. Pamula, M. G. Pollack, and R. B. Fair, "Clinical diagnostics on human whole blood, plasma, serum, urine, saliva, sweat, and tears on a digital microfluidic platform", *Proc. μTAS*, pp. 1287-1290, 2003.
30. V. Srinivasan, "*A Digital Microfluidic Lab-on-a-Chip for Clinical Diagnostic Applications*", PhD Thesis, Duke University, 2005.
31. S. Senturia, "Microfabricated structures for the measurement of mechanical properties and adhesion of thin films", *Proc. Int. Conf. Sold-State Sensors and Actuators (Transducers)*, pp. 11-16, 1987.
32. G. K. Fedder and Q. Jing, "A hierarchical circuit-level design methodology for microelectromechinal system", *IEEE Trans. Circuits and Systems II,* vol. 46, pp.1309-1315, 1999.
33. S. K. De and N. R Aluru, "Physical and reduced-order dynamic analysis of MEMS", *Proc. IEEE/ACM Int. Conf. Computer Aided Design*, pp. 270-273, 2003.
34. T. Mukherjee and G. K. Fedder, "Design methodology for mixed-domain systems-on-a-chip [MEMS design]", *Proc. IEEE VLSI System Level Design*, pp. 96 - 101, 1998.
35. B. Kahng, I. Mandoiu, S. Reda, X. Xu, and A.Z. Zelikovsky, "Evaluation of placement techniques for DNA probe array layout", *Proc. IEEE/ACM Int. Conf. Computer Aided Design*, pp. 262-269, 2003.
36. N. Chatterjee and N.R. Aluru, "Combined circuit/device modeling and simulation of integrated microfluidic systems", *Journal of Microelectromechanical Systems*, vol. 14, pp. 81-95, 2005.
37. B. Shapiro, H. Moon, R. Garrell, and C. J. Kim, "Modeling of electrowetted surface tension for addressable microfluidic systems: dominant physical effects, material dependences, and limiting phenomena" *Proc. IEEE Conf. MEMS*, pp. 201- 205, 2003.
38. J. Zeng and F. T. Korsmeyer, "Principles of droplet electrohydrodynamics for lab-on-a-chip", Lab on a Chip, vol. 4, pp. 265-277, 2004.
39. CoventorWare™, http://www.coventor.com.
40. F. Su and K. Chakrabarty, "Architectural-level synthesis of digital microfluidics-based biochips", *Proc. IEEE International Conference on CAD*, pp. 223-228, 2004.
41. F. Su and K. Chakrabarty, "Design of fault-tolerant and dynamically-reconfigurable microfluidic biochips", accepted for publication in *Proc. Design, Automation and Test in Europe (DATE) Conference*, 2005.
42. F. Su, S. Ozev and K. Chakrabarty, "Test planning and test resource optimization for droplet-based microfluidic systems", *Proc. European Test Symposium*, pp. 72-77, 2004.

43. F. Su, S. Ozev and K. Chakrabarty, "Concurrent testing of droplet-based microfluidic systems for multiplexed biomedical assays", *Proc. IEEE International Test Conference*, pp. 883-892, 2004.
44. G. De Micheli, *Synthesis and optimization of digital circuits.* New York: McGraw-Hill, 1994.
45. R. Camposano, "Behavioral synthesis", *Proc. IEEE/ACM Design Automation Conference*, pp.33-34, 1996.
46. Kolpekwar and R. D. Blanton, "Development of a MEMS testing methodology", *Proc. International Test Conference*, pp. 923-93, 1997.
47. N. Deb and R. D. Blanton, "Analysis of failure sources in surface-micromachined MEMS", *Proc. International Test Conference*, pp. 739-749, 2000.
48. H. G. Kerkhoff, "Testing philosophy behind the micro analysis system", *Proc. SPIE: Design, Test and Microfabrication of MEMS and MOEMS,* vol. 3680, pp.78-83, 1999.
49. H. G. Kerkhoff and H. P. A. Hendriks, "Fault modeling and fault simulation in mixed micro-fluidic microelectronic Systems", *Journal of Electronic Testing: Theory and Applications*, vol. 17, pp. 427-437, 2001.
50. H. G. Kerkhoff and M. Acar, "Testable design and testing of micro-electro-fluidic arrays", *Proc. IEEE VLSI Test Symposium*, pp. 403-409, 2003.
51. A. Jee and F. J. Ferguson, "Carafe: An inductive fault analysis tool for CMOS VLSI circuits", *Proc. IEEE VLSI Test Symposium*, pp. 92-98, 1993.
52. F. Su, S. Ozev and K. Chakrabarty, "Testing of droplet-based microelectrofluidic systems", *Proc. IEEE International Test Conference*, pp. 1192-1200, 2003.
53. F. Su, S. Ozev and K. Chakrabarty, "Ensuring the operational health of droplet-based microelectrofluidic biosensor systems", *IEEE Sensors Journal*, vol. 5, pp. 763-773, 2005.
54. T. Mukherjee, "MEMS design and verification", *Proc. IEEE International Test Conference*, pp. 681-690, 2003.

Chapter 2

MODELING AND SIMULATION OF ELECTRIFIED DROPLETS AND ITS APPLICATION TO COMPUTER-AIDED DESIGN OF DIGITAL MICROFLUIDICS

Jun Zeng
Coventor, Inc., 625 Mount Auburn Street, Cambridge, MA 02138, USA.
jun.zeng@coventor.com.

Abstract: Digital microfluidics is the second-generation lab-on-a-chip architecture based upon micromanipulation of droplets via a programmed external electric field by an individually addressable electrode array. Dielectrophoresis (DEP) and electrowetting-on-dielectric (EWOD) are of the dominant operating principles. The microfluidic mechanics of manipulating electrified droplets are complex and not entirely understood. In this article, we present a numerical simulation method based on droplet electrohydrodynamics (EHD). First we show a systematic validation study comparing the simulation solution with both analytical and experimental data, quantitatively and qualitatively, and in both steady state and transient time sequences. Such comparison exhibits excellent agreement. Simulations are then used to illustrate its application to computer-aided design of both EWOD-driven and DEP-driven digital microfluidics.

Key words: biochips; dielectrophoresis; droplet; electrohydrodynamics; electrowetting; microfluidics; simulation.

1. INTRODUCTION

Simulation-based computer-aided design (CAD) for lab-on-a-chip holds the promise of accelerating the design process from concept to volume production. Simulation permits the prediction of performance prior to

K. Chakrabarty and J. Zeng (eds.),
Design Automation Methods and Tools for Microfluidics-Based Biochips, 31–52.

building a device, supports the troubleshooting of device designs during development, and enables critical evaluations of failure mechanisms after a device has entered the manufacturing stage. Simulation can also expose physical properties that are otherwise difficult to measure through experiments.

Simulation is gaining acceptance among lab-on-a-chip developers as means to investigate device physics and to optimize device performance. However, the multi-disciplinary nature of lab-on-a-chip's operating principle gives rise to serious challenges to the integration of simulation-based CAD into standard lab-on-a-chip design practice. Computational prototyping of lab-on-a-chip demands simulation engines that can deliver coupled solutions across different domains of science. Computer-aided design of digital microfluidics is such an example.

Digital microfluidics[1] is the second generation lab-on-a-chip architecture based upon micromanipulation of discrete fluid particles (droplets). In the operation of a digital microfluidics, minute amounts of chemical sample are drawn from individual sample reservoirs in the form of metered droplets. These droplets are then delivered to a reaction chamber where multiple droplets may reside simultaneously. When droplets containing different chemical samples arrive at the same location, the droplets will merge into one droplet and a chemical reaction can occur. Chemical reactions can be detected, categorized, and reported. Hierarchical reactions can be achieved by merging droplets of intermediate reactions. A larger droplet may be split into smaller ones for parallel manipulation or detection.

Compared to the first generation channel-based lab-on-a-chip that operates under conditions of continuous flow, digital microfluidics enables reconfigurability and scalability such that complex procedures can be built up through combining and reusing a finite set of basic instructions within hierarchical chip architecture. The basic operations required in digital microfluidics include droplet generation, or separating a liquid stream into discrete droplets; droplet translocation; droplet fusion, or merging multiple droplets into one; and droplet fission, or dividing one droplet into smaller ones. The electrohydrodynamic (EHD) forces generated by the presence of an electric field are utilized to accomplish this set of operations. An individually addressable electrode array can generate the electric field to produce these forces in a predetermined manner; dielectrophoresis (DEP) and electrowetting-on-dielectric (EWOD) are of the dominant operating principles. Fig. 2-1 illustrates example digital microfluidics chip architecture.

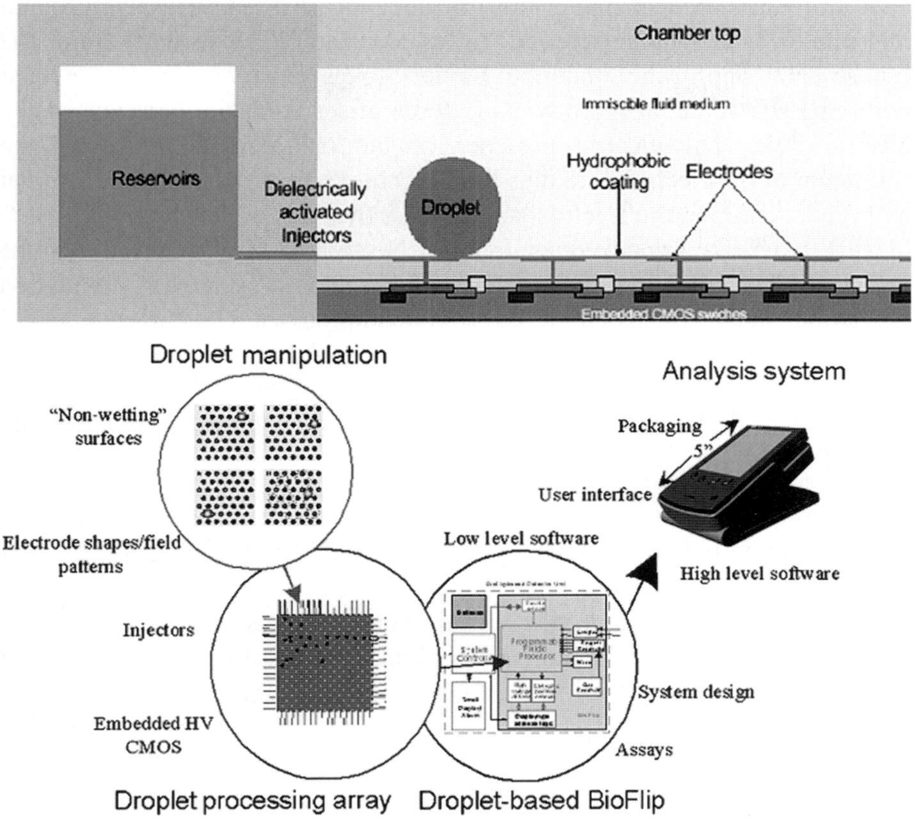

Figure 2-1. Example of a digital microfluidics architecture. (Top) Elevation view of the chip (PFP) architecture. (Below) System architecture.

Analytical means that were broadly adopted in design of channel-based lab-on-a-chip are much less applicable to digital microfluidics. Let's take DEP as example (This argument is also applicable to that driven by EWOD). It is possible to derive an integral DEP force exerted on a spherical and rigid particle in an analytical form and use that to predict the particle motion under an electric field. However, a droplet in a typical DEP digital microfluidics chip does not hold a spherical shape, but undergoes complicated deformation, even to the extent of topological change. This makes an analytical expression of the DEP force either not useful or, at best, not completely trustworthy. Therefore, simulation-oriented design practice for digital microfluidics has to rely on detailed simulation solutions. Detailed EHD simulation of electrified droplets and its application to computer-aided design of digital microfluidics is the theme of this article.

A unified theoretical framework for modeling DEP and EWOD in digital microfluidics, hereafter referred to as droplet EHD,[2] stems from the fundamental approach[3,4] that the Navier-Stokes equations are augmented with an EHD force, a hydrodynamic force arises from the presence of the electric field. The electric presence of the operating fluid affects the formation of the electric field thus the EHD force. In other words, the droplet hydrodynamics and the electrostatic field are two-way coupled.

This article is organized as follows. Section 2 briefly introduces the droplet EHD. Following that, section 3 focuses on a systematic simulation validation study composed of three validation cases. Sections 4 and 5 analyze EWOD and DEP, respectively. A simulation of on-chip droplet-based chemistry is presented in section 6, to illustrate the essence of the digital microfluidics, and the outlook of the virtual prototyping of digital microfluidics.

2. DROPLET ELECTROHYDRODYNAMICS

The theoretical basis of droplet electrohydrodynamics was elaborated in[2]. When a fluid is exposed to an electric field, an EHD force arises acting in concert with other hydrodynamic forces to dictate the fluid motion. This EHD force density $\boldsymbol{f}^e$ exerted on this fluid can be expressed as[5]

$$\boldsymbol{f}^e = \rho_e \boldsymbol{E} - \sum_{i=1}^{m} \alpha_i \nabla\left(\frac{\partial W}{\partial \alpha_i}\right) \tag{2.1}$$

where ρ_e is the volumetric density of the free charge, $\boldsymbol{E}$ is the electric field, $\alpha_1, \alpha_2 \ldots \alpha_m$ are material properties of this fluid, and W is the volumetric density of the electroquasistatic energy,

$$W = \int_0^{\boldsymbol{D}} \boldsymbol{E}(\alpha_1 \ldots \alpha_m, \boldsymbol{D}')\ d\boldsymbol{D}' \tag{2.2}$$

where $\boldsymbol{D}$ is the electric displacement. In Eq. (2.1), the first term is the Coulombic force density, representing the contribution of the free charge to the EHD force. The second term is the DEP force density, representing the polarization effect.

The governing equations for droplet EHD have two components. The hydrodynamic component is dictated by the Navier-Stokes equations and the interfacial stress boundary condition unique to the presence of droplets. The electric component is described by a truncated version of Maxwell's equations under the electro-quasistatic assumption. These two components are coupled through Eq. (2.1), the EHD force expression.

In this article, these droplet EHD governing equations are solved by the multi-physics simulation software CoventorWare® (Coventor, Inc., Cambridge, Massachusetts, USA); and specifically, FLOW-3D® (Flow Science Inc., Santa Fe, New Mexico, USA), which is embedded in CoventorWare as its multiphase-flow hydrodynamics solution engine.

3. VALIDATION STUDY

3.1 Pellat's Experiment

Pellat's classical experiment[6] demonstrates the phenomenon using two vertical, parallel-plate electrodes dipped into a pool of dielectric fluid having permittivity ε. Upon application of an electric field $\boldsymbol{E}$, the liquid rises to a relative height H against the gravitational acceleration $\boldsymbol{g}$. For a liquid having mass density ρ that is large compared with that of the ambient gas, $H = (\varepsilon-\varepsilon_0) E^2/2\rho g$, where ε_0 is the permittivity of the vacuum.

Bond number $Bo = \Delta\rho g W^2/\gamma$ quantifies the ratio of the gravitational impact and that of surface tension force, where $\Delta\rho$ is the density difference of the liquid and air, γ is the surface tension coefficient, and W is the gap dimension (Fig. 2-2(a)). A typical macroscopic gap dimension W results a fairly large Bond number. In other words, the surface tension force is insignificant in this problem. Thus, according to dimensional analysis, one obtains

$$H^{nd} = func\ (\varepsilon^{nd}, U^{nd}), \tag{2.3}$$

where H^{nd}, ε^{nd} and U^{nd} are non-dimensionalized liquid height (H), permittivity (ε) and liquid velocity (U), respectively. The non-dimensional parameters for length, permittivity and fluid velocity are ε_0, W and U_{ref}, where

$$U_{ref} = \sqrt{\Delta\rho g W^3 / \varepsilon_0}\ . \tag{2.4}$$

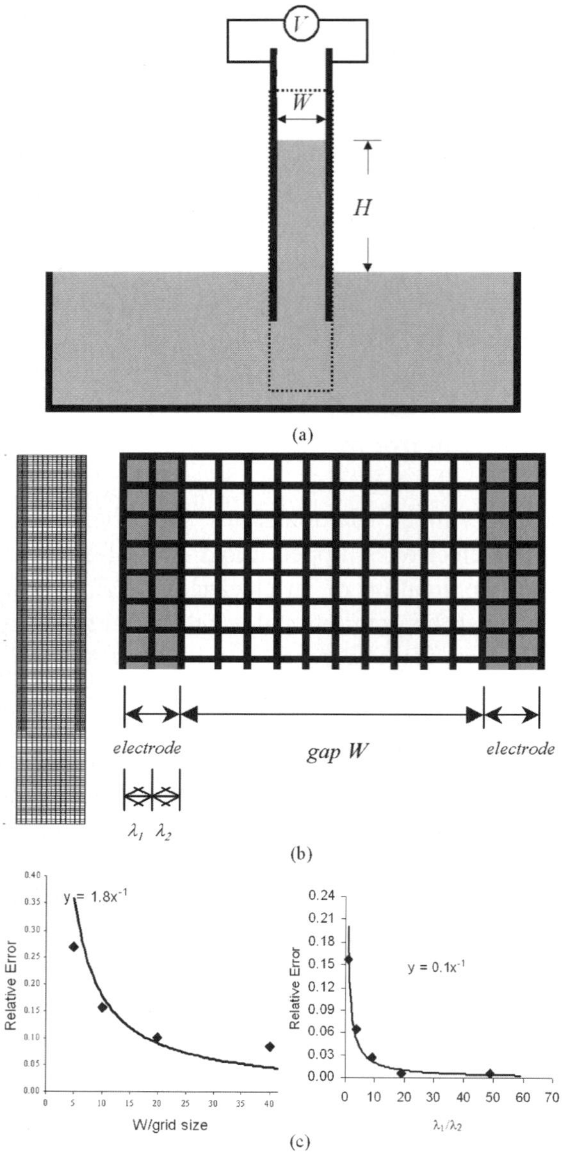

Figure 2-2. Simulation validation against the Pellat's experiment. (a) Illustration of the apparatus. The computational domain is within the dashed lines. (b) Numerical grids used in simulation. Left shows the uniform grids with cell size equal to *0.1W*. Portions of the grids are zoomed up shown at right. The electrode is covered by two computational cells. (c) Grid convergence study. The *y*-axis records the relative error of liquid column height *H* obtained from simulations against theoretical prediction. Left show numerical results obtained via uniform grids. The *x*-axis plots reciprocals of non-dimensional grid size. Without increasing the number of numerical cells used, rather biasing the electrode cells close to the fluids, a much higher numerical accuracy can be achieved, shown at right.

Equation (2.3) indicates that a value pair (ε^{nd}, U^{nd}) uniquely determines one Pellat's experiment. Numerical experiments presented below are with respect to one instance of Pellat's experiment with the value pair (*80, 0.3*).

Simulation results presented in Fig. 2-2(c) shows the grid convergence study. The left figure shows a set of simulations where uniform grids of different resolutions are used. This figure clearly indicates that the numerical solution does converge to the theoretical prediction however the relative error remains noticeable.

One source of the numerical error is the coarse grids used to discretize the electrodes. The voltage of the electrode is applied to the center of the computational cell therefore the effective gap size in the calculation is in fact ($W + \lambda_2$), where λ_2 is the size of the electrode cell right next to the fluids (Fig. 2-2(b)). For instance, for case that the grid size is equal to $0.1W$, as shown in Fig. 2-2(b), the effective gap size is $1.1W$. This weakens the electrical field by *9%*. This alone will contribute to the relative numerical error of the height of the liquid column by *17%*.

A second set of simulations is carried out to test the effect of the discretization of the electrodes. The grids shown in Fig. 2-2(b) are used as the base case: each electrode is discretized by two cells with non-dimensional size λ_1 and λ_2, and there are ten uniform grids covering the gap W. Without increasing total numerical cells used in simulation, we simply modify λ_2 ($\lambda_1 = 0.2 - \lambda_2$) and record the simulation accuracy. The simulation results are shown at right of Fig. 2-2(c). This set of simulations show that high numerical accuracy can be achieved by tightening up the size of the numerical cell in the electrode next to the fluids. Noticeably, optimal bias is achieved when λ_2 is between *0.01* and *0.02*, which delivers relative error of from *2%* to a fraction of *1%*.

This numerical exercise, first of all, has successfully demonstrated the correctness and the accuracy of the simulation code; furthermore, it has come up a guideline for discretization of computational domain, more specifically, the non-dimensional grid size and biasing that can deliver high fidelity simulation solutions.

3.2 Melcher-Taylor Experiment

The Melcher-Taylor experiment[3] illustrates the presence of an electric field excites a cellular convective liquid flow. The device is shown in Fig. 2-3(a). A shallow, slightly conducting liquid fills an insulating container to depth *b*. An electrode (at left) extends over the interface, is canted, and reaches a height *a* at the extreme right. The length *l* is much larger than *a* and *b*. A voltage difference of *V* is applied between these two electrodes.

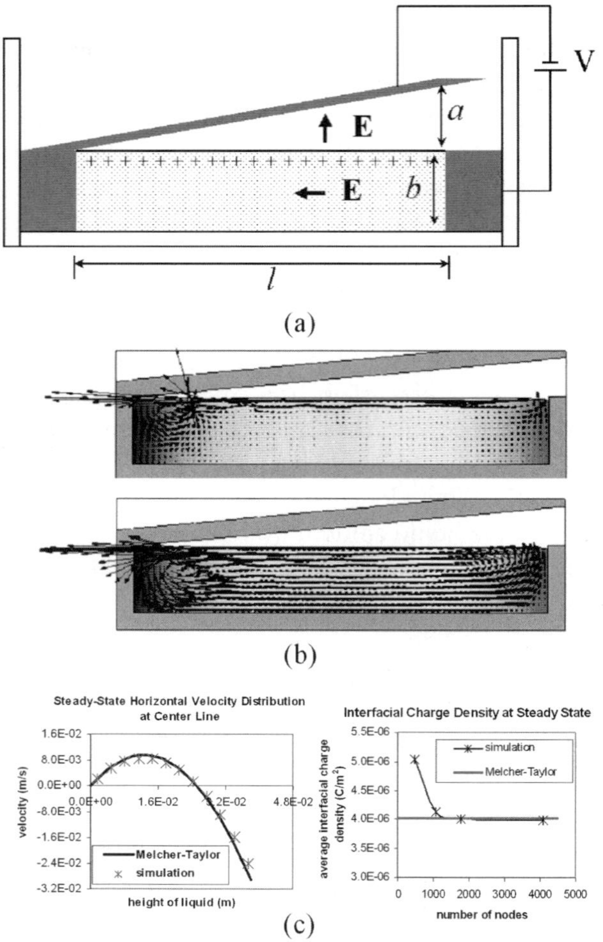

Figure 2-3. Simulation validation against the Melcher-Taylor experiment. (a) Illustration of the apparatus. The induced interfacial charge acts in concert with the electric field resulting in a counterclockwise cellular convection. (b) Numerical simulation of the development of the cellular convective liquid flow. The interfacial shear flow generated by EHD force induces a cellular convective flow in the liquid bulk. The gray scale shows electric potential; the vector plots the liquid velocity. (Above) Interfacial shear flow (time is at 0.20τ). (Below) Developed cellular convective flow in the bulk (time is at 5.2τ). τ is the time unit. (c) Quantitative validation of simulation against Melcher-Taylor theory. Left shows the steady-state horizontal velocity distribution along the centerline. 1794 nodes are used in the simulation. Right shows the steady-state average interfacial charge density plotted versus mesh resolution. The dashed line is the Melcher-Taylor steady-state solution.

The liquid is leaky dielectric, in other words, it possesses both finite electric conductivity κ and finite permittivity ε. Consequently the interfacial boundary conditions include an interfacial charge accumulation equation. The presence of the interfacial charge sustains the discontinuity of the electric field at the interface.[7]

The transient process of interfacial charge accumulation, electric field development, and the excitation of the liquid flow have been simulated,[7] shown in Fig. 2-3. The right figure in Fig. 2-3(c) shows a mesh convergence study: the average interfacial surface charge density at steady state as a function of mesh resolution. A good convergence behavior is observed. The left figure in Fig. 2-3(c) shows an excellent agreement of the simulation results with the Melcher-Taylor analytical solution. Fig. 2-3(b) shows how the flow that was originally present only at the interface propagates into the bulk and results a cellular convective flow.

3.3 Formation of Taylor Cone

The electric shear stress at the interface can be used to create a liquid jet. As illustrated in Fig. 2-4, a constant electric potential difference V is maintained between the conducting cylindrical needle and the metal plate, which is separated by a distance L. The metal plate is beyond the top boundary of the images therefore not shown. A semi-insulating liquid flows through the needle.

The electric shear stress augmented by a small hydrodynamic pressure overhead in liquid creates a Taylor cone, which is a capillary jet with a cone-shaped base narrowing down to a fine liquid filament of dimension much smaller than that of the needle nozzle. The transient process of the development of a Taylor cone has been simulated.[8] Fig. 2-4 presents the comparison between the simulation and experiment results. Fig. 2-4(a) shows a fully developed Taylor cone. Fig. 2-4(b) shows the snapshots of the transient process of Taylor cone formation. The images obtained from simulations and experiments are qualitatively in agreement.

4. ELECTROWETTING ON DIELECTRIC

EWOD is one of the two operating principles for electrically controlled digital microfluidics. Experimental and theoretical research on EWOD has been carried out extensively.[9-16] EWOD connotes a configuration where a thin layer of insulating solid material is inserted in between a droplet and an

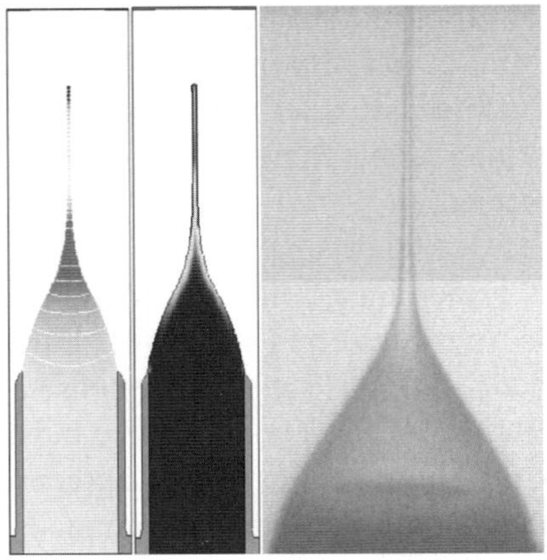

(a)

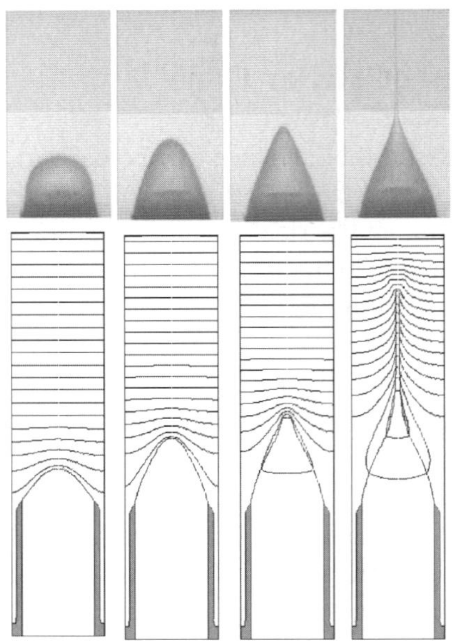

(b)

Figure 2-4. Simulation validation against the experiment of Taylor cone formation. (a) Fully developed Taylor cone. At right is a micrograph from experiment. Left two images are obtained from simulation, gray scale in the first image indicates the electric potential; gray scale in the second image indicates the interfacial charge density. (b) Transient sequence of the Taylor cone formation. Experimental images are shown at the top row. Simulation results are shown at the bottom row. Their correspondence is indicated by the vertical alignment.

electrode. Upon application of an electric field, free charges will be present at the interface between the droplet and the solid, which gives rise of an electric force. This electric force acts on the tri-phase contact line and causes the contact angle reduction that is usually observed in experiments. A programmed electric field can create a strength disparity of this wetting force around the tri-phase contact line to make droplets move.

EWOD force acts on the tri-phase contact line and results the contact angle reduction,

$$\Delta\theta = \theta_0 - acos(cos\theta_0 + f^{EWOD}) \tag{2.5}$$

where θ_0 is the contact angle measured when the electric field is absent, and $\Delta\theta$ is the magnitude of the contact angle reduction due to EWOD. f^{EWOD} is the EWOD force, an EHD line force density acting on the tri-phase contact rim originating from this EWOD configuration. Since $\Delta\theta$ is measurable experimentally, it is commonly used to describe the effect of EWOD and the strength of f^{EWOD}.

EWOD chip relies on creating contact angle disparity along the tri-phase contact rim to manipulate droplets. The magnitude of the contact angle disparity, thus the powerfulness of the EWOD chip, is constrained by the occurrence of the contact angle saturation.

When the applied voltage V is smaller than certain ceiling value V_c, the contact angle decreases with the increase of V. The experimental measurements of $\theta(V)$, the contact angle θ as function of the applied voltage V, is conformable to the droplet EHD prediction (equation 30 of [2]). However, when V exceeds V_c, θ abruptly ceases from further decreasing and stays at $\theta(V_c)$, deviating from the theoretical prediction. This is called the contact angle saturation. The experimental measurement of $\theta(V)$ in figure 4 of [17] shows the abrupt occurrence of the contact angle saturation once V exceeds the ceiling voltage V_c.

The physical origin of the contact angle saturation is under active debate[17,18] and is not understood at this moment. Therefore the modeling practice has to be somewhat empirical. One modeling approach accounting for the contact angle saturation is to extract the saturation point from experiments. The EWOD is governed by droplet EHD when the saturation point is not reached; the contact angle reduction halts when the operating condition reaches or is beyond the saturation point.

In addition, EWOD chip requires droplets being in direct contact with the surface of the solid substrate, the reaction surface. The smoothness of the reaction surface affects the occurrence and the speed of the droplet translocation hence the performance of the EWOD chip. In order to incorporate this effect into EWOD chip simulation, two extreme cases are implemented, that is, the no-slip condition modeling a very rough reaction surface that the liquid velocity at the reaction surface is set to zero, and the free-slip condition reflecting a perfect smooth surface that no tangential stresses are present at the droplet-surface contact. An additional weight

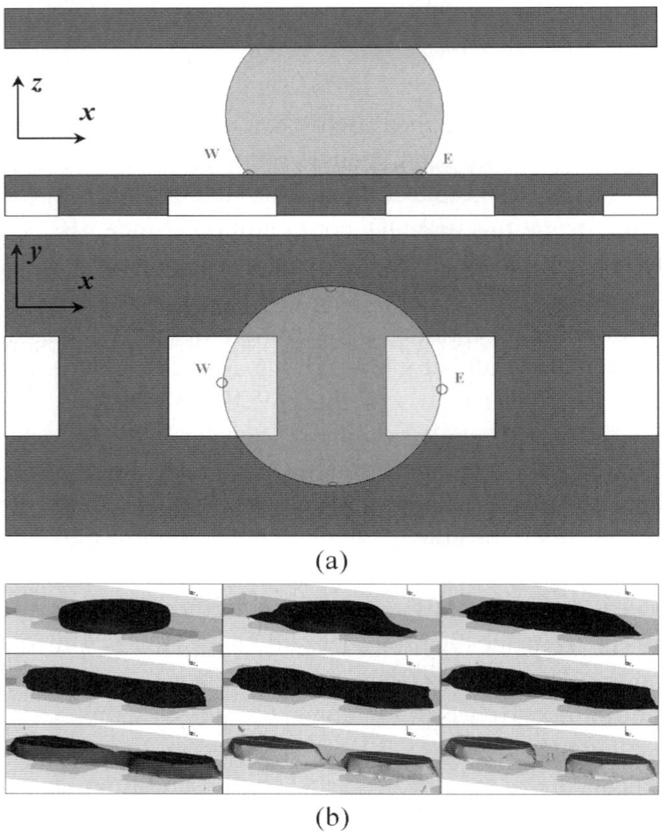

Figure 2-5. Droplet fission on an EWOD-driven lab-on-a-chip. (a) Device configuration. All four electrodes embedded in the insulating material are *ON* electrodes, *100μm* wide and *100μm* apart. The thickness of the insulating coating is *5μm*. (b) Simulation solution of the transient sequence of the droplet fission process. The snapshots are at a *75μs* time interval. Initially (without the presence of the electric field), this water-based droplet of *1μL* is of a "pancake" shape maintaining a contact angle of *117°*. Upon application of *70V* to all four electrodes, the reduction of the contact angle elongates the droplet in the *x* direction, shrinking the *yz*-plane cross-section at the center of the droplet, which eventually breaks the droplet into two parts. (Satellite droplets can also be observed.)

parameter is created to simulate the contact between the droplet and the reaction surface that is of partial-slip, *i.e.*, in between the no-slip condition and free-slip condition.[19] This weight parameter is empirical and requires calibration with experiments. Simulations presented in this article assume free-slip surface condition.

An individually addressable electrode array can be used to shape an electric field surrounding a droplet hence create a spatial variation of $\boldsymbol{f}^{\mathrm{EWOD}}$ to accomplish droplet generation, translocation, fission, and fusion. Fig. 2-5 shows a simulation of a transient process of droplet fission, that is, one droplet is cut into two smaller ones by EWOD. As shown in Fig. 2-5(a), the electrodes are aligned along the x direction, and a droplet initially is centered in between two neighboring electrodes. Upon application of a voltage to all the electrodes, a spatial disparity of $\boldsymbol{f}^{\mathrm{EWOD}}$ is created. It may be observed that the contact angle at the tri-phase contact point closer to the electrodes (the vicinity of points *W* and *E*) is smaller than that at the tri-phase contact point further from the electrodes (the vicinity of points *N* and *S*). Consequently, as shown in Fig. 2-5(b), the droplet is elongated in the x direction at both sides (along *W-E* plane), and simultaneously the *y-z* cross-section at the center of the droplet (on *N-S* plane) is reduced. Eventually the cross-section in the *N-S* plane reduces to a point and two droplets are created to conclude the fission process.

5. DIELECTROPHORESIS

When a neutral particle (droplets are fluid particles) is suspended in a dielectric fluid medium and is exposed to an electric field, this particle is subject to a DEP force.[20] Such force can be applied to manipulate and discriminate particles and has inspired DEP-based electrically controllable trapping, focusing, translation,[21] fractionation[22] and characterization of particulate mineral, chemical and biological analytes within fluid suspending media.[23] Of particular interest are fractionation,[24-26] sorting,[27,28] trapping/positioning[29-32] and characterization[33-36] of biological cells. Digital microfluidics based on manipulation of droplets by DEP has been developed.[37]

It has been a common theoretical practice to approximate the DEP force on the particle as a lumped function of the surrounding electric field $\boldsymbol{E}$ [38-40]

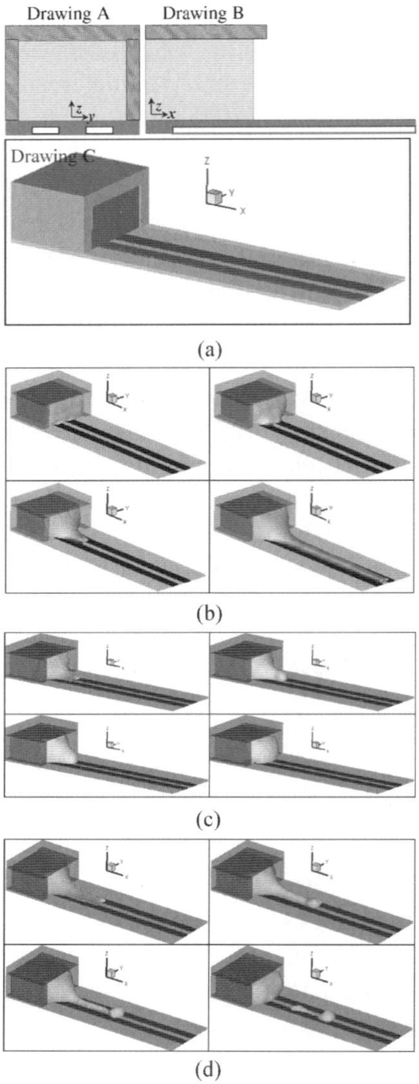

Figure 2-6. Droplet generation on a DEP-driven lab-on-a-chip. (a) Injector configuration. The injector nozzle, $60\mu m$ in height and $120\mu m$ in width, is connected with the chemical sample reservoir. Two parallel and co-planar electrodes, $20\mu m$ wide and $20\mu m$ apart, are embedded in an insulating layer. The thickness of the insulating coating is $2\mu m$. The sample liquid is water based. The contact angle at the tri-phase contact line is 117° in the absence of the electric field. A high frequency AC voltage $400V$ (rms) is applied between the two electrodes at time $t=0+$. (b) Simulation of a DEP finger formation. (c) Simulation of a failed attempt of droplet generation. The voltage difference between the two electrodes is reset to zero at time $t=26\mu sec$. The images are in sequence corresponding to time starting at $t=26\mu sec$ with a $16\mu sec$ interval. (d) Simulation of droplet-on-demand. The voltage difference between the two electrodes is reset to zero at time $t=50\mu sec$. The images are in sequence corresponding to time starting at $t=50\mu sec$ with a $20\mu sec$ interval.

$$\vec{F} = 2\pi\varepsilon_m r^3 \,\mathrm{Re}[f_{CM}]\nabla\vec{E}^2_{rms} \tag{2.6}$$

where r is the radius of the particle, subscript *rms* stands for root-mean-square, ε stands for material permittivity, subscript m denotes surrounding medium. f_{CM} is the Clausius-Mossotti factor,

$$f_{CM} = \frac{\varepsilon_p^* - \varepsilon_m^*}{\varepsilon_p^* + 2\varepsilon_m^*} \tag{2.7}$$

where $\varepsilon^* = \varepsilon - j(\sigma/\omega)$, j is the imaginary unity, σ is the conductivity, ω is the angular frequency of the field; and the subscript p stands for properties of the particle. With this lumped DEP force expression, much of the additional effort was devoted to analyzing the electrostatic field utilizing Green's theorem[41-43] or numerical means such as the Finite Element Method.[44-46]

However, such a lumped DEP force expression assumes the suspended particle always be spherical, sufficiently far away from electrodes such that its presence has little impact on "far field" calculations, and the flow of suspending liquid around the particle and any possible circulating flow inside the particle induced by DEP is negligible. Such assumptions are not valid for DEP driven digital microfluidics. The droplets used in DEP chip are comparable to the electrode pad in size, and are placed close to the electrode array, that is, the presence (deformation and translocation) of the droplets are expected to affect the electric field. Furthermore, the droplets used in DEP chip are not hard particles: a circulating flow may be generated inside the droplet due to the electrical tangential stress at the interface; the droplet may also deform from spherical shape due to the non-uniform normal stress imbalance over the spherical droplet surface; both fluids are subject to translational forces due to field non-uniformity. To faithfully analyze the dynamic behavior of droplets on a DEP driven digital microfluidics chip, one has to rely on a detailed simulation modeling approach based on droplet EHD.[2]

Figure 2-6 shows the application of the droplet EHD simulation to the droplet formation process on a DEP driven digital microfluidics chip. Fig. 2-6(a) shows the device configuration. The injector nozzle, an insulating material, is connected with the liquid reservoir. The substrate is also an insulator. Two identical, co-planar, and rectangular electrodes are embedded inside the substrate. Their length is much larger than their breadth and thickness. The electrodes are placed symmetrically with respect to the injector nozzle opening. When the electric field is absent, the tri-phase contact is hydrophobic with a contact angle θ_0. A small hydrodynamic pressure overhead is applied to the reservoir, working against the interfacial tension force, such that the liquid interface stops at the opening of the injector nozzle. Upon application of an AC voltage difference between the two electrodes, liquid is drawn out of the injector nozzle to form a liquid column on the substrate. This liquid column is referred to as a DEP finger.[47] A simulation of the transient process of DEP finger formation is shown in Fig. 2-6(b). The DEP finger is drawn along the x direction on top of the two electrodes. The DEP forces suppress hydrodynamic instability so that a very long slim finger can be formed and sustained without a break-up of the liquid column. Fig. 2-6(b) shows that a finger $420\mu m$ long is obtained at time $t=125\mu sec$.

Removing the DEP forces by resetting the voltage difference between the two electrodes to zero will trigger the hydrodynamic instability[48] thus can be used as mechanism to release individual droplets. The timing of turning off the electric field is one of the most critical design parameters for successful DEP-driven droplet generation, and simulations can be used to narrow down the design space and even pinpoint an optimal design. Fig. 2-6(c) shows a simulation where the electric voltage is reset to zero at $t=26\mu sec$. Upon removal of the electric field, the DEP finger starts to retreat and eventually is pulled back completely inside the injector nozzle and no discrete droplet is released. Fig. 2-6(d) shows a simulation where the electric field is removed at time $t=50\mu sec$. It is observed that the liquid column forms a droplet-like head with a thin tail (the third image) and eventually a single droplet is released along with satellite droplets from the break-up of the tail (the forth image). This indicates, this injector design and its operating condition can potentially achieve drop-on-demand.

6. OUTLOOK: VIRTUAL PROTOTYPING OF DIGITAL MICROFLUIDICS

A simulation example of a two-stage chemical reaction on a droplet-based DEP-driven lab-on-a-chip is presented here to illustrate the feasibility of virtual prototyping of digital microfluidics, and to conclude this article.

Figure 2-7(a) illustrates the set-up. An electrode array is coated with an insulating material, on the top of which initially are three droplets of uniform size containing different chemical compounds. Electrodes *E1*, *E2*, … *E6* can be addressed individually, switching between two voltage states *OFF* and *ON*. At time $t=0+$, *E4* is turned *ON*. At time $t=\tau$, *E4* is turned *OFF*. At $t=2\tau$,

E3 is turned *ON*. Such a sequence will cause a desired droplet translocation and droplet fusion. A chemical reaction is described by

$$\frac{\partial C_i}{\partial t} = \sum_{j=1}^{n} \alpha_j^{(i)} \prod_{k=1}^{n} C_k^{\gamma_k^{(j,i)}} \tag{2.8}$$

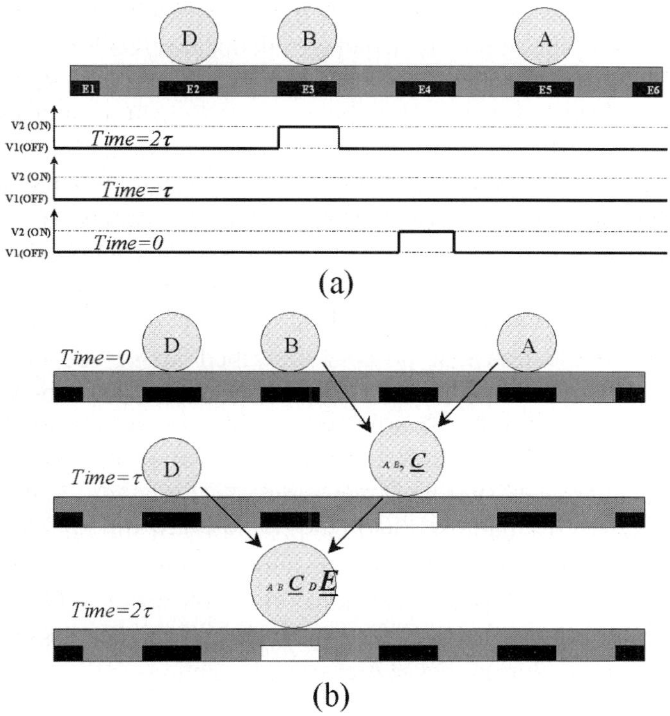

Figure 2-7. Two-stage chemical reaction on a droplet-based DEP-driven lab-on-a-chip. (a) Configuration. An array of electrodes labeled as *E1*, *E2*, ... is embedded in an insulator. Initially three equal-sized droplets are placed on the top of electrodes *E2*, *E3* and *E5* containing chemical compounds *A*, *B* and *C*. The electrodes of the array can be addressed individually, switching between two voltage states *V1* (*OFF*) and *V2* (*ON*). (b) Flow of operation. The *ON* electrodes are colored in white, and the *OFF* electrodes are colored in black. The droplets are labeled by their chemical contents. Chemicals produced by reactions are underlined. The arrows indicate the direction of the droplet fusion process. At time $t=0+$, Electrode *E4 ON* induces fusion of droplets *A* and *B* and triggers the reaction between *A* and *B* producing chemical *C*. At time $t=2\tau$, *E3 ON* induces fusion of droplets *D* and *ABC* producing chemical *E*.

where C_i stands for the volumetric concentration of chemical compound *i*, *i=1,2,…,n.*, and $\alpha_j^{(i)}$ and $\gamma_k^{(j,i)}$ are constants that represent intrinsic properties of a chemical reaction. In this simulation, five different chemical compounds are in play, namely *A*, *B*, *C*, *D* and *E*, with volumetric concentration C_1, C_2, C_3, C_4 and C_5, respectively. Initially only chemicals *A*, *B* and *D* exist. A two-stage chemical reaction is defined according to Eq. (2.8): (1) C_1, C_2 and C_3 are interdependent; and (2) C_3, C_4 and C_5 are interdependent. That is, chemicals *A* and *B* react with each other to produce chemical *C*, then chemicals *C* and *D* react to produce chemical *E*. The existence of chemical *E* indicates the completion of this two-stage chemical reaction.

Figure 2-7(b) illustrates the lab-on-a-chip operation sequence. The droplets are labeled by their chemical content. From time $t=0+$ to τ, DEP fusion of droplets *A* and *B* occur at *E4*. The internal circulation inside droplet *AB* promotes the first stage of the chemical reaction producing *C*. *E4* is turned *OFF* once the fusion of droplets *A* and *B* is accomplished. At time $t=2\tau$, *E3* is turned *ON* and fusion of droplets *ABC* and *D* occurs at *E3*. The mixing of chemicals inside droplet *ABCD* enables the reaction between chemicals *C* and *D* to produce chemical *E*.

Snapshots of the transient process described above are obtained from simulation and shown in Fig. 2-8. Fig. 2-8(a) shows the two-stage droplet fusion. The gray scale in Fig. 2-8(b) indicates the concentration of individual chemical compounds. The dark color in the last snapshot of Fig. 2-8(b) indicates that chemical E has been generated.

This simulation has demonstrated the possibility of full degree of freedom control over droplets on the two-dimensional reaction surface by using an individually addressable two-dimensional electrode array. When droplets are used as carriers of biochemical agents, such control enables programmed chemical reactions, that is, desired chemical reactions will occur at desired sites at desired times – the essence of digital microfluidics. This simulation has also demonstrated the feasibility and powerfulness of the simulation based virtual prototyping for digital microfluidics design.

ACKNOWLEDGEMENTS

This work was supported in part by The Defense Advanced Research Projects Agency (DARPA) under contract DAAD10-00-1-0515 from the Army Research Office to the University of Texas M. D. Anderson Cancer Center.

Professor Peter R. C. Gascoyne of M. D. Anderson Cancer Center provided drawings shown in Fig. 2-1. Dr. Daniel Sobek of Agilent Technologies Inc. provided the experimental images shown in Fig. 2-4.

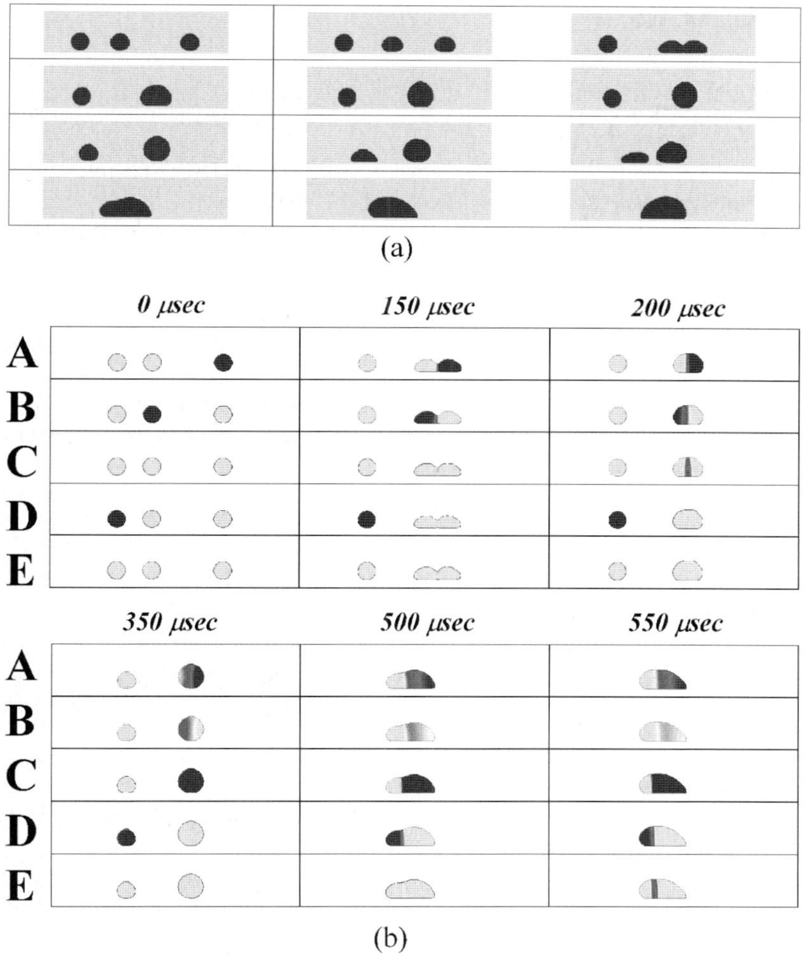

Figure 2-8. Two-stage chemical reaction on a droplet-based DEP-driven lab-on-a-chip. The electrodes are *30μm* wide and *30μm* apart. Initially three equal-sized droplets of *14nL* are placed on the top of electrodes *E2*, *E3* and *E5*. Time unit τ is equal to *150μsec*. (a) Simulation of two-stage droplet fusion. Snapshots are at *50μsec* time interval. (b) Simulation of chemical diffusion and reaction. The gray scale indicates the volumetric density of concentration of chemicals *A*, *B*, *C*, *D* and *E*, respectively.

This chapter was based on an invited presentation at SPIE Optics East 2004 Symposium, Philadelphia, PA, 25–28 October 2004, published in SPIE Proceedings of Lab-on-a-Chip: Platforms, Devices, and Applications, Volume 5591, 125–142.

REFERENCES

1. http://www.tutorgig.com/ed/Digital_microfluidics.
2. J. Zeng and F. Korsmeyer, Principles of droplet electrohydrodynamics for lab-on-a-chip, *Lab. Chip.*, **4**, 265–277 (2004).
3. J. R. Melcher and G. I. Taylor, Electrohydrodynamics: a review of the role of interfacial shear stresses, *Annu. Rev. of Fluid Mech.*, **1**, 111–146 (1969).
4. D. A. Saville, Electrohydrodynamics: The Taylor-Melcher leaky-dielectric model, *Annu. Rev. Fluid Mech.*, **29**, 27-64 (1997).
5. J. R. Melcher, *Continuum Electromechanics*, Section 3.7 (The MIT Press, 1981).
6. H. Pellat and C. R. Seances, *Acad. Sci., Paris*, 119, 675 (1894), see: T. B. Jones and J. R. Melcher, Dynamics of electromechanical flow structures, *Physics of Fluids*, **16**(3), 393-400 (1973).
7. J. Zeng, D. Sobek and F. T. Korsmeyer, Electro-hydrodynamic modeling of electrospray ionization: CAD for a μFluidic device – mass spectrometer interface, *Transduers'03 Digest of Technical Papers*, 1275-1278 (2003).
8. D. Sobek, J. Cai, H. Yin and J. Zeng, Fundamental study of Taylor cone dynamics of nano-electrosprays, *52nd American Society for Mass Spectrometry Conference* (Nashville, TN, May 23-27, 2004).
9. J. L. Jackel, S. Hackwood and G. Beni, Electrowetting optical switch, *Appl. Phys. Lett.*, **40**(1), 4-5 (1982).
10. H. J. J. Verheijen and M. W. J. Prins, "Contact angles and wetting velocity measured electrically", *Review of Scientific Instruments*, **70**(9), 3668-3673 (1999).
11. R. Digilov, Charge-induced modification of contact angle: the secondary electrocapillary effect, *Langmuir*, **16**, 6719-6723 (2000).
12. C. Quilliet and B. Berge, Electrowetting: a recent outbreak, *Current Opinion in Colloid & Interface Science*, **6**, 34 (2001).
13. M. G. Pollack, A. D. Shenderov, R. B. Fair, Electrowetting-based actuation of droplets for integrated microfluidics, *Lab. Chip.*, **2**, 96-101 (2002).
14. S. K. Cho, H. Moon and C-J Kim, Creating, transporting, cutting, and merging liquid droplets by electrowetting-based actuation for digital microfluidic circuits, *Journal of microelectromechanical systems*, **12**(1), 70-80 (2003).
15. K.-L. Wang and T. B. Jones, Electrowetting dynamics of microfluidic actuation, *Langmuir*, **21**, 4211-4217 (2005).
16. T. B. Jones, On the relationship of dielectrophoresis and electrowetting, *Langmuir*, **18**, 4437-4443 (2002).
17. H. J. J. Verheijen and M. W. J. Prins, Reversible electrowetting and trapping of charge: model and experiments, *Langmuir*, **15**, 6616-6620 (1999).
18. B. Shapiro, H. Moon, R. L. Garrell, C-J. Kim, Equilibrium Behavior of Sessile Drops under Surface Tension, Applied External Fields, and Material Variations, *Journal of Applied Physics*, **93**(9), 5794-5811 (2003).
19. A. B. Basset, *A Treatise on Hydrodynamics* (Cambridge University Press, 1888).
20. H. A. Pohl, *Dielectrophoreisis: The behavior of neutral matter in nonuniform electric fields*, (Cambridge University Press, Cambridge, 1978).
21. A. Desai, S. W. Lee and Y. C. Tai, A MEMS electrostatic particle transportation system, *MEMS 1998* (1998).
22. X.-B. Wang, J. Vykoukal, F. F. Becker and P R. C. Gascoyne P. R. C., Separation of polystyrene microbeads using dielectrophoretic/gravitational field-flow-fractionation, *Biophysical J.*, **74**, 289-2701 (1998).

23. P. R. C. Gascoyne and J. Vykoukal, Particle separation by dielectrophoresis, *Electrophoresis*, **23**, 1973-1983 (2002).
24. J. Yang, Y. Huang, X.-B. Wang, F. F. Becker and P. R. C. Gascoyne, Differential analysis of human Leukocytes by dielectrophoretic field-flow-fractionation, *Biophysical J.*, **78**, 2680-2689 (2000).
25. X.-B. Wang, J. Yang, Y. Huang, J. Vykoukal, F. F. Becker and P. R. C. Gascoyne, Cell separation by dielectrophoretic field-flow-fractionation, *Analytical Chemistry*, **72**(4), 832-839 (2000).
26. J. Xu, L. Wu, M. Huang, W. Yang, J. Cheng, X.-B. Wang, Dielectrophoretic separation and transportation of cells and bioparticles on microfabricated chips, *Micro Total Analysis Systems 2001* (2001).
27. F. F. Becker, X.-B. Wang, Y. Huang, R. Pethig, J. Vykoukal and P. R. C. Gascoyne, Removal of human leukaemia cells from blood using interdigitated microelectrodes, *J. Phys. D: Appl. Phys.*, **27**, 2659-2662 (1994).
28. P. R. C. Gascoyne, X.-B. Wang, Y. Huang and F. F. Becker, Dielectrophoretic separation of cancer cells from blood, *IEEE Transactions on Industry Applications*, **33**(3), 670-678 (1997).
29. J. Suehiro and R. Pethig, The dielectrophoretic movement and positioning of a biological cell using a three-dimensinonal grid electrode system, *J. Phys. D: Appl. Phys.*, **31**, 3298-3305 (1998).
30. P. R. C. Gascoyne, *Physiology, Pathobiology, Technology, and Clinical Applications*, E. P. Diamandis, editor 499-502 (AACC Press, New York, 2002).
31. J. Voldman, R. A. Braff, M. Toner, M. L. Gray and M. A. Schmidt, Holding forces of single-particle dielectrophoretic traps, *Biophysical J.*, 80, 531-541 (2001).
32. T. Heida, W. L. C. Rutten and E. Marani, Dielectrophoretic trapping of dissociated fetal cortical rat neurons, *IEEE Trans. Biomed. Eng.*, **48**, 921-30 (2001).
33. X.-B. Wang, Y. Huang, P. R. C. Gascoyne and F. F. Becker, Dielectrophoretic manipulation of particles, *IEEE Transactions on Industry Applications*, **33**(3), 660-669 (1997).
34. Y. Huang, X.-B. Wang, R. Holzel, F. F. Becker and P. R. C. Gascoyne, Electrorotational studies of the cytoplasmic dielectric properties of Friend murine erythroleukaemia cells, *Phys. Med. Biol.*, **40**, 1789-1806 (1995).
35. J. Yang, Y. Huang, X. Wang, X.-B. Wang, F. F. Becker and P. R. C. Gascoyne, Dielectric properties of human Leukocyte subpopulations determined by electrorotation as a cell separation criterion, *Biophysical J.*, **76**, 3307-3314 (1999).
36. P. R. C. Gascoyne, J. Noshari, F. F. Becker and R. Pethig, Use of dielectrophoretic collection spectra for characterizing differences between normal ansd cancerous cells, *IEEE Transactions on Industry Applications*, **30**(4), 829-834 (1994).
37. J. Vykoukal, J. Schwartz, F. F. Becker and P. R. C. Gascoyne, A programmable dielectric fluid processor for droplet-based chemistry, *Micro Total Analysis Systems 2001*, 72-74 (2001).
38. T. B. Jones and G. W. Bliss, Bubble dielectrophoresis, *Journal of Applied Physics*, **48**(4), 1412-1417 (1977).
39. L. Benguigui and I. J. Lin, The dielectrophoresis force, *Am. J. Phys.*, **54**(5), 447-450 (1986).
40. X.-B. Wang, Y. Huang, F. F. Becker and P. R. C. Gascoyne, A unified theory of dielectrophoresis and traveling wave dielectrophoresis, *J. Phys. D: Appl. Phys.*, **27**, 1571-1574 (1994).

41. X. Wang, X.-B. Wang, F. F. Becker and P. R. C. Gascoyne, A theoretical method of electrical field analysis for dielectrophoretic electrode arrays using Green's theorem, *J. Phys. D: Appl. Phys.,* **29**,1649-1660 (1996).
42. D. S. Clague and E. K. Wheeler, Dielectrophoretic manipulation of macromolecules: The electric field, *Physical Review E.*, **64**, 26605/1-26605/8 (2001).
43. M. Washizu and T. B. Jones, Generalized multipolar dielectrophoretic force and electrorotational torque calculation, *J. of Electrostatics*, **38**, 199-211 (1996).
44. N. G. Green, A. Ramos and H. Morgan, Numerical solution of the dielectrophoretic and traveling wave forces for interdigitated electrode arrays using the finite element method, *J. Electrostatics*, **56**, 235-254 (2002).
45. T. J. Snyder, J. B. Schneider and J. N. Chung, Dielectrophoresis with application to boiling heat transfer in microgravity. I. Numerical analysis, *J. of Applied Physics*, **89**(7), 4076-4083 (2001).
46. T. Heida, W. L. C. Rutten and E. Marani, Understanding dielectrophoretic trapping of neuronal cells: modeling electric field, electrode-liquid interface and fluid flow, *J. Phys. D: Appl. Phys.,* **35**, 1592-1602 (2002).
47. T. B. Jones, M. Gunji, M. Washizu and M. J. Feldman, Dielectrophoretic liquid actuation and nanodroplet formation, *Journal of Applied Physics*, **89**, 1441-1448 (2001).
48. P. G. Drazin and W. H. Reid, *Hydrodynamic stability* (Cambridge University Press, 1981).

Chapter 3

MODELING, SIMULATION AND OPTIMIZATION OF ELECTROWETTING*

Jan Lienemann, Andreas Greiner, and Jan G. Korvink
Lab of Simulation, Department of Microsystems Engineering (IMTEK)
University of Freiburg, Georges-Köhler-Allee 103, 79110 Freiburg, Germany
lieneman@imtek.de

Abstract: Electrowetting is an elegant method to realize the motion, dispensing, splitting and mixing of single droplets in a microfluidic system without the need for any mechanical – and fault-prone – components. By only applying an electric voltage, the interfacial energy of the fluid/solid interface is altered and the contact line of the droplet is changed. However, since the droplet shape is usually heavily distorted, it is difficult to estimate the droplet shape during the process. Further, it is often necessary to know if a process, e.g., droplet splitting on a given geometry, is possible at all, and what can be done to increase the system's reliability. It is thus important to use computer simulations to gain understanding about the behavior of a droplet for a given electrode geometry and voltage curve. Special care must be exercised when considering surface tension effects. We present computer simulations done with the *Surface Evolver* program and a template library combined with a graphical user interface which facilitates standard tasks in the simulation of electrowetting arrays.

Keywords: Electrowetting arrays, biochip microfluidics, simulation, surface tension, droplet pumping

1. INTRODUCTION

Microfluidics is currently one of the fields of microsystem engineering with the largest market opportunities. Reproducible parallel batch fabrication of large numbers of low cost devices is ideal for the varied disposable devices

*This work is supported by the Commission of the European Communities under contract number G5RD-CT-2002-00744, Competitive and Sustainable Growth Program, Micrometer Scale Patterning of Protein & DNA chips, MICROPROTEIN, and by an operating grant of the University of Freiburg.

K. Chakrabarty and J. Zeng (eds.),
Design Automation Methods and Tools for Microfluidics-Based Biochips, 53–84.

dictated by contamination concerns in biology and medicine. Design of such devices will need to focus on exploiting device scaling while optimizing for reliability and lifetime. In the world of microsystems, where all dimensions are downscaled by several orders of magnitude, surface and edge effects become more important as the size shrinks. For example, a certain amount of water will form a droplet, the shape of which is barely influenced by gravity; further, the influence of electrostatic forces increases, while the effect of inertia decreases.

The displacement of fluid volume is a fundamental design issue in microfluidic devices. A variety of micropumps [Laser and Santiago, 2004] have been proposed that use movable mechanical parts like membranes for displacing fluid volumes or spotting droplets. They mostly operate with a continuous stream of fluid after being primed at the start. This paper focuses on an alternate fluid displacement mechanism: electrowetting. Electrowetting works without mechanical parts; the only moving mass is the fluid itself. It is technologically much easier as well: The manufacturing process requires only one step to pattern a metallic layer, whereas other micropumps require a number of lithographic steps and complex etching procedures.

1.1 Electrowetting Setup and Devices

The underlying idea of electrowetting is to change the wetting properties of a liquid on a substrate. By applying an electric voltage, surfaces can be switched between a wetting and non-wetting state. If the substrate is only partially wetting, the liquid seeks to cover *this* part to minimize its energy. A phase boundary between liquid and surrounding air would thus be shifted towards the wettable spot, and a fluid motion can be observed. This spatial control of wetting is accomplished by applying the voltage only on certain parts of the substrate – it is partitioned into an array of controllable spots by an assembly of electrodes.

One possible application to biochips is the switching between flow channels. Channel based biochips [Reyes et al., 2002, Auroux et al., 2002] are typically configured at design time. In contrast, the use of an electrode array that controls wettability offers the possibility for reconfiguration of the "virtual" fluid channel at runtime [Pollack et al., 2000, Ding et al., 2002]. One can imagine the device to be like the field programmable gate arrays used in microelectronics. Here, fluidic gates can allow the fluid meniscus to traverse a certain spot in the channel, inhibit the motion, or alter the fluid path [Tkaczyk et al., 2003]. It is even possible to omit preprocessed channels at all and form virtual channels by a suitable actuation of an assembly of electrodes. Since the effect mostly acts at phase boundaries, these devices usually operate with a quantized flow of single droplets instead of a continuous flow.

Figure 3-1. The four main operations of a microfluidic electrowetting array: Droplet creation (1), droplet motion (2), droplet splitting (3) and droplet merging (4).

Figure 3-1 shows an illustration of a possible electrowetting electrode array. By using the electrowetting effect, the droplets are moved from one electrode to another. Four main operations need to be possible for the technology to be useful for fluid processing:

1 **Creation:** to take a certain amount of liquid from a reservoir to form droplets of a given size.

2 **Transport:** to move the droplet along a path to or from other functional components like detectors, catalytic converters, supplies and waste outlets.

3 **Splitting:** to split a droplet into smaller parts for parallel processing.

4 **Merging/Mixing:** to merge droplets and mix their contents. This can be achieved by diffusion aided by periodic actuation.

Virtual reaction vials can be formed at a single spot of the array. Possible applications are arrayed bioassays and custom combinatorial synthesis of, e.g., deoxyribonucleic-acid (DNA) probes. But there are also other applications beyond the scope of biochips, e.g., for computer displays [Hayes and Feenstra, 2003] or adaptive lenses [Berge and Peseux, 2000], each with their own requirements.

1.2 Device Design

The shapes of droplets during this process can often become quite distorted and difficult to estimate. Computer simulations give insight in the driving forces leading to motions of a droplet. Calculated energy curves give hints to help the designer understand what happens energetically, and show optimization potential to increase the speed or reliability of the motion. They can also show if a process, like splitting, is possible at all for a given configuration, and which parameters need to be tuned to allow for a reliable operation.

This enables the designer to experiment without the need to wait for possibly expensive prototypes; simulations hence speed up development cycles and allow for a lower time to market which is crucial in such innovative fields as bio-microelectromechanical systems (BioMEMS).

Some design goals which a simulation could help to achieve are the following:

- **Fluid processing algorithms.** An actuation scheme must be found to achieve each of the four operations listed above, and its parameters must be tuned. It is important to estimate the droplet geometry to design the size of electrodes and the amplitude of electric voltages.
- **Fluid process flow.** Many single operations must be combined to form the complete process sequence. The start and end of one sequence must fit, and if parallel processing is wanted, the operations must not interfere.
- **Reliability.** It is of utmost importance to determine the reliability and the limits of a certain design, and under which circumstances the success of an operation is not sensitive to parameters which are difficult to control. The design should be made such that independence from those is achieved. Further, the influence of tolerances should be quantified.
- **External constraints.** Since the device will be connected to external equipment, other constraints might have to be considered, such as power consumption, processing time, external forces, etc.

For the electrowetting setup as treated in this paper, most design goals are strongly influenced by geometrical quantities. The droplet shape and the device setup play an important role: electrical fields and substrate geometries influence the motion of the droplet. Also the electrowetting actuation voltage curve $V(t)$ determines the result.

For the droplet transport – also called pumping – some questions that will be asked are:

- Which droplet volumes can be transported with a given electrode setup?
- How should the electrodes be shaped to allow transport with maximal speed at minimal actuation voltages for a large range of droplet volumes?
- How accurate are the processes?
- Is there a voltage limit where, e.g., complete wetting of a surface occurs?
- If the liquid is transported in channels, how would it be possible to fill a larger chamber? Are there optimal "flange" shapes?
- To understand the motion and be able to optimize, it is important to know how the potential energy distribution a droplet sees during the process is influenced by the setup.

For droplet splitting, both droplet shapes and actuation parameters are of interest; an optical engineer would be interested to extract the geometry of a droplet to determine the focal length of a droplet lens; if fluorescent markers are inside the droplet, one would be interested in knowing the thickness variation of the droplet to calibrate the light output.

1.3 Computer Simulation Aided Design

Experiments can answer many of those questions, and for the given setups are fairly easy to perform and hence quite satisfying. But there are some limitations involved. Due to the small size of some features (dielectric layer thickness, electrode fine structure), facilities for production and measurement of prototypes must be available. Especially if cost is an issue, experiments should be prepared using estimates of the results. Optimizations with a large number of evaluations might be easier narrowed down by computer simulation; otherwise, we have found it fairly costly to make quantitative and qualitative experimental comparisons when it comes to, e.g., finding optimal electrode shapes. Further, effects of changes can be estimated without interferences and contaminations in a simulation. Finally, one can also estimate the response of the system to inputs which are difficult to reliably apply in a laboratory setup, giving potential to perform thought experiments. This motivated the development of a modeling tool with which we could quickly perform what-if calculations.

However, also simulations have their limits. First, they are always based on a model. The amount of detail, the number of physical effects involved, the validity of assumptions and simplifications determine the accuracy of such a model. Material and geometrical data need to be obtained, and solver and discretization parameters must be chosen correctly. The possible resolution of details depends on the speed of the implementation. In conclusion, simulations should not replace experiments but complement and accompany them.

In this paper, we present a tool which is very effective in helping to understand the process of electrowetting. A number of approaches are possible. One could implement a simulation coupling at least the electrostatics and fluidic domain and providing treatment of the droplet shape by, e.g., a levelset or volume-of-fluid method, which is the way to go if a design should be characterized before the production of prototypes.

Another possibility is to view the droplet motion in its quasi-static limit, which is the approach we have used in our model. The goals of the simulation tool are to provide a fast methodology to compute the change of the shape of a droplet subject to electrowetting and to investigate the effect of this change. It is not meant as a full computational fluid dynamics tool as presented in [Zeng, 2004]; the questions asked to a CFD tool are different from the ones we want to answer here. Our simulation is based on the energy equilibrium of surface tensions. Dynamic effects and the fluidic transport process within the moving droplet are excluded from the simulation, thus yielding results in the quasi-static limit of very slow motion and long time. This was motivated by a) the complete overshadowing of inertial forces by electrostatic forces (Re = 0.01) and b) the conservative nature of quasi-static computations. In short, we clearly see whether a droplet can get "stuck" in a local minimum and hence block a fluidic path. While this kind of simulation makes no statement about the exact time response as CFD, it still provides general physical insight as shown in the electrode fine structure optimization further on in this paper. There, the potential energy curves of certain device variations give strong hints on the performance, and the effect of modifications is much more visible and detectable than in other simulation and experimental approaches. This potential energy is the main quantity in the Surface Evolver and thus easily accessible, allowing us to obtain a fairly accurate picture of, e.g., the energy saddle point that gives rise to droplet splitting.

The modification of the geometry of, e.g., an electrode is a matter of a few seconds work, and allows to quickly perform parameter variation studies, the results of which aid in developing compact models of the droplet operations.

A further design decision is the representation of the model in a computer program. We use the *Surface Evolver* program for our simulation, which explicitly represents the fluid surface by a mesh consisting of triangles. The spatial constraints that the Surface Evolver makes available for use with nodes, edges and faces are a useful modeling tool with which it is possible to simplify and hence speed up the computation of droplet motion. Due to the exclusion of internal fluid transport, the number of equations is strongly reduced; there is also no need for a boundary element treatment of the droplet interior. This simplified model allows for a fast integration with a much lower CPU time compared to full CFD simulations. For example, the droplet splitting (section 6.2) needed only a few minutes for solving. Further, there is no need to store a three-dimensional (3D) grid but only a two-dimensional (2D) surface. This makes this approach well suited for optimization loops. However, these decisions also have some disadvantages which will be discussed later. Table 3-1 gives a short comparison of the two options.

In the following section, we describe the basic physical effects and the typical setup of electrowetting devices, and we discusses the application of the Young-Lippmann equation to electrowetting on dielectrics for biochip applications as already formulated in the literature. We then present our simulation methodology along with a discussion of its limits. The methodology was integrated into a template library for the Surface Evolver program with a graphical user interface, addressing a number of specific problems of the design of electrowetting arrays. Finally, we present results of these simulations, which answer some of the questions posed above.

Table 3-1. Comparison between full computational fluid dynamics simulation (CFD) and our quasi-static approach (QS). A "+" in this table means that the method is appropriate or good, a "-" means that there might be some difficulties, and a "+/-" means a limited suitability.

	CFD	QS
CPU time and memory	-	+
Potential energy landscape	-	+
Compact models	+/-	+
Optimization	+/-	+
Interaction with solver	+/-	+
Design for "worst case"	+/-	+
Surface representation	implicit	explicit
Surface recovery	+/-	+
Topological changes	+	-
Inertia and damping	+	-
Fluidic transport	+	-
Overshooting	+	-
Transient behavior	+	-

2. WETTING AND ELECTROWETTING

Electrowetting is a method to alter the wetting properties of a surface. A voltage is applied, and an electrostatic field is built up. The energy stored in this field can be formulated as equivalent interfacial tension and thus related to the liquid-gas-substrate contact angle of the droplet. This leads to a deformation of the droplet shape, which can be used for the fluidic operations described above. In this section, we describe wetting in general and the influence of an applied electric potential.

Moving droplets by an applied electric potential (without additional energy transducers like piezos or electrostatic actuators) can be achieved by two main effects: Dielectrophoresis and Electrocapillarity, and can be described under the framework of electrohydrodynamic forces [Zeng, 2004]. The setup we will consider in this article is a electrocapillarity approach called "Electrowetting on dielectrics" (EWOD), which modulates the wetting properties of a substrate via electrostatic energy.

2.1 Wetting on Surfaces

An amount of a single phase liquid L forms a spherical droplet if no external influences are present. This is the configuration with the smallest surface area for a given volume. The effect which leads to this minimization of surface area is the *surface tension*, measured as energy per area [de Gennes, 1985, Israelachvili, 1991]. The reason for this property of the droplet surface is the differing environment of a liquid molecule. A molecule feels forces from neighboring molecules: *Van der Waals* forces for non-polar molecules, the Keesom interaction (orientation effect) for polar molecules and the Debye interaction (induction effect) for a polar and a non-polar molecule. In the interior of the liquid, these forces equilibrate in the time average, while on the surface only one half of the surrounding is contained in the liquid interior.

After normalization to the surface area, the sum of these forces gives the the *Laplace pressure* [de Gennes, 1985]

$$\Delta p = 2\sigma H \tag{3.1}$$

with Δp the pressure difference, σ the surface tension, and H the local *mean curvature* of the surface. If the droplet is not surrounded by vacuum but by another medium V (for **v**apor phase), the surface tension is replaced by the *interfacial tension* γ_{LV}, since now the second medium also causes a force on a liquid molecule. The energy of such a surface A can be calculated with the surface integral

$$E = \int_A \gamma_{LV} \mathrm{d}A'. \tag{3.2}$$

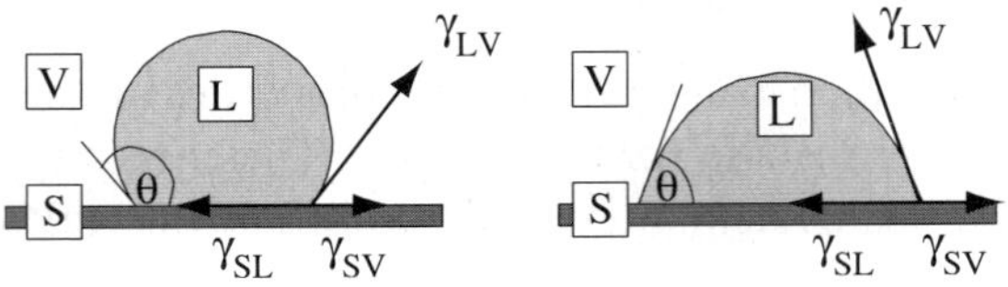

Figure 3-2. A droplet on a substrate. Left: hydrophobic surface. Right: hydrophilic surface.

Now consider a droplet sitting on a surface S (Fig. 3-2). The droplet is in contact with two materials: The vapor phase and the substrate. Now, three different interfacial energies interact: γ_{LV} for the liquid/vapor interface, γ_{SL} for the substrate/liquid interface and γ_{SV} for the substrate/vapor interface. The line where the three phases meet is called the *contact line*, and the angle of the liquid phase is called the *contact angle* θ. It can be calculated by considering the variation of the free energy F due to a virtual displacement of the contact line (Fig. 3-3):

$$\delta F = \gamma_{SL} 2\pi r \delta x - \gamma_{SV} 2\pi r \delta x + \gamma_{LV} 2\pi r \delta x \cos\theta \tag{3.3}$$

where r is the radius of the contact line.

Equilibrium and thus an energy minimum is reached when $\delta F/\delta A = 0$. This leads to the Young equation

$$\cos\theta = \frac{\gamma_{SV} - \gamma_{SL}}{\gamma_{LV}}. \tag{3.4}$$

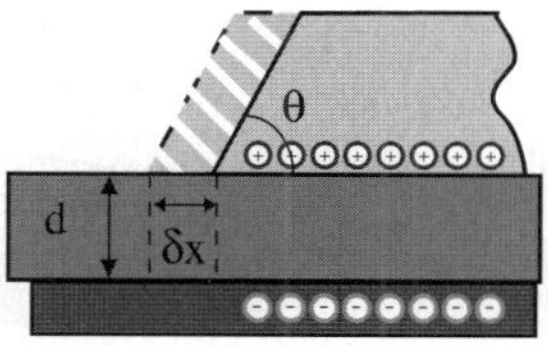

Figure 3-3. Schematic picture of the virtual displacement of the contact line.

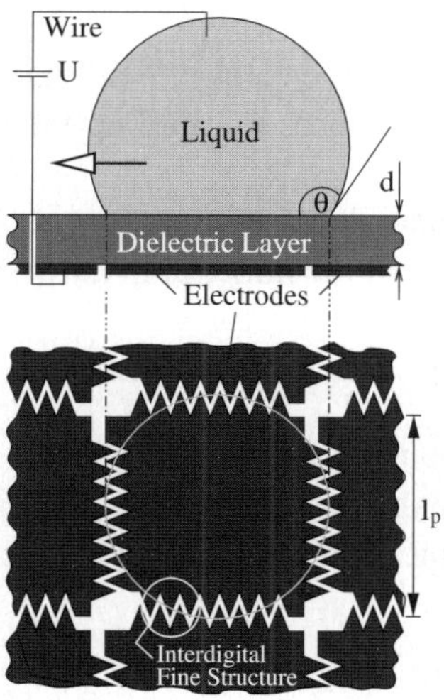

Figure 3-4. Typical setup of an electrowetting device. The contact angle θ is lowered if a voltage U is applied.

2.2 Electrowetting

The typical setup of an electrowetting device is shown in Fig. 3-4. The system consists of a dielectric layer of thickness d with metal electrodes below, while a droplet of conducting liquid (electrolyte) is situated on the upper exposed surface. It is essential that the dielectric layer is a good insulator with no pinholes, and that ions cannot easily be trapped inside the layer; this would inhibit the correct functioning of the device. In this particular setup, the droplet is contacted with a wire as shown in Fig. 3-4, further possibilities are discussed later.

By applying an electric voltage U between the electrode and the droplet, charge is accumulated as in a capacitor. This decreases the interfacial tension between the droplet and the dielectric layer due to the stored electrostatic energy, leading to a change of the contact angle of the droplet (Fig. 3-5) [Lippmann, 1875]. The variation of free energy (3.3) then reads [Vallet et al., 1996, Verheijen and Prins, 1999]:

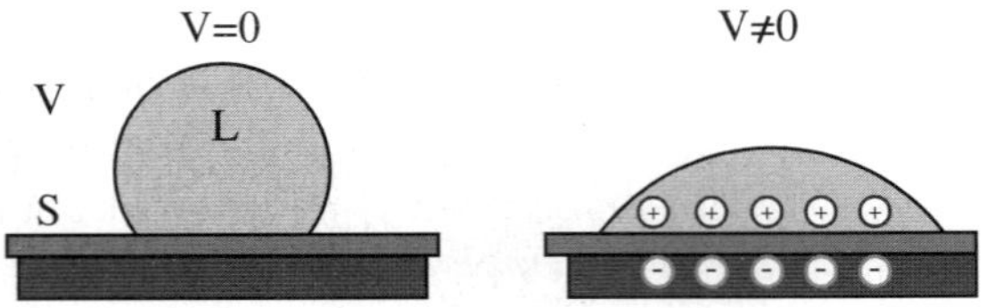

Figure 3-5. A droplet changing its contact angle due to electrowetting.

$$\delta F = \gamma_{SL} 2\pi r \delta x - \gamma_{SV} 2\pi r \delta x + \gamma_{LV} 2\pi r \delta x \cos\theta + \delta U - \delta W_B \tag{3.5}$$

where U is the energy stored in the electric field in the dielectric layer and W_B the work done by the voltage source to build up the potential between droplet and electrode.

The energy stored in a capacitor with large area A, small plate distance d and relative dielectric constant ε_r of the material in between for a voltage V is given as

$$U = \frac{1}{2} C V^2 = \frac{1}{2} \frac{\varepsilon_r \varepsilon_0 A}{d} V^2, \tag{3.6}$$

where ε_0 is the dielectric constant of vacuum.

Now we assume that the droplet changes its area by $\delta A = 2\pi r \delta x$ because of movement of the contact line. Then the energy of the electric field changes by

$$\frac{\delta U}{\delta A} = \frac{1}{2} \frac{\varepsilon_r \varepsilon_0}{d} V^2. \tag{3.7}$$

The additional energy is fed into the system by the voltage source, so that

$$\frac{\delta W_B}{\delta A} = \frac{\varepsilon_r \varepsilon_0}{d} V^2. \tag{3.8}$$

$\delta U / \delta A$ and $\delta W_B / \delta A$ can be combined to an electrowetting term $\gamma_{EW} = \delta W_B / \delta A - \delta U / \delta A$, whereupon (3.5) reads

$$\frac{\delta F}{\delta A} = \gamma_{SL} - \gamma_{SV} + \gamma_{LV} \cos\theta - \gamma_{EW} \tag{3.9}$$

with

$$\gamma_{EW} = \frac{1}{2} \frac{\varepsilon_r \varepsilon_0}{d} V^2. \tag{3.10}$$

The Young equation (3.4) then becomes

$$\cos\theta = \frac{\gamma_{SV} - \gamma_{SL} + \frac{1}{2} \frac{\varepsilon_r \varepsilon_0}{d} V^2}{\gamma_{LV}}. \tag{3.11}$$

This can be modeled as an equivalent interfacial tension of the liquid to the substrate, i.e.,

$$\gamma_{SL}(V) = \gamma_{SL}(0) - \frac{1}{2} \frac{\varepsilon_r \varepsilon_0}{d} V^2 \tag{3.12}$$

on those parts of the contact area where it overlaps with the respective electrode.

By applying a voltage to an adjacent electrode pad, and provided that the contact interface overlaps this second electrode, a droplet seeks to increase its contact area on that pad at the cost of the area on the current pad. Therefore, a motion to the next electrode takes place. Subsequent application of this algorithm allows to transport the droplet over a larger distance. By moving two droplets to the same spot, mixing can be achieved. Splitting requires more complicated actuation schemes, which can benefit from proper design tools. An analysis of droplet splitting was presented in [Cho et al., 2001, Cho et al., 2003].

2.3 Electrowetting Devices

The setup described above requires tracking of the droplet and moving the wire accordingly. Further, the wire also distorts the droplet shape, impeding use in optical applications. Therefore, some more advanced setups are used, as demonstrated in Fig. 3-6.

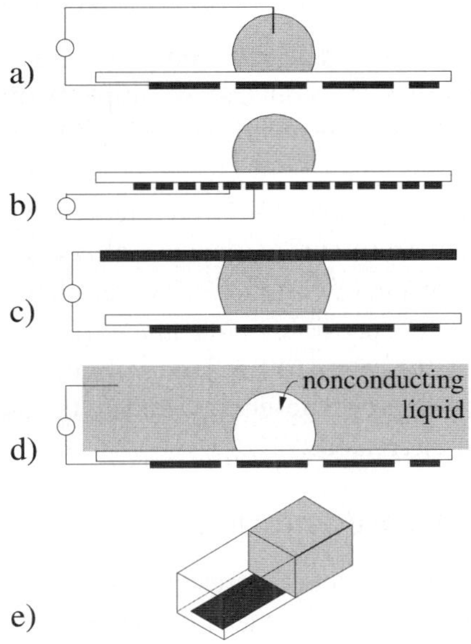

Figure 3-6. Different actuation setups for electrowetting. a) Droplet contacted by wire; b) two capacitive contacts; c) confined droplet (electrode is complete upper substrate); d) inverted setup; e) liquid in a channel.

Figure 3-6a shows the classical setup with one wire to provide an ohmic contact and the capacitive coupling on the substrate. It is also possible to operate the droplet with two capacitive contacts (Fig. 3-6b); the contact line must overlap with two electrodes, between which the voltage is applied. The droplet then wets both electrodes. In this case, only half of the applied voltage is available for each pad, since the electric field passes the dielectric layer twice.

Another solution is to use a conductive plate instead of the wire, such that the droplet is confined between two substrates (Fig. 3-6c). This also facilitates the splitting of droplets [Cho et al., 2003], since the Laplace pressure of the droplet surface is lower. Using a transparent conductor like indium tin oxide (ITO), optical monitoring is still possible.

Especially for optical purposes, it is useful to invert the setup (Fig. 3-6d): A non-conducting liquid is immersed into an electrolyte; the voltage is applied between the surrounding medium and the substrate. The main advantage is that the droplet shape is not distorted by an electrode, and setups with radial symmetry are easy to build, so that good adaptive lenses can be created.

Finally, electrowetting can also be used to pump liquid in a channel; besides moving droplets in capillaries, one possible use is the priming of a microfluidic device, to avoid bubbles of air clogging the system.

3. SIMULATION WITH THE Surface Evolver

An analytical solution to these equations is only possible for very simple cases; in general, numerical models are required. We implemented the electrowetting model with the Surface Evolver, a powerful program for the numerical modeling of minimal surfaces.

The Surface Evolver by K. A. Brakke is an interactive program for the study of surface shapes arising from surface tension effects and other energies. It "evolves" the surface to an energy minimum by a gradient descent or conjugate gradients minimization. It is possible to introduce spatial constraints as well as global surface integral constraints like a fixed volume [Brakke, 1992, Brakke, 2003]. By formulating appropriate energy terms, the effect of non-uniform surface tensions can be integrated.

3.1 Numerical Representation

In the Surface Evolver, the droplet is represented by its bounding facets, which are flat triangles defined by three vertices (points in the Euclidean $\mathbb{R}^3$ space) and three connecting edges.

The basic operation for the evolution of the surface is the iteration step which moves the vertices along the energy gradient. The actual displacement is the product of the energy slope of the respective degree of freedom and a global *scale factor*, which can be specified by the user or optimized by the Surface Evolver. An additional quantity correcting motion enforces global quantity constraints.

For a facet with edges s_0 and s_1, the facet energy due to surface tension γ can be calculated by

$$E = \frac{\gamma}{2} |s_0 \times s_1| . \tag{3.13}$$

It is straightforward to show that the gradient $g_i = \partial E/\partial x_i$ of the first edge s_0 is then

$$g_{s_0} = \frac{\gamma}{2} \frac{s_1 \times (s_0 \times s_1)}{|s_0 \times s_1|} . \tag{3.14}$$

Summing up all gradient parts of the adjacent faces yields the total free energy gradient of the vector motion [Brakke, 1992].

3.2 Substrate-Liquid Interfaces

The interface of the droplet to air is modeled by a triangle mesh as described. However, for the interface to the substrate, a mesh is inappropriate for a number of reasons: First, on those parts of the interface with constant interfacial tension, there is no gradient for the vertices sitting on the interface; this could lead to numerical problems and mesh degradation. Second, to model a varying interfacial tension as needed for electrowetting, the surface energy of the triangles would have to be updated whenever the triangle changes its position. Finally, it would be a waste of resources since there is a very elegant way to solve this problem: Instead of an explicit representation of the interface between droplet and substrate, the energy is added to the total energy by transforming the surface integral (3.2) into a line integral over the surface boundary [Brakke, 2003, Lienemann, 2002, Lienemann et al., 2004c]. This boundary is represented by the edges of the triangles at the contact line. With the Green-Gauss theorem, we have

$$\int_A \gamma_{SL} \vec{n} \mathrm{d}\vec{A}' = \int_{\partial A} \vec{g} \mathrm{d}\vec{l} \tag{3.15}$$

with normal vector $\vec{n}$ and $\gamma_{SL}\vec{n} = \nabla \times \vec{g}$.

Since on the bottom surface $\mathrm{d}\vec{A}' = \vec{k}\mathrm{d}A'$, where $\vec{k}$ is the unit vector in the z direction, we require a $\vec{g}$ such that the third component of its rotation, $f_z = \partial g_y/\partial x - \partial g_x/\partial y$, is equal to the interfacial tension γ_{SL}. Choosing $g_x = 0$, we get

$$\vec{g} = \begin{pmatrix} 0 \\ \int_x \gamma_{SL}(x', y)\mathrm{d}x' \\ 0 \end{pmatrix} . \tag{3.16}$$

On the top (confined droplet), the sign is inverted.

This approach can also handle spatially varying interfacial tensions.

If only the interface on the bottom is replaced, the volume calculation is left to the Surface Evolver. For a confined droplet, the removed interface at the top must also be manually integrated in the volume calculation, as shown in [Lienemann, 2002, Lienemann et al., 2002, Brakke, 2003].

3.3 Electrowetting

Electrowetting effects are modeled by using equation (3.12) and setting up $\gamma_{SL}(x, y)$ such that, above the electrode, the second (electrowetting) term is switched on, and is left zero otherwise. Multiple electrodes with different voltages can be treated analogously. $\gamma_{SL}(x, y)$ is then integrated according to (3.16) and written to the Surface Evolver script file. With the use of parameters, the voltage can be changed during runtime.

Typically, the edges of the electrodes for this kind of electrowetting pump feature spikes reaching into the adjacent electrode. The reason for this arrangement is that the dynamics at the start of the droplet motion is essentially determined by the shape of the potential energy curve at the adjacent electrode, and thus, by the drag force on the contact line. For a flat electrode edge, the interfacial tension is likewise approximately flat with a transition at the pad boundary (Fig. 3-7). According to (3.2), this results in a flat potential curve as long as the contact line does not touch the actuated pad, and thus, a zero force.

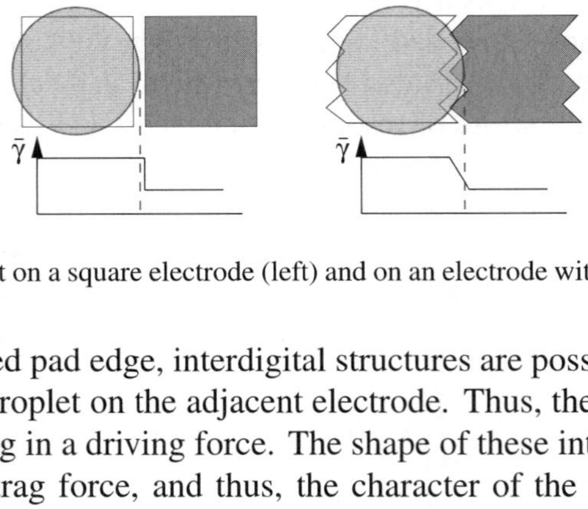

Figure 3-7. Droplet on a square electrode (left) and on an electrode with a jagged edge (right).

With the jagged pad edge, interdigital structures are possible which are also in touch with a droplet on the adjacent electrode. Thus, there exists an energy gradient, resulting in a driving force. The shape of these interdigital structures determines the drag force, and thus, the character of the initial motion. By optimizing its shape, it is possible to account for different droplet sizes and chemical contaminations on the substrate. Those contaminations can lead to a contact angle hysteresis [de Gennes, 1985] and even inhibit the motion of the droplet.

These shapes are not implemented in detail, because to resolve a jagged electrode shape in all its complexity would require a very fine mesh resolution of the contact line; further mesh degeneracy and instabilities were observed in numerical experiments. Instead, we assume that a spikes' size is small enough so that its effect can be averaged along the edge direction [Lienemann et al., 2003]:

$$\overline{\gamma}(x) = \int_{y_1}^{y_2} \frac{\gamma(x, y')\mathrm{d}y'}{(y_2 - y_1)} \tag{3.17}$$

4. EDEW, A TOOL FOR ELECTROWETTING CHIP DESIGN

To implement a simulation, some experience in writing of Surface Evolver script files is required to specify the model along with constraints and surface energies. Writing new models cat so slow down the design process where ready-made solutions for standard problems could be used. We therefore provide a tool to simplify this process: A script template library is provided along with a user friendly graphical user interface (GUI) for all relevant model parameters. For experienced users, direct interaction with the Surface Evolver remains still possible. The frontend is written in *Java* for portability reasons. Figure 3-8 shows the main components of the program: The panel on the left allows entering parameters for the template library. Then, after starting the simulation, the control window (top right) opens, which allows interactive control of the simulation process.

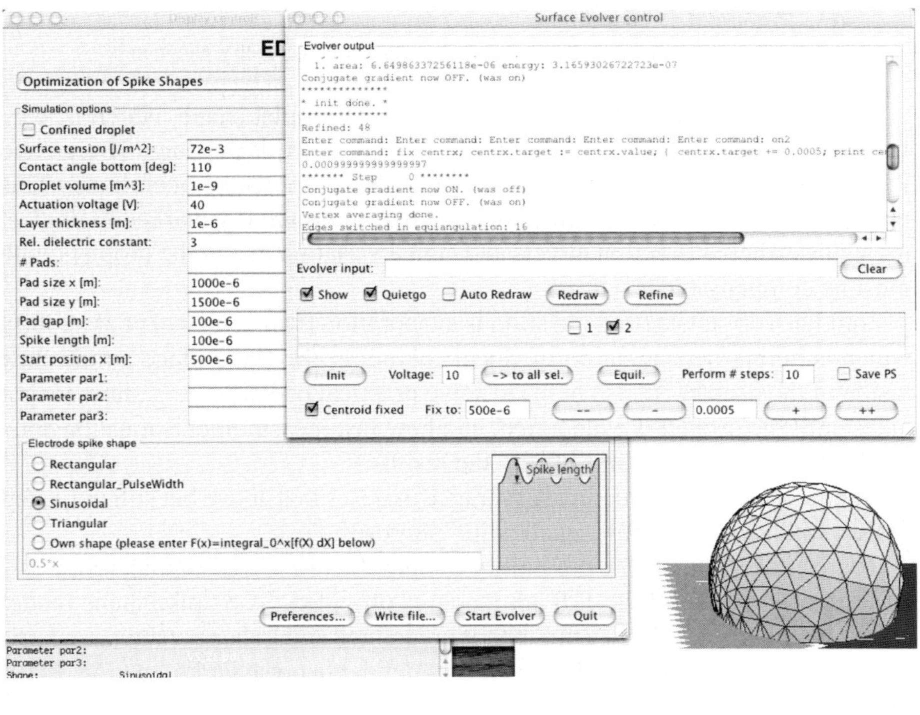

Figure 3-8. EDEW user interface. Left: Simulation parameters; Top: Surface Evolver control window; Bottom: Graphics window (provided by the Surface Evolver).

Each template set provides its own parameter and control panel. Currently, three models are implemented; extending the library is easily done by extending the `Simulation` Java class. Details of the available models and the Java class are provided in the user manuals [Lienemann et al., 2004a, Lienemann et al., 2004b].

The first model (1DPath) provides a line of electrode pads both for confined and non-confined droplets (Fig. 3-9). It allows one to test basic operations of an electrowetting array like moving, dispensing, merging and splitting. Since the topology of the droplet remains unchanged during the simulation, splitting and merging is detected by the designer using the graphical output.

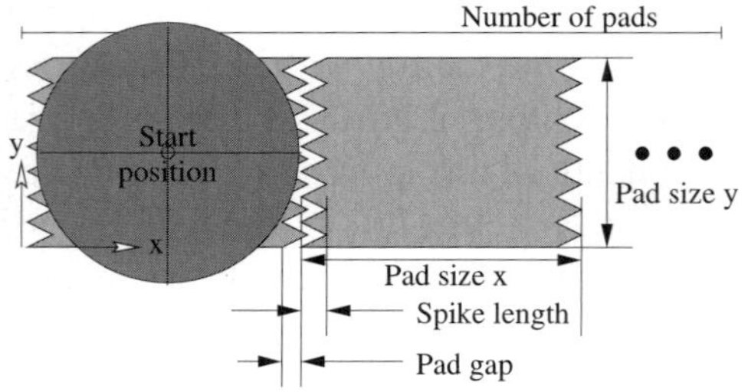

Figure 3-9. The 1DPath model and the adjustable geometry parameters.

This model is also useful to explore the exact droplet shape, which is, e.g., valuable for estimation of optical properties for micro-lens design [Berge and Peseux, 2000], and how the shape behaves during the basic operations. This may also be critical for the optical detection of fluorescently marked biological molecules where refraction effects and signal variation due to the droplet depth must be compensated.

Another very interesting question is evaporation [van den Doel et al., 2000], which is highly dependent on the local curvature; this can induce a fluid flow inside of the droplet. It also helps to solve practical questions, e.g., how long it takes until the droplet is evaporated, and how fast measurements must be done until the decrease of volume affects the results.

The electrode edge structure is averaged as described above but still indicated in the graphical output for visualization purposes.

To give the designer the possibility to optimize these interdigital structures, an extended version of the 1DPath model is provided. The SpikeShape model allows to either select from a number of predefined spike shapes (sinusoidal, triangular, rectangular, rectangular with user definable pulse width, see Fig. 3-10) or define additional shapes.

For the latter, two steps are necessary:

1 Find the function for $\overline{\gamma}_{EW}(x)$ and normalize its support to the interval $[0, 1]$ such that the normalized new function $f(x)$ fulfills $f(0) = 0$ and $f(1) = 1$.

2 According to (3.16), find the integral $F(x) = \int_0^x f(x')\mathrm{d}x'$.

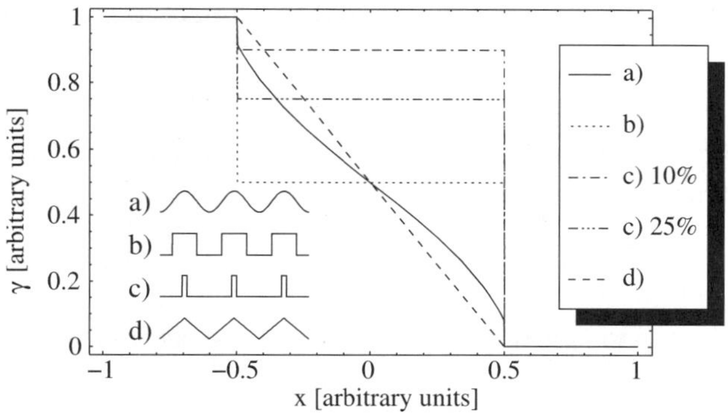

Figure 3-10. Variation of the interfacial tension $\overline{\gamma}(x)$ at the pad edge for different shapes.

These functions can also feature parameters, which can even be changed during runtime.

It is possible to operate the model in a free motion mode, where the droplet moves only according to the electrowetting forces, or in a constrained mode, which is the recommended mode for spike shape optimization: Here, the centroid of the droplet is forced to a given position, resulting in an potential energy over centroid curve. Based on this curve, a dynamic model can be extracted that also allows the estimation of inertial effects.

A third model simulates a liquid meniscus in a rectangular channel. The mesh consists only of the meniscus area; the liquid volume is modeled through surface integrals and constraints. For all four channel walls different properties and voltages can be specified. This model can be interesting if electrowetting serves for priming a fluidic structure by placing electrodes on one of the walls, e.g., for estimating the minimum voltage for wetting the complete channel wall. To avoid numerical problems, the voltages should be changed in small steps.

5. LIMITS

As already indicated above, the chosen approach results in a number of limitations. In this section, we discuss the consequences for the use of the presented model.

Inertia and damping. The model evolves the droplet shape and position to a point of minimal potential energy. The trajectory of the degrees of freedoms is not necessarily the path a fully dynamic simulation would take. This also means that inertia and damping effects are not included in the model. However, under some assumptions, there is nevertheless a close relation to a more complete simulation due to the way the Surface Evolver calculates the motion of the mesh vertices. Their motion is proportional to the (negative) energy gradient, i.e., the resulting force f acting on the vertex, subject to constraints [Brakke, 1992]:

$$x_{n+1} = x_n + d * f(x), \tag{3.18}$$

where d is the scale factor chosen by the Surface Evolver. By using $\dot{x} \approx (x_{n+1} - x_n)/d$, linearizing and reordering,

$$\dot{x} + Kx = f_{\text{ext}}. \tag{3.19}$$

Now let us have a look at a mass/damper/spring system subject to an external force:

$$f_I + f_D + f_S = f_{\text{ext}}, \tag{3.20}$$

where $f_I = M\ddot{x}$ is the reaction force of the inertial mass M subject to acceleration, $f_D = C\dot{x}$ is the damping force of the system, $f_S = Kx$ is the reaction force of the stiffness K, and f_{ext} is the external force. x may also be vector-valued; M, C and K then turn into matrices.

The external force is balanced by the inertial, damping, and stiffness force. The work applied by the external force is converted into kinetic energy, potential energy and dissipation by the damping. At the beginning of the motion, energy mainly goes into the acceleration of the mass (kinetic energy), which can drive the system beyond the equilibrium point where $Kx = f_{\text{ext}}$, leading to an oscillation. This is true if the ratio of the damping force over the inertia force is small enough.

For a massless or strongly damped system where the damping force is much higher than the inertial force, $f_I \approx 0$ and the remaining ordinary differential equation (ODE) reads

$$f_D + f_S = f_{\text{ext}}, \tag{3.21}$$

or,

$$C\dot{x} + Kx = f_{\text{ext}}. \tag{3.22}$$

With $C = 1$, this is the same formula as for the Surface Evolver evolution step except for the provision of constraints [Brakke, 1992] and the timestep.

The result is a damped motion – similar to what can be seen from movies of droplets moved by electrowetting [Duke University Digital Microfluidics Research Group, 2004], which is indicated by the scale effects discussed in the introduction. Damping was also found to be important in the context of droplet vibrations [Prosperetti, 1980].

Still, this damping should not considered to be the real damping of the physical system, which is influenced by the fluid motion and other friction effects, but the equilibrium position after long time is still the same. The main trait of such a damped system is the absence of overshooting effects which can push the system to a state which is not reachable in the quasi-static limit. One example is a droplet which is accelerated and moved to an electrode which is much larger than the droplet. In a full dynamic model, the droplet may end up further in the interior of the electrode. Droplet splitting is another example, where inertia may lead to an augmented droplet motion. The numerical experiment must therefore be carefully checked if it is necessary to include such effects – analogous to a RF switch consisting of two beams which are attracted to each other by electrostatic actuation: In one case, one would like to find a minimal voltage where the switch will close independent from squeeze film damping and from the applied voltage curve which may be distorted by parasitic line capacities; in another, one wants to find the maximum voltage one can apply with a given curve such that no switching occurs. It is the first case where the main value of a quasi-static simulation lies: Even if due to a slow actuation the inertia is not as high as expected, there is still the wanted effect, and the massless system gives a conservative design rule for these circumstances.

Energy dissipation. There is no information on energy dissipation by damping. Therefore, the energy needed for a strongly damped process cannot be calculated by this simulation. We assume that the voltage source is capable of delivering all the energy needed to reach the equilibrium state.

Peripheral electric field. The electrostatic energy is calculated only below the droplet/substrate interface. Peripheral electric fields and the electric field in the air are not considered. However, the contribution of the air to the energy is small due to the fact that the field strength is smaller due to the lower dielectric constant (usually a factor of 2 or 3) and the longer length of the electric flux lines (the potential difference remains constant). Whereas the thickness of the dielectric layer is in the micrometer range or even below, the lengths of aerial flux lines are in millimeter dimensions. More errors could come from the region near the contact line, both from the contribution of the peripheral field inside the dielectric layer as well as in the surrounding air. We have performed a finite element simulation to investigate the effect of this on the calculation of the modified interfacial tension.

The simulation shows the region close the the contact line of a droplet. We assume a potential of 1 V at the droplet boundary and 0 V at the bottom electrode. The dielectric layer with a relative permittivity of 2 is 1 μm thick. The result (Fig. 3-11) shows that, except for a small region around the contact line, the electric energy density is close to the assumed values of 0 away from the interface and 8.85 J/m^3 just below it.

Near the contact line singularity, there is a small region where large values of the electrostatic energy are observed; nevertheless, this region is small compared to the remainder of the droplet. In conclusion, we observe a distortion of the electric field only at a region in the order of the size of the layer thickness, which is small compared to the droplet dimensions.

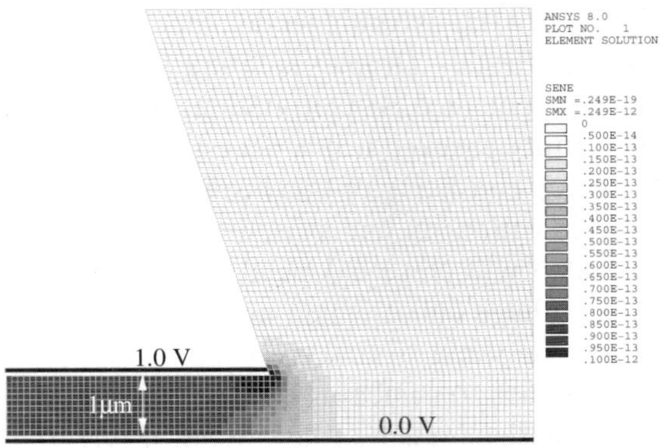

Figure 3-11. FEM solution of the electrostatic energy near the contact line. The indicated values must be multiplied by 10^{14} J/m^3 to obtain the energy density.

Further, since only the energy difference of two systems (or the energy gradient) is important, we expect an influence of these distortions only if the length of the contact line part on the electrode or its curvature experience a large change. This happens, e.g., on that point where it intersects with the electrode boundary; still, the change of effective diameter is small for a small dielectric layer thickness and a large droplet. The distortions inside and outside of the contact area also partly balance each other.

Charge trapping. If the dielectric layer is penetrable by charged particles and the voltage is applied for a certain time, charges may be trapped inside [Verheijen and Prins, 1999]. This is often seen as one reason for the so called contact angle saturation, where the contact angle does not change any more if the voltage is increased above a certain limit; further, it impedes the reversibility of the interfacial tension change, leading to contact angle hysteresis. This could be modeled by an additional voltage contribution, such that turning the voltage "off" means setting it to a finite value which models the trapped charges.

Charged biomolecules. A further distortion of the process can come from large charged molecules – or molecules with a nonuniform charge distribution, which distort the Helmholtz layer of the droplet and modify the capacity of the droplet/electrode system. This could also cause a contact angle hysteresis, if the molecules remain attached to the substrate. This can be modeled by an additional "off" voltage as discussed above and by a modified layer thickness. However, these approaches need further experimental validation.

Topological changes. Droplet splitting and merging is not fully implemented in the model, manual inspection remains necessary. This is due to the explicit surface representation; with a levelset or volume of fluid approach, this is only a minor issue. However, in these methods, the determination of the contact line and the surface reconstruction is more difficult, which is important for, e.g., optical applications. For droplet merging, on the other hand, it is easy to see from the graphical output whether the operation was successful and whether the droplets touch. Droplet splitting is more difficult to see, it occurs when the liquid bridge connecting the two parts collapses to a line, or even overlap and interpenetration occurs. Due to the implementation of the energy calculations, this singularity poses no numerical problems.

6. RESULTS

In this section, we show the results of a number of simulations performed with our model. All of them with the exception of the curved channel and the tube model can be performed with the EDEW tool; however the pinch-off simulation requires some manual input.

6.1 Droplet Motion

Figure 3-12 shows the simulation of a non-confined droplet moved by electrowetting with the material and operation data of Tab. 3-2. There is no other external force to the droplet except for the change of interfacial energy.

Figure 3-12. Simulation results for moving droplet: a) after actuation of electrode; b) moved to second pad, electrode actuated; c) relaxed after grounding electrode.

At the beginning of the motion (a), the change of the hydrophobic to hydrophilic behavior of the pad is clearly visible at the contact line on the actuated electrode. The droplet then moves without external influences, only because of the change in interfacial energy, to the next pad. After turning off the voltage, the droplet relaxes to its initial state. Another simulation where the droplet

Table 3-2. Parameters for the simulation in Fig. 3-12.

Surface tension	$72\,J/m^2$
Contact angle bottom	110 °
Droplet volume	1 nl
Actuation voltage	40 V
Layer thickness	1 μm
Rel. dielectric constant	3

was not overlapping the adjacent electrode in the start shows no motion. This simulation can be used as a first validation against experiments.

6.2 Droplet Splitting

This simulation shows the successful splitting of a confined droplet. We repeat the experiment in [Cho et al., 2003] using the values in Tab. 3-3. We place the droplet off center so that unbalanced splitting occurs as is sometimes seen in experiments. Another simulation with a centered droplet (not shown) resulted in an even partition.

Table 3-3. Parameters for the simulation in Fig. 3-13.

Surface tension	$72\,J/m^2$
Contact angle	120 °
Vertical spacing between substrates	80 μm
Droplet volume	62.8 pl
Actuation voltage	25 V
Layer thickness	0.1 μm
Rel. dielectr. const	2

The procedure for splitting is as follows:

1 Spread the droplet over a number of electrodes (e.g. 3) by activating all of them.

2 Switch off electrodes in the center of the droplet. While the outer active electrodes still attract the droplet, the central inactive electrode repels the droplet due to its natural hydrophobicity. If the parameters are well chosen, the droplet splits and two single droplets, each with half the volume, remain.

We stop the simulation just before topological changes occur due to pinch-off, resulting in the shape shown in Fig. 3-13. The computation time for this simulation was about 3.5 minutes on an AMD Athlon 64 3000+ (1.8 GHz), the surface is discretized using about 1000 vertices.

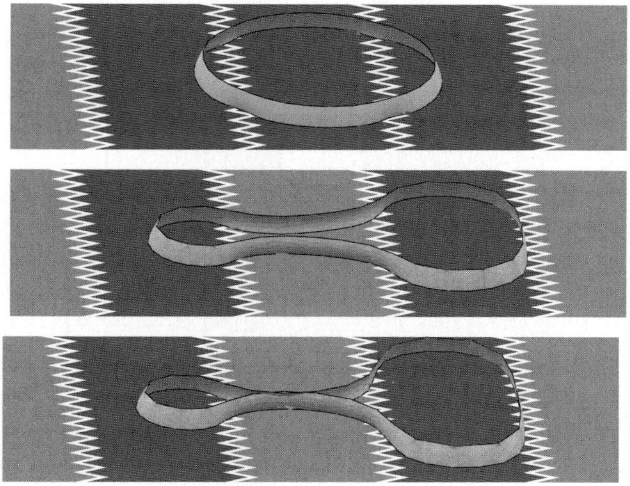

Figure 3-13. Splitting of a droplet by electrowetting. The dark electrodes are actuated with a voltage of 25 V.

6.3 Rising Fluid in Tube

This example shows a liquid column rising in a cylindrical tube due to capillary action. The capillary forces are balanced by gravity in the direction of the tube:

$$F_c = F_g \tag{3.23}$$

$$2\pi r\gamma = \pi r^2 \varrho g h \tag{3.24}$$

$$\Rightarrow \quad h = \frac{2\gamma}{r\varrho g}, \tag{3.25}$$

where F_c, F_g are capillary and gravitational force, respectively, r is the tube radius, γ the interfacial tension to the wall of the tube, ϱ is the fluid density, g the gravity constant and h the height of the meniscus.

The interfacial tension to the wall of the tube can be varied by electrowetting. Since an analytical solution is available, we can use this example as a verification for our approach.

Figure 3-14 shows a comparison between the analytical result and the Surface Evolver result (height average of meniscus vertices), yielding a very close match between the two.

6.4 Pinch-Off in Confined Setup

This simulation considers the case of a confined droplet losing volume, e.g., by evaporation. A failure of such a setup can occur because of two geometrical effects. The first danger is that the droplet becomes smaller than the electrode size. If it is then sitting in the interior of the electrode, with no overlap with an

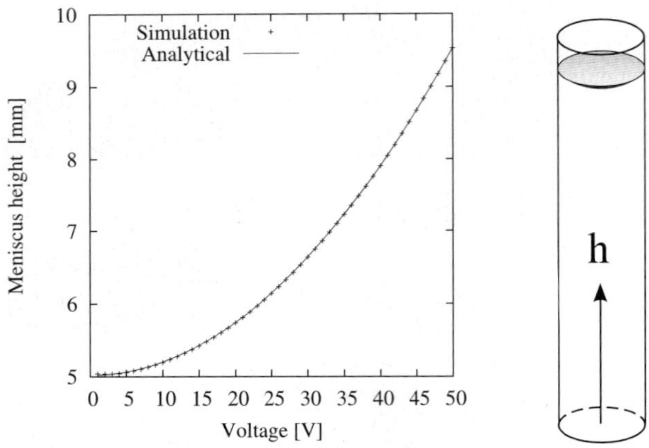

Figure 3-14. Height of a liquid column in a tube subject to electrowetting.

adjacent electrode, it is not possible any more to move the droplet away from this spot (see Figs. 3-16, 3-17). This problem can be easily tackled by making the electrodes smaller than the considered "worst case" droplet volume, such that even in that case transport remains possible. But since the confined setup only works properly as long as the droplet is in contact with both substrates, also pinch-off must be avoided at all circumstances. Assuming a contact angle θ at the substrate and a distance of h between top and bottom covers, we can calculate that the sufficient volume, where contact is always guaranteed, is:

$$V \geq \pi h^3 \left(\frac{1}{1 - \cos\theta} - \frac{1}{3} \right). \tag{3.26}$$

If the contact angles on the substrates differ, the smaller of the two must be used, since a smaller contact angle decreases the height of the droplet and thus is the more critical part.

Fortunately, there is a safety margin between the theoretical value and the actual pinch-off. As can be seen in Fig. 3-15, the shape of the evaporating droplet just before pinch-off is almost cylindrical near the hydrophobic part. This corresponds to a local energy minimum, which traps the surface in this shape. A further decrease in droplet volume finally results in the system leaving the local minimum. However, once the droplet has detached, recovery is impossible.

This margin is clearly visible in Fig. 3-15, with minimal volume where the contact angles of both substrates are equal.

6.5 Channels

When electrowetting is performed in channels, there is an additional constraint to the droplet motion: The surfaces of the channel walls heavily influence the

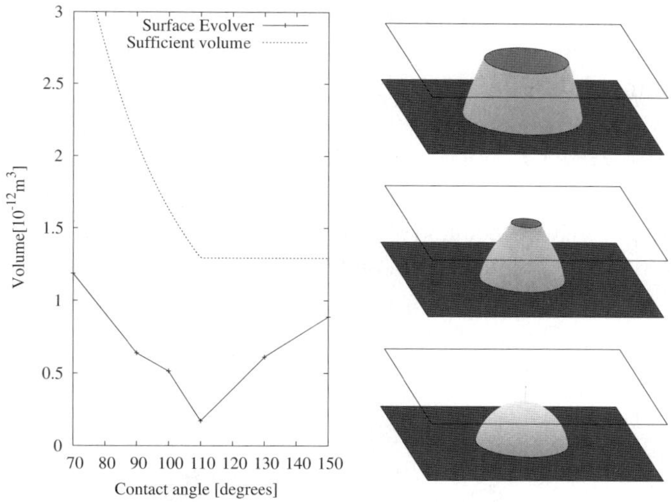

Figure 3-15. Minimal transportable volume of a droplet in a sandwich structure. Left: Simulated minimal volume compared to sufficient transport condition for a plate distance of 100 μm and a constant contact angle of 110° on one plate. Right: Development of the droplet shape with decreasing volume.

droplet shape and thus the balance of surface tension and interfacial energies. This becomes especially important if the channel changes its cross section or ends at a larger reservoir: The fluid might get stuck, because a large force is necessary to modify the surface. Figure 3-16 shows a series of pictures of a liquid meniscus in such a channel with a varying cross section. The fluid itself is not discretized, but included by surface integral transformations similar to (3.16). The voltage on the meniscus is increased from left to right, but still the meniscus stops at a certain point, and more voltage is needed for a further shift.

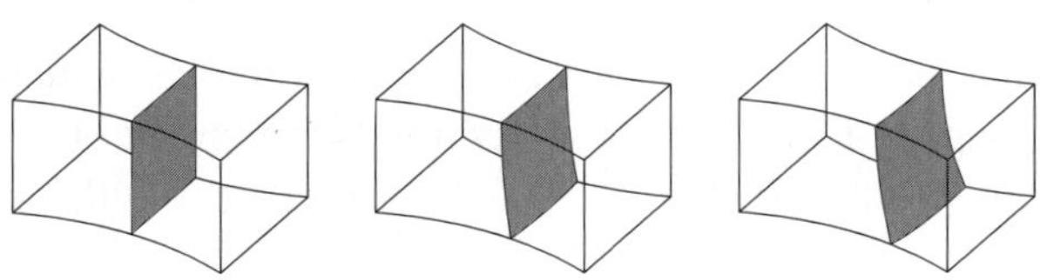

Figure 3-16. Liquid meniscus in a curved channel for different voltages.

However, for a straight channel as implemented in the EDEW model library, we observed that at a certain voltage we get a large increase in the proceeding of the contact line in the channel; its position increases further and further. Figure 3-17 shows the different states of the meniscus for the system given in Tab. 3-4: Figure 3-17a shows the equilibrium state for zero voltage. The other two graphs show the meniscus for a voltage of 86 V. This is not the equilibrium state; since complete wetting occurs for this value, the contact line proceeds further and further into the channel, until the finite resolution of the mesh leads to numerical instabilities.

Table 3-4. Parameters for the simulation in Fig. 3-17.

Surface tension	$72\,\mathrm{J/m^2}$
Contact angle	$110\,^\circ$
Channel width and height	$100\,\mu\mathrm{m}$
Layer thickness	$1\,\mu\mathrm{m}$
Rel. dielectr. const	3

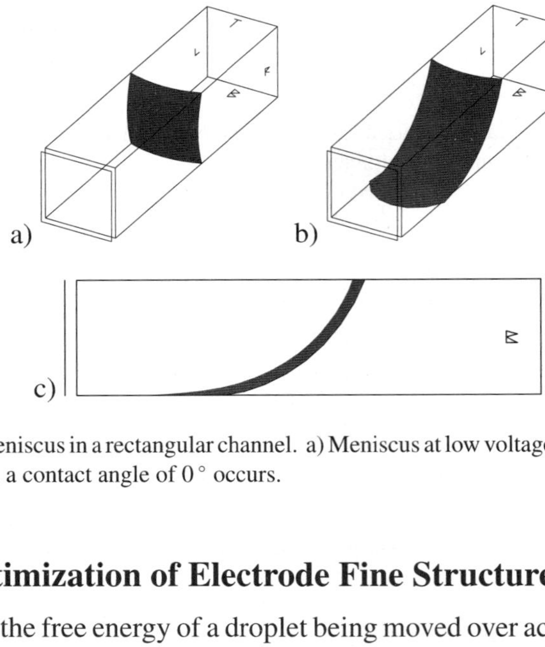

Figure 3-17. Meniscus in a rectangular channel. a) Meniscus at low voltage. b) and c) Meniscus at higher voltage; a contact angle of 0 ° occurs.

6.6 Optimization of Electrode Fine Structure

We calculated the free energy of a droplet being moved over actuated electrodes with different shapes of interdigital structures [Lienemann et al., 2003]. We studied the shapes shown in Fig. 3-10 for a structure length of $100\,\mu\mathrm{m}$ and $400\,\mu\mathrm{m}$. The parameters of the model are shown in Tab. 3-5.

Initially, the droplet resides next to the pad to which the voltage is applied such that it does not touch the pad edge structure of the actuated pad at all. We assume that only one pad is actuated at a time. We further assume that the motion happens on a much larger time scale than the fluidic relaxation of the droplet, i.e., the fluid shape follows the movement adiabatically.

The droplet is then moved manually onto the pad. For every simulation step, the energy minimum for the droplet surface is calculated, with the constraint that the centroid of the droplet is fixed at a given location. The surface energy is evaluated and is plotted versus the centroid position.

Table 3-5. Parameters for the electrode fine structure optimization.

Surface tension	$72\,\mathrm{J/m^2}$
Contact angle bottom	110 °
Droplet volume	2 µl
Actuation voltage	33 V
Layer thickness	1 µm
Rel. dielectric constant	2.1

We compare the results to a geometric model, for which the following assumptions have been made:

- The liquid-air interface does not contribute to the energy change, i.e., its area is approximately constant
- The base radius of the contact line does not change
- The contact line always forms a circle (see Fig. 3-18)

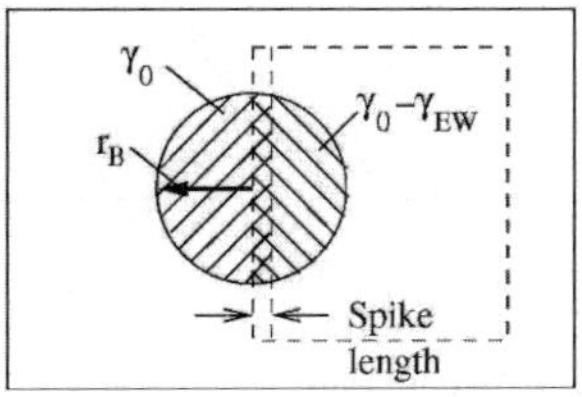

Figure 3-18. Schematic drawing of the geometric model.

The potential energy change can then be calculated by evaluating

$$\Delta E(x_c) = \int_{-r_B}^{r_B} 2\sqrt{r_B^2 - \xi^2}\,\overline{\gamma}(\xi + x_c)\,\mathrm{d}\xi, \tag{3.27}$$

where r_B is the radius of the contact line and x_c is the position of the center of the contact line. The radius of the droplet base for a contact angle θ can be calculated with

$$r_B = \sin\theta \cdot \sqrt[3]{\frac{3V}{\pi(1-\cos\theta)^2(2+\cos\theta)}}. \tag{3.28}$$

The results of the Surface Evolver model are shown in Fig. 3-19 and 3-21. White circles indicate where the contact line arrives at the interdigital edge structure and where it arrives on the bulk pad.

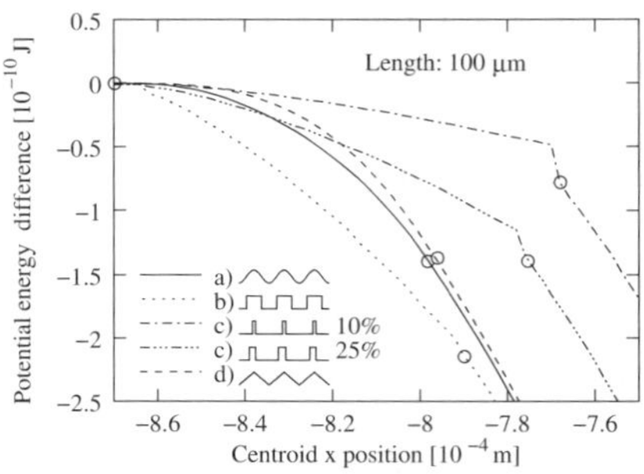

Figure 3-19. Potential energy for different pad edge shapes with a length of 100 µm.

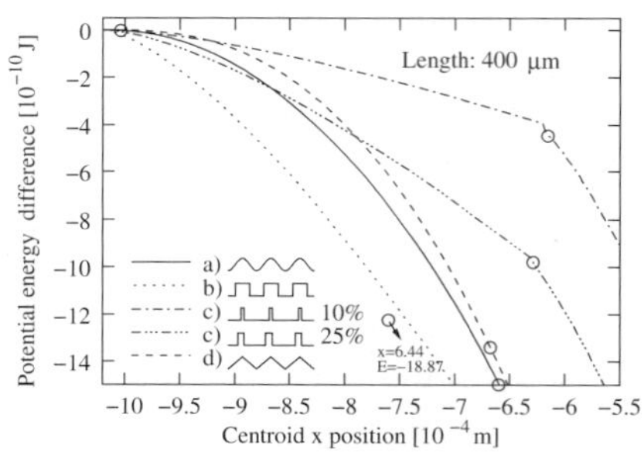

Figure 3-20. Potential energy for different pad edge shapes with a length of 400 µm. The shape of the curves is similar to Fig. 3-19.

6.6.1 Influence of the spike shape

The difference of the potential energy for the different shapes is clearly visible. The rectangular shape shows a very steep energy descent from the beginning, which indicates rapid acceleration.

The triangular shape shows a very shallow decrease and thus a vanishing energy gradient at the beginning, even the 10% spike shape performs better. However, the curve recovers very fast, and in the long run, the curves of b) and d) coincide.

The sinusoidal shape lies in between. The energy gradient is larger than for the triangular shape at the beginning, but lower after half of the structure is passed.

The spike shapes show a very low energy gradient. We also see that the energy curve is shifted to the right, because the structure does not cover half

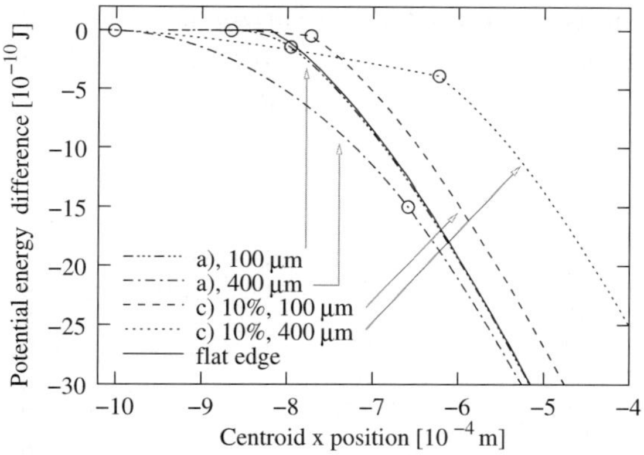

Figure 3-21. Potential energy for different spike lengths. The curves for sinusoidal shapes coincide after an initial energy difference; the curves for the c) shapes show a clear shift to the right.

of the area as for the other examples, but only 10% and 25%, respectively. In contrast, the overall energy decrease is equal once the complete contact line of the droplet has passed the structure.

A rectangular shape seems to be optimal with respect to the acceleration of the droplet; however, since the adjacent interdigital edge structures would touch at a pulse ratio of 50%, the fabrication of this ideal case is challenging and expensive. But since a smaller pulse ratio would impair the performance of the structure – as is visible for the c) shapes – either the sinusoidal shape or a mix of the triangular and the rectangular shape should be preferred.

6.6.2 Influence of the spike length

The curves for different spike lengths show good congruence for different sizes; the length does not affect the shape of the energy curve deformation, only its extent (Fig. 3-21). The contact line above the structure moves faster than the remaining part of the droplet, thus the effective structure length is smaller than the true length. The overall energy decrease is independent of the spike structure. Again, the c) shapes show a large shift towards positive x values.

Since the initial energy gradient becomes lower the larger the spikes are, there is a tradeoff between a large size to reach small drops and a small size for a large gradient.

6.6.3 Comparison with geometric model

Figure 3-22 shows the potential energy difference calculated with the geometric model. The curves are in excellent agreement with Fig. 3-20, showing the same features for the different shapes. For the droplet further on the pad, the curves were found to slightly diverge from the Surface Evolver curves; however, the curve shapes remains identical.

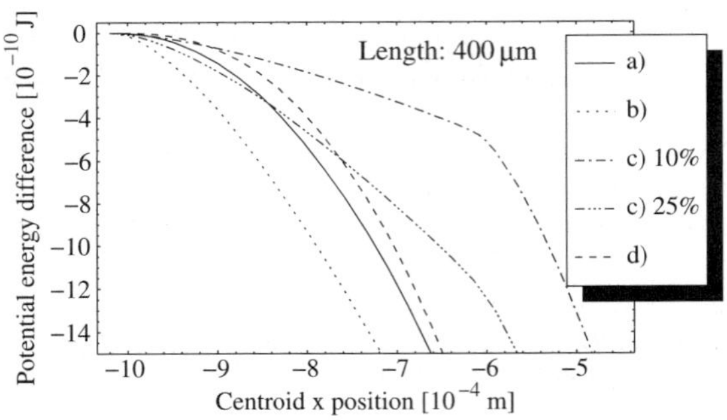

Figure 3-22. Potential energy for different pad edge shapes with length 400 μm, calculated with the geometric model.

7. CONCLUSIONS

We have presented a modeling and simulation methodology for electrowetting effects, which enables the designer to calculate droplet shapes and provides insight into the energy configuration of electrowetting arrays, which is useful for the dimensioning and layout of biochips. A method for the calculation of the fine structure of the electrodes was presented and applied to the optimization of spike shapes for interdigital edge structures, which help to make the electrowetting process more reliable. The comparison with an analytic model confirms the resulting energy curves.

In all, the Surface Evolver simulation does much more than merely simulate the motion of electrowetted droplets, for it enables us to obtain a clear picture of the potential energy landscape for a specific electrode setup together with a moving droplet. In this way, we can go back and reshape the electrodes until the obtained energy landscape is of a configuration that allows controlled behavior of the "gadget" we are implementing with the electrode, be it a mover, splitter or merger.

These simulations were integrated into a user-friendly simulation tool based on the Surface Evolver code. A template library provides ready-made scripts, so that in most cases the simulation can be performed without the need for manual script input. The tool is available from `http://www.imtek.de/simulation/microprotein`.

REFERENCES

Auroux, Pierre-Alian, Iossifidis, Dimitri, Reyes, Darwin R., and Manz, Andreas (2002). Micro total analysis systems. 2. analytical standard operations and applications.2 *Analytical Chemistry*, 74(12):2637–2652.

Berge, B. and Peseux, J. (2000). Variable focal lens controlled by an external voltage: An application of electrowetting. *The European Physical Journal E*, 3(2):159–163.

Brakke, Kenneth A. (1992). The surface evolver. *Experimental Mathematics*, 1(2):141–165.

Brakke, Kenneth A. (2003). *Surface Evolver Manual, Version 2.20*. Susquehanna University, Selinsgrove, PA 17870.

Cho, Sung Kwon, Moon, Hyejin, Fowler, Jesse, and Kim, Chang-Jin (2001). Splitting a liquid droplet for electrowetting-based microfluidics. In *Proceedings of the ASME International Mechanical Engineering Congress and Exposition*, number IMECE2001/MEMS-23831, New York, NY. ASME.

Cho, Sung Kwon, Moon, Hyejin, and Kim, Chang-Jin (2003). Creating, transporting, cutting, and merging liquid droplets by electrowetting-based actuation for digital microfluidic circuits. *J. Microelectromech. Syst.*, 12(1):70–80.

de Gennes, P. G. (1985). Wetting: Statics and dynamics. *Reviews of Modern Physics*, 57(3):827–863.

Ding, Jie, Chakrabarty, Krishnendu, and Fair, Richard B. (2002). Scheduling of microfluidic operations for reconfigurable two-dimensional electrowetting arrays. *IEEE Trans. Circuits Syst.*, 20(12):1463–1468.

Duke University Digital Microfluidics Research Group (2004). Digital microfluidics by electrowetting. http://www.ee.duke.edu/research/microfluidics.

Hayes, Robert A. and Feenstra, B. J. (2003). Video-speed electronic paper based on electrowetting. *Nature*, 425(6956):383–385.

Israelachvili, Jacob (1991). *Intermolecular and Surface Forces*. Academic Press, 2nd edition.

Laser, D. J. and Santiago, J. G. (2004). A review of micropumps. *Journal of Micromechanics and Microengineering*, 14(6):R35–R64.

Lienemann, Jan (2002). Modeling and simulation of the fluidic controlled self-assembly of micro parts. Diplomarbeit, University of Freiburg – IMTEK, Freiburg, Germany.

Lienemann, Jan, Greiner, Andreas, and Korvink, Jan G. (2002). Surface tension defects in micro-fluidic self-alignment. In *Symposium on Design, Test, Integration and Packaging of MEMS/MOEMS DTIP 2002*, pages 55–63, Cannes-Mandelieu, France.

Lienemann, Jan, Greiner, Andreas, and Korvink, Jan G. (2003). Electrode shapes for electrowetting arrays. In *Proc. Nanotech 2003*, volume 1, pages 94–97, Cambridge, USA. NSTI.

Lienemann, Jan, Greiner, Andreas, and Korvink, Jan G. (2004a). *EDEW Version 1.0, A simulation tool for fluid handling by electrowetting effects*. University of Freiburg – IMTEK, Georges Köhler Allee 103, D-79110 Freiburg, Germany.

Lienemann, Jan, Greiner, Andreas, and Korvink, Jan G. (2004b). *EDEW Version 2.0, A simulation and optimization tool for fluid handling by electrowetting effects*. University of Freiburg – IMTEK, Georges Köhler Allee 103, D-79110 Freiburg, Germany.

Lienemann, Jan, Greiner, Andreas, Korvink, Jan G., Xiong, Xiaorong, Hanein, Yael, and Böhringer, Karl F. (2004c). Modelling, simulation and experimentation of a promising new packaging technology – parallel fluidic self-assembly of micro devices. *Sensors Update*, 13:3–43.

Lippmann, M. G. (1875). Relations entre les phenomenes electriques et capillaires. *Ann. Chim. Phys.*, 5(11):494–549.

Pollack, Michael G., Fair, Richard B., and Shenderov, Alexander D. (2000). Electrowetting-based actuation of liquid droplets for microfluidic applications. *Applied Physics Letters*, 77(11):1725–1726.

Prosperetti, Andrea (1980). Free oscillations of drops and bubbles: the initial-value problem. *Journal of Fluid Mechanics*, 100(2):333–347.

Reyes, Darwin R., Iossifidis, Dimitri, Auroux, Pierre-Alian, and Manz, Andreas (2002). Micro total analysis systems. 1. introduction, theory, and technology. *Analytical Chemistry*, 74(12):2623–2636.

Tkaczyk, AlanH., Huh, Dongeun, Bahng, JoongHwan, Chang, Yu, Wei, Hsien-Hung, Kurabayashi, Katsuo, Grotberg, James B., Kim, Chang-Jin, and Takayama, Shuichi (2003). Fluidic switching of high-speed air-liquid two-phase flows using electrowetting-on-dielectric. In *Proceedings of the 7th International Conference on Miniaturized Chemical and Biochemical Analysis Systems*, pages 461–464, Squaw Valley, California, USA.

Vallet, M., Berge, B., and Vovelle, L. (1996). Electrowetting of water and aqueous solutions on poly(ethylene terephthalate) insulating films. *Polymer*, 37(12):2465–2470.

van den Doel, L. R., van Vliet, L. J., Hjelt, K. T., Vellekoop, M. J., Gromball, F., Korvink, J. G., and Young, I. T. (2000). Nanometer-scale height measurements in micromachined picoliter vials based on interference fringe analysis. In Sanfeliu, A., Villanueva, J.J., Vanrell, M., Alquezar, R., Huang, T., and Serr, J., editors, *Proceedings of the 15th International Conference on Pattern Recognition*, volume 3 of *Image, Speech, and Signal Processing*, pages 57–62, Barcelona, Spain. IEEE, IEEE Computer Society Press.

Verheijen, H. J. J. and Prins, M. W. J. (1999). Reversible electrowetting and trapping of charge: model and experiments. *Langmuir*, 15(20):6616–6620.

Zeng, Jun (2004). Electrohydrodynamic modeling and simulation and its application to digital microfluidics. In Smith, Linda A. and Sob, Daniel, editors, *Lab-on-a-Chip: Platforms, Devices, and Applications*, volume 5591 of *Proceedings of the SPIE*, pages 125–142. SPIE.

Chapter 4

ALGORITHMS IN FASTSTOKES AND ITS APPLICATION TO MICROMACHINED DEVICE SIMULATION

Xin Wang, Joe Kanapka, Wenjing Ye, Narayan Aluru, and Jacob White[†]

Synopsys Inc., Mathworks Inc., Georgia Institute of Technology, University of Illinois at Urbana-Champaign, and Massachusetts Institute of Technology

Abstract: For a wide variety of micromachined devices, designers need accurate analysis of fluid drag forces for complicated three-dimensional problems. In this paper we describe FastStokes, a recently developed three-dimensional fluid analysis program. FastStokes rapidly computes drag forces on complicated structures by solving an integral formulation of the Stokes equation using a precorrected-FFT accelerated boundary-element method. The specializations of the precorrected-FFT algorithm to the Stokes flow problem are described, and computational results are presented. Timing results are used to demonstrate that FastStokes scales almost linearly with problem complexity, can easily analyze structures as complicated as an entire comb drive in under an hour, and can produce results that accurately match measured data.

Key words: FastStokes, Stokes flow, BEM, MEMS, fluid, simulation.

[†] This work was supported by grants from the DARPA Composite CAD program, NSF, the Singapore-MIT Alliance, the Semiconductor Research Program, and Analog Devices Inc.
Xin Wang is with the Synopsys Inc., 700 E Middlefield Rd., Mountain View, CA 94043. (email: wangxin@synopsys.com)
Joe Kanapka is with the MathWorks, Inc., 3 Apple Hill Drive, Natick, MA 01760.
Wenjing Ye is with the Woodruff School of Mechanical Engineering, Georgia Institute of Technology, Atlanta, GA 30339
N. R. Aluru is with the Beckman Institute for Advanced Science and Technology, the Department of Mechanical and Industrial Engineering, the Department of Electrical and Computer Engineering, and the Bioengineering Department, University of Illinois at Urbana-Champaign, Urbana, IL 61801 USA.
Jacob White is with the Department of Electrical Engineering and Computer Science, Massachusetts Institute of Technology, Cambridge, MA 02139 (email: white@mit.edu)

K. Chakrabarty and J. Zeng (eds.),
Design Automation Methods and Tools for Microfluidics-Based Biochips, 85–107.

1. INTRODUCTION

Nearly all the micromachined devices being developed for biological applications manipulate gas or liquid, and for many of these devices, performance optimization depends critically on understanding fluid forces in very complicated three dimensional geometries[1]. Although general finite-volume and finite-element fluid flow analysis programs can perform these analyses, these programs are too time-consuming to be used for design optimization, particularly for very complicated geometries. In the case of micromachined devices, faster approaches can be developed by noting that the fluid flow is primarily Stokes flow and the quantities of interest are typically drag forces on bodies or structures in the fluid.

In this paper we describe the algorithms used in FastStokes, a very fast fluid analysis program useful for extracting surface forces on very complicated micromachined devices. The FastStokes program is based on solving an integral formulation of Stokes equation using a specialized accelerated boundary-element method (BEM). We describe the approach by first reviewing background material on the integral formulation of the Stokes equation, the standard BEM discretization, and the precorrected-FFT (PFFT) accelerated iterative algorithm for solving the BEM equations. In sections 3 and 4 we present the main contributions of the paper, starting with a description of the specializations of the PFFT algorithm to Stokes flow problem in Section 3. In Section 4, we discuss the singularity in the BEM operators and present a modified Krylov subspace algorithm for addressing the singularity. In Section 5, we provide several numerical examples, and compare computational results with experimental results to demonstrate the accuracy and efficiency of FastStokes.

2. BACKGROUND

The FastStokes program numerically solves the incompressible Stokes equation. In this section we describe the incompressible Stokes equation, which can be derived by assuming a small Reynolds number in the incompressible Navier-Stokes equation, and give an integral formulation. We also describe the basic BEM method for discretizing integral equations, and show how it is as applied to the integral form of Stokes equation. Finally, we use a simple electrostatics example to give a brief presentation of the PFFT-accelerated iterative method for solving BEM equations.

2.1 Stokes Integral Equations

Fluid flow in which viscous forces dominate over inertial forces is referred to as a Stokes flow or a creeping flow. The flow in many micromachined devices, such as air-packaged actuators or liquid-handling mixers, pumps and valves, is Stokes flow, as can be seen by examining the associated Reynolds number. The Reynolds number is defined as $\mathrm{Re} = UL/\nu$, where U is the velocity, L is the characteristic length, and ν is the kinematic viscosity of the fluid. Since $\mathrm{Re} \propto \text{inertia force}/\text{viscous force}$, Reynolds number is frequently used in experiments to determine whether viscous or inertial forces dominate. Because the feature size of many microelectromechanical system (MEMS) devices is on the order of micrometer, the UL product is small even if the movable parts oscillate at a reasonably high frequency. For example, consider an air-packaged resonator oscillating at 10 KHz with an oscillation amplitude of 1 μm. Using an air kinematic viscosity[2] at 300K of $\nu = 1.566 * 10^{-5} m^2/s$, the Reynolds number is $\mathrm{Re} = 2\pi * 10^4 * 10^{-6} * 10^{-6} / 1.566 * 10^{-5} \approx 0.004$, and therefore the inertial force can be neglected. For structures in liquids, the kinematic viscosity is much lower because of the higher density of liquid, but the operating frequencies (or structure velocities) are commensurately lower than those in air, and so Reynolds number remains low. Applying the small Reynolds number assumption to the Navier-Stokes equations yields the steady Stokes equations:

$$\begin{cases} -\nabla P + \mu \nabla^2 \vec{u} = 0 \\ \nabla \bullet \vec{u} = 0 \end{cases} \tag{4.1}$$

where $\vec{u}$ is the vector velocity, P is the pressure, and μ is the dynamic viscosity of the fluid. The surface velocities and forces satisfy an integral relation

$$\begin{aligned} u_i(\vec{x}) = &\sum_{j=1}^{3} \int G_{ij}(\vec{x},\vec{y}) f_j(\vec{y}) ds(\vec{y}) \\ &+ \sum_{j=1}^{3} \sum_{k=1}^{3} \int T_{ijk}(\vec{x},\vec{y}) u_j(\vec{y}) n_k(\vec{y}) ds(\vec{y}) \\ &\qquad i = 1,2,3 \end{aligned} \tag{4.2}$$

where the domain of integration for the surface integrals is the union of the surfaces of the structures embedded in the fluid, $\vec{x}$ is any field point, $\vec{y}$ is a point on a structural surface, $u_i(\vec{x})$ and $f_i(\vec{x})$, i=1,2,3, are the x-, y- or z-directed surface velocities and surface forces, respectively, and $\vec{n}$ is the surface outward normal. The Green's functions[3] are

$$G_{ij} = -\frac{1}{8\pi\mu}\left(\frac{\delta_j^i}{r} + \frac{\hat{x}_i\hat{x}_j}{r^3}\right)$$

$$T_{ijk} = -\frac{3}{4\pi}\frac{\hat{x}_i\hat{x}_j\hat{x}_k}{r^5} \tag{4.3}$$

$$r = |\vec{x} - \vec{y}|, \quad \hat{x}_i = x_i - y_i$$

For micromachined devices, the fluid-embedded structures are either stationary and rigid, or are deforming more slowly than the fluid response time and can be treated as quasistatically rigid. For rigid bodies, the surface integral with kernel T_{ijk} in Eq. (4.2) is zero[3], greatly simplifying the integral equation. The FastStokes program makes use of this rigid body assumption and solves the simplified integral equation

$$u_i(\vec{x}) = \sum_{j=1}^{3} \int G_{ij}(\vec{x},\vec{y}) f_j(\vec{y}) ds(\vec{y}), \qquad i = 1,2,3 \tag{4.4}$$

Integral equation (4.4) can be used to determine traction and pressure forces on the surface of a fluid-embedded structure given velocity boundary conditions. The form of Eq. (4.4) is also referred to as the single layer integral equation representation of the Stokes flow problem[3].

2.2 Discretization

In order to compute traction and pressure forces using Eq. (4.4), FastStokes uses a BEM scheme in which the integral equation is first discretized by subdividing the surface into flat panels. In particular, the FastStokes program reads an input file which contains data describing a surface mesh discretization comprised of flat triangle or quadrilateral panels. The primary reason for using flat panel discretizations is that they are easy to generate, but flat panels are also particularly well suited to micromachined structures as they usually have nearly flat surfaces.

After surface discretization, a piece-wise constant collocation method is applied to solving the integral equation in (4.4). This collocation method is based on assuming panel force densities are constant on each panel, and that when these panel force densities are used as the traction forces in Eq. (4.4), they produce velocities at panel centroids that exactly match given velocity boundary conditions. This discretized form of the velocity integral equation is then

$$u_i(\vec{x}_l) = \sum_{k=1}^{\text{number of panels}} \sum_{j=1}^{3} f_j(\vec{y}_{centroid}) \int_{panel\,k} G_{ij}(\vec{x}_l, \vec{y}) ds_k(\vec{y}), \qquad i = 1,2,3 \tag{4.5}$$

where $\vec{x}_l$ is the centroid of the l_{th} panel. The panel integrals in Eq. (4.5) can be evaluated analytically, at least for the case of polygonal flat panels, using an extension of the approach presented by Newman[4]. The technique is described in detail in the appendix.

Equation (4.5) can be written in matrix-vector form as

$$\begin{bmatrix} \mathbf{U}_1 \\ \mathbf{U}_2 \\ \mathbf{U}_3 \end{bmatrix} = \begin{bmatrix} \mathbf{G}_{11} & \mathbf{G}_{12} & \mathbf{G}_{13} \\ \mathbf{G}_{21} & \mathbf{G}_{22} & \mathbf{G}_{23} \\ \mathbf{G}_{31} & \mathbf{G}_{32} & \mathbf{G}_{33} \end{bmatrix} \begin{bmatrix} \mathbf{F}_1 \\ \mathbf{F}_2 \\ \mathbf{F}_3 \end{bmatrix} \tag{4.6}$$

or

$$\mathbf{U} = \mathbf{GF} \tag{4.7}$$

where $\mathbf{U}_1, \mathbf{U}_2, \text{and}\, \mathbf{U}_3$ are the x, y and z components of the known panel centroid velocity vectors, $\mathbf{F}_1, \mathbf{F}_2, \text{and}\, \mathbf{F}_3$ are the x, y and z components of the unknown piece-wise constant panel surface force density vectors, and $\mathbf{G}$ is the matrix form of the G_{ij} integral operator. Given the surface velocities, which automatically satisfy the continuity equation if all the surfaces are boundaries of rigid bodies, Eq. (4.6) can be used to compute surface forces. Finding methods for efficiently solving Eq. (4.6) is the key to developing a fast solver since the $\mathbf{G}$ matrix is not only dense but also singular. We discuss solving the dense matrix problem in the next subsection; the singularity problem is discussed in Section 4.

2.3 PFFT Algorithm

It is well-known that traditional BEM methods are too slow for large problems because they generate dense linear systems that are expensive to form and solve. Very efficient techniques for handling these linear systems were developed during the past two decades, by combining sparsification techniques with rapidly converging preconditioned iterative schemes, as first proposed by Rokhlin[5] and first used for accelerating BEM in general three-dimensional problems by Nabors and White[6]. As will be described below, such methods avoid explicitly forming Eq. (4.6), and can be used to reduce the cost of solving Eq. (4.6) from $O(n^3)$ to nearly $O(n)$ operations.

The basic idea behind these accelerated BEM methods follows from first considering applying an iterative method, like the Krylov-subspace based GMRES algorithm[7], to solving a dense system like the one in Eq. (4.6). When applied to solving a generic linear system, $Ax = b$, the qth iteration of a Krylov subspace method constructs an approximate solution by selecting the best weighted combination of the vectors in a *qth* order Krylov subspace

$$\{b, Ab, A^2 b, \dots, A^q b\}.$$

The *(q+1)th* order Krylov-subspace can be computed from the qth order subspace by multiplying a vector by the matrix A, and since that matrix is dense in the BEM case, computing matrix-vector products is $O(n^2)$ operations and dominants the cost of iteratively solving BEM equations.

For the case of many BEM matrices, the matrix-vector products can be computed approximately in $O(n)$ or $O(n\log(n))$ time by exploiting the fact that the matrices associated with BEM have certain properties. For example, nearby panels can be clustered together when evaluating their contributions to the potential at distant collocation points. This multiresolution idea is exploited in methods based on the Fast Multiple algorithm[8]. Alternatively, the near convolutional structure of the underlying integral equation can be exploited using the PFFT algorithm[9]. The FastStokes program uses a modification of the PFFT algorithm, which is described in more detail below.

The basic PFFT algorithm is easily illustrated using the single-variable electrostatic problem as an example. In the simpler electrostatic problem, the integral equation and its discretized form are

$$V(\vec{x}) = \frac{1}{4\pi\varepsilon}\int \frac{1}{\|\vec{x}-\vec{y}\|} q'(\vec{y})ds(\vec{y})$$
$$V_i = \sum_{j=1}^{\text{number of panels}} \Pi_{ij} q_j \qquad \text{or } \mathbf{V} = \mathbf{\Pi q} \tag{4.8}$$

with

$$\Pi_{ij} = \frac{1}{4\pi\varepsilon}\frac{1}{area(\text{panel j})}\int_{\text{panel j}} \frac{1}{\|\vec{x}-\vec{y}\|} ds(\vec{y})$$

where V_i is the voltage at the centroid of the ith panel, and $\mathbf{V}$ is the voltage vector. The charge density q' is assumed to be constant over each panel, and q_j is used to denote the net charge on the jth panel. The matrix element Π_{ij} is the potential at collocation point i due to unit net charge on panel j.

Given a voltage vector $\mathbf{V}$, consider computing the charge vector $\mathbf{q}$ by solving $\mathbf{V} = \mathbf{\Pi q}$ using the GMRES algorithm mentioned above. Since GMRES is a Krylov subspace method, it will be necessary to compute many matrix-vector products with the dense matrix $\mathbf{\Pi}$. The PFFT algorithm can be used to reduce the cost of computing matrix-vector products by separating the panel interactions into nearby and far field interactions. Then, the cost-dominant and smoother far field interactions are computed very rapidly by projecting and interpolating from an underlying uniform grid, and then resolving the grid interactions sing multidimensional FFT's. Note that we say the far-field interaction is smoother because the kernel $1/\|\vec{x}-\vec{y}\|$ varies much more slowly in space when the source point $\vec{y}$ is far from the field point $\vec{x}$. Nearby interactions have very rapid spatial variations, so they are computed directly using an accurate kernel integration algorithm to avoid large numerical errors.

The four major steps of the PFFT algorithm are listed below and are also pictorially illustrated in Fig. 4-1:

1. Project the panel charges onto the FFT grid $\mathbf{q}^{\mathbf{grid}} = \mathbf{W}^{\mathbf{projection}}\mathbf{q}^{\mathbf{panel}}$;
2. Compute grid voltages due to grid charges using the FFT's. This step can be expressed as $\mathbf{V}^{\mathbf{grid}} = \mathbf{ifft}\left(\mathbf{fft}(\mathbf{\Pi}^{\mathbf{grid}})\cdot \mathbf{fft}(\mathbf{q}^{\mathbf{grid}})\right)$;
3. Interpolate the grid voltages back to panel voltages. $\mathbf{V}^{\mathbf{panel}} = \mathbf{W}^{\mathbf{interpolation}}\mathbf{V}^{\mathbf{grid}}$;
4. Directly compute nearby interactions and use the results to replace the inaccurate nearby parts of the voltages calculated from the grid.

In addition to the modification of the transform part of the PFFT algorithm, there are other optimizations helpful for Stokes problems. The Stokes integral equation has three velocity components and three force components, but only six independent kernels, as $G_{ij} = G_{ji}$. This is a helpful observation because the FFTs of the grid kernels $\widetilde{G}_{jk}^{\text{gird}} = FFT(G_{jk}^{\text{gird}})$ are calculated once and then stored so that they can be used repeatedly for each matrix-vector product. The projection and interpolation matrices are also stored, and if polynomial projection is used then these matrices are coordinate independent and only one set is needed[10].

4. NULL SPACE OF THE SINGULAR INTEGRAL OPERATORS AND THE MODIFIED GMRES

The fact that only the derivative of pressure arises explicitly in the Stokes equation implies that any constant pressure can be added to the solution of the Stokes equation, and therefore the equation does not have a unique solution. This constant-pressure zero-velocity solution is a "singular mode" or a null space vector that does not affect the total forces on a single rigid body, but the singularity can impact the results produced by a numerical procedure. One approach presented by Tausch[11] to eliminating the null space is to add an addition operator to integral equation that maps the Stokes flow operator's null space to its defect in the range. Below we describe an alternative, one that removes the null-space using a modification of the GMRES iterative matrix solution algorithm. Our approach is not as general as the technique of Tausch[11], but it fits with the fast solver methodology and guarantees a null-space free solution independent of discretization or sparsification errors. Note that the null-space free solution is only useful for computing total body forces. To correctly compute the detailed force distribution, the null space contributions must be determined by solving an additional pressure matching equation[12].

Constant pressure force on the surface of a rigid body generates zero net body force or torque, and therefore zero velocity. Consider such a force, denoted f_j, acting on a rigid body. This force acts only in the surface normal direction of the rigid body and has a constant magnitude, so it is a multiple of the surface normal vector for the rigid body, denoted n_j. And since f_j generates zero velocity, it must follow that

$$\int_{surface} G_{ij} n_j ds = 0 \tag{4.11}$$

and therefore n_j is in the null space of the integral equation. In general, a problem with m independent bodies will have m independent null space vectors that correspond to being equal to the surface normal on one body and zero on the others. The discretization of an m-body system generates a system equation $\mathbf{U} = \mathbf{GF}$, where $\mathbf{G}$ is now the discrete form of the integral operator with an m-dimension null space given by the outward-normal vectors of the m objects in the system.

If a Krylov-subspace based method is applied to solve $\mathbf{U} = \mathbf{GF}$, then removing the null space of the $\mathbf{G}$ matrix can be performed by removing the null space from every Krylov-subspace vector since the final solution is in the Krylov-subspace.

$$Krylov\ \ Subspace = [U, GU, G^2U, G^3U, ...] \tag{4.12}$$

A simple approach to removing the null space is to remove the orthogonal projection on to the null space from every matrix-vector computed in the Krylov-subspace algorithm. Such an approach guarantees that the null space vectors will not contaminate any orthogonalization being performed on the Krylov-subspace. This is important because contamination of the Krylov subspace by the null space can interfere with convergence. Thus, this modification not only generates a null-space-free solution, but also makes the Krylov-subspace algorithm converge faster. To demonstrate this phenomenon, the GMRES algorithm was applied to solving a system of the form of Eq. (6) generated from a complicated fluid analysis. The convergence of GMRES with and without the null-space-remover is shown in Fig. 4-3, and demonstrates that without the null-space remover, the GMRES algorithm stalls.

It is worth noting that when a velocity vector associated with rigid body motion forms the right-hand side of Eq. (4.6), that velocity vector must satisfy a divergence-free condition. This implies the velocity vector is orthogonal to the integral equation null space. Orthogonality should guarantee that the associated Krylov subspace is also null-space free, and the null-space remover should be unnecessary. However, since the PFFT algorithm is used to compute approximates to matrix-vector products with G, the null space can easily appear and contaminate the subspace. Therefore, the null-space remover substantially enhances robustness.

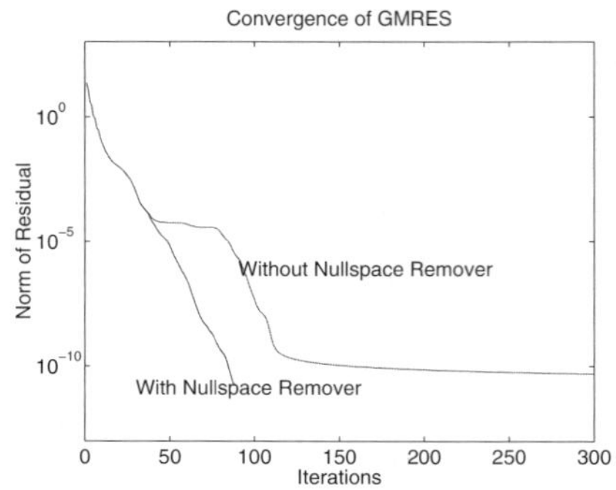

Figure 4-3. Convergence of the modified GMRES algorithm.

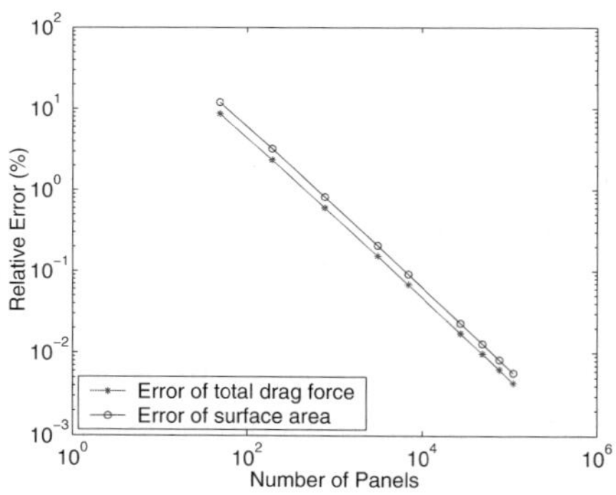

Figure 4-4. Percentage relative error of the sphere vs. the number of panels.

5. SIMULATION EXAMPLES

We present three simulation examples in this section to show the effectiveness of the steady incompressible FastStokes solver. The first simple sphere example demonstrates that the fast solver does not interfere with the convergence of the discretization method. The second and third examples, a comb drive and a micromirror, are used to demonstrate that the FastStokes program generates drag results that correlate surprisingly well with measured data.

5.1 A Translating Sphere

For the simple spherical geometry, an analytical solution of the Stokes equation exists. Given the radius of the sphere R_0 and a constant velocity $\vec{U}$, the drag force on the sphere is:

$$\vec{F} = 6\pi\mu R_0 \vec{U} \tag{4.13}$$

For this computational experiment, it is assumed that $\mu = 1, R_0 = 1, U_x = 1, U_y = U_z = 0$, and FastStokes is used to calculate the X-direction drag forces numerically. The red line in Fig. 4-4 shows the percentage relative error, and clearly indicates that the error decreases from approximately 2% for one hundred panels to 0.004% for 100,000 panels, and that decrease is a straight line when viewed on a log-log plot. The blue line in Fig. 4-4 shows the error of total surface area due to the flat panel discretization. Note the blue line is parallel and very close to the red line. This is because the error of the flow calculation is mainly due to the geometrical error of using a flat panel discretization, and this geometrical error is reflected by the error of total surface area. The CPU times of using the $O(n\log(n))$ FastStokes solver and the traditional $O(n^3)$ Gaussian Elimination method (LU decomposition) are compared in Fig. 4-5. If 5,000 panels are used, FastStokes is about 3,000 times faster than Gaussian Elimination. The memory used by Gaussian Elimination is $O(n^2)$ while that of FastStokes is much less (about $O(n) \sim O(n^{1.5})$); the comparisons are shown in Fig. 4-6. A 500-Mhz dual-processor computer running Alpha-Linux system is used for the simulations.

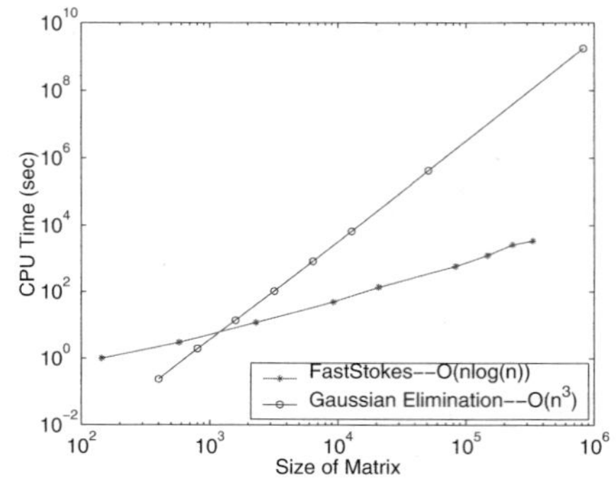

Figure 4-5. CPU times of FastStokes and Gauss Elimination.

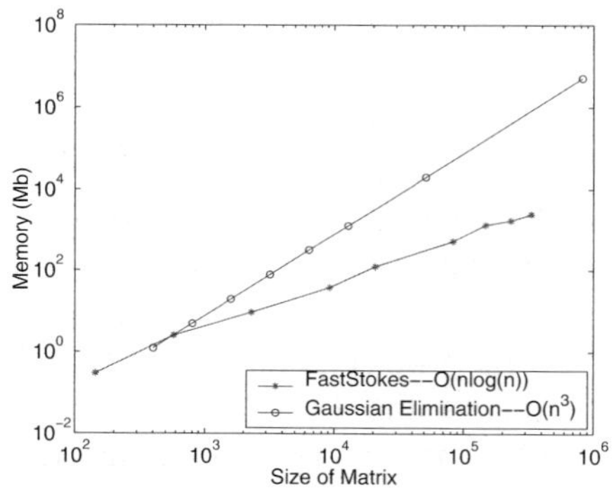

Figure 4-6. Memory usages of FastStokes and Gauss Elimination.

5.2 Comb-Drive Resonator

A lateral comb-drive resonator is shown in Fig. 4-7. The test structure was fabricated using the MUMPS process at MCNC (now Cronos Integrated Microsystems Inc., Research Triangle Park, NC). The dimensions of the resonator are given in Table 4-1. The movable comb-drive was set into motion in air at atmospheric pressure using an electrical stimulus to one static comb-drive. The magnitude and angle of the resulting motions were measured using the computer microvision technology[13]. The measured resonant frequency of the lateral motion is 19.2 kHz and the quality factor is 27.

Table 4-1. Resonator dimensions.

	Dimensions (μm)
Finger gap	2.88
Finger length	40.05
Finger overlap	19.44
Tether length	151
Tether width	1.1
Thickness	1.96
Substrate gap	2

A discretization using 16544 panels is shown in Fig. 4-8. The lateral direction surface force solution is shown in Fig. 4-9. Using the rigid-body assumption and a second order spring-mass-damper system as a macro-model, we calculate the damping coefficient b from the FastStokes result and then further calculate the quality factor Q, i.e.,

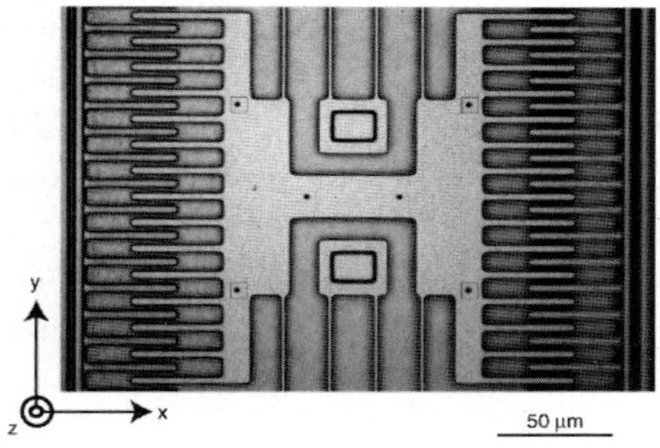

Figure 4-7. SEM of a lateral resonator.

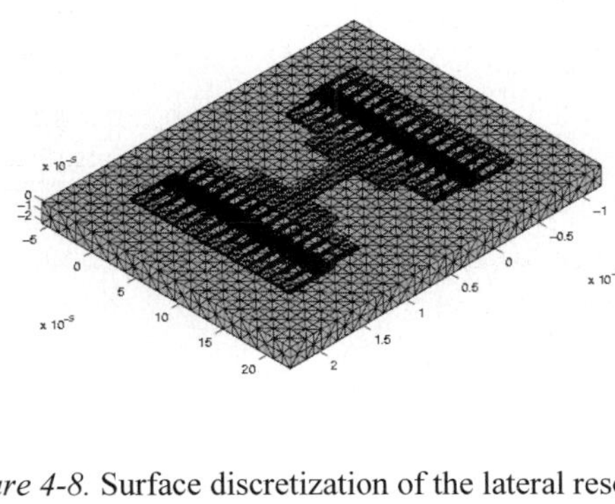

Figure 4-8. Surface discretization of the lateral resonator.

$$m_{eff}\ddot{x} + b\dot{x} + kx = F_{electro-static}$$
$$Q = \frac{\sqrt{km_{eff}}}{b} \quad (4.14)$$

$$m_{eff} = m_m + \frac{12}{35}m_b + \frac{1}{4}m_t = 5.61\times10^{-11}kg$$

where m_m and m_b are the masses of the movable comb-drive and the beam respectively[13], and m_t is the mass of the connecting truss. Stiffness k can be calculated from the resonance frequency and the effective mass using $k = (2\pi f_0)^2 m_{eff} = 0.816N/m$. The simulation result is compared with experimental result in Table 4-2. The steady incompressible FastStokes solver gave a numerical solution that is very close to the experiment results, while simple approaches such as using the Couette flow model failed. The convergence is shown in Fig. 4-10; the solution is accurate even if a coarse mesh with 4868 quadrilateral panels is used. The CPU time is shown in Fig. 4-11, a very fine discretization with 59,280 panels takes a little more than an hour's time.

Table 4-2. Comb-drive resonator simulation and measurement results.

	Q
Couette Flow	58.9
FastStokes	29.8
Experiment	27

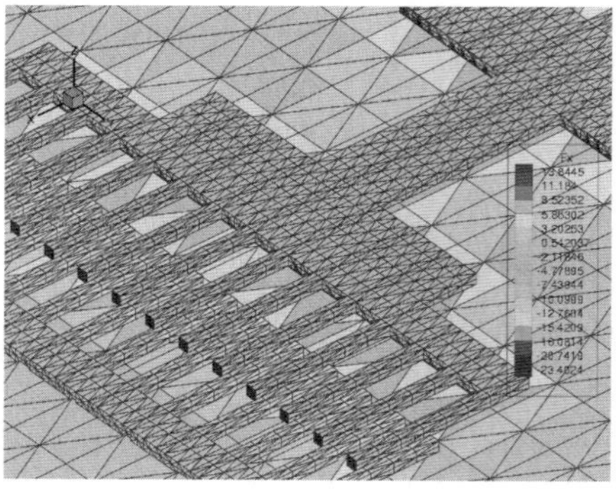

Figure 4-9. Detailed drag force on a lateral resonator using the incompressible Stokes model.

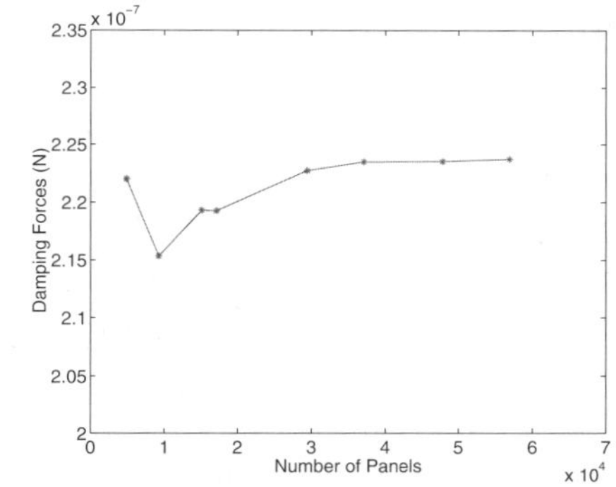

Figure 4-10. Convergence of the drag forces of comb-drive resonator simulation.

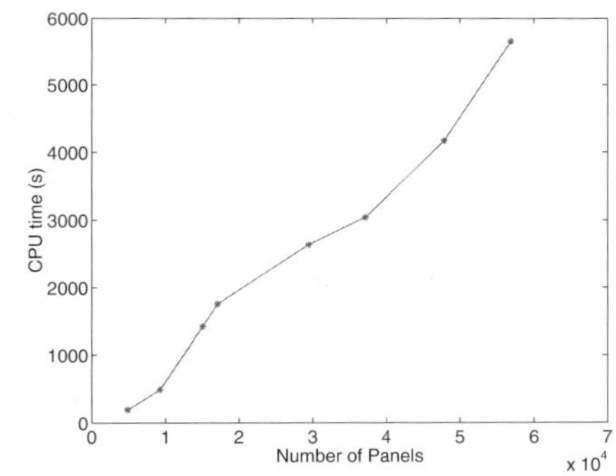

Figure 4-11. CPU times of comb-drive resonator simulation.

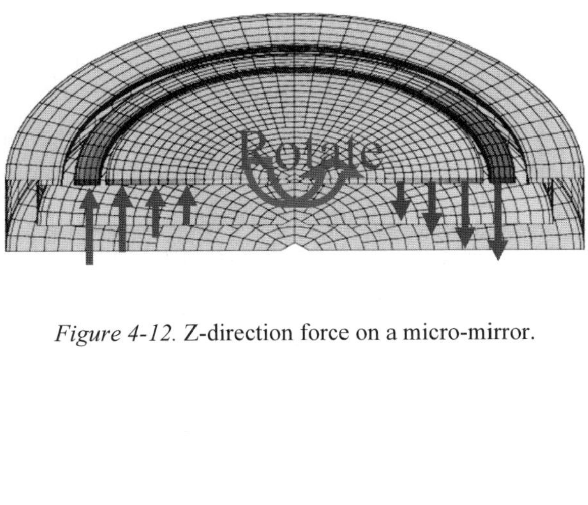

Figure 4-12. Z-direction force on a micro-mirror.

5.3 Micro-Mirror

An electrostatically actuated micro-mirror is simulated using FastStokes[12]. The micro-mirror is fabricated and tested in the Micromachined Product Division of Analog Devices Inc. (Cambridge, MA). The air-packaged micro-mirror is the critical part of an optical switch, and its dynamic performance is strongly affected by the viscous drag forces. Testing data have showed that the mirror is heavily damped with a quality factor around 2 in certain designs. Two major modes, the "mirror only" rotation mode and "mirror + gimbal" rotation mode, are simulated here. Table 4-3 compares the simulation results and experimental results of two different designs.

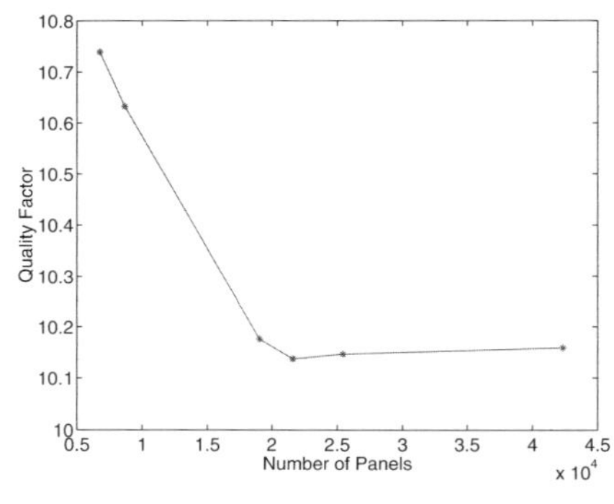

Figure 4-13. Convergence of the micro-mirror simulation.

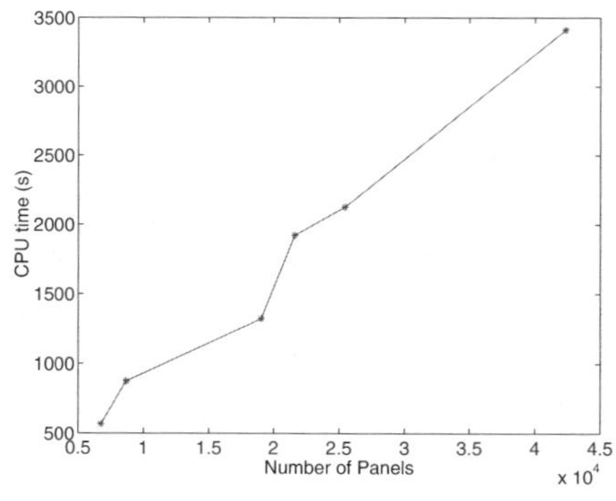

Figure 4-14. CPU times of the Micro-mirror simulation.

Table 4-3. Quality factors of the Micro-mirror simulations and measurements.

		Measured Q	Simulated Q	Error (%)
Mirror 1	Mirror +gimbal	2.31	2.36	2.16
	Mirror	3.45	3.14	8.99
Mirror 2	Mirror +gimbal	4.27	4.69	9.84
	Mirror	10.63	10.16	4.42

The simulated and measured quality factors match within 10%. Again, the small differences prove the accuracy of the FastStokes program. Fig. 4-12 shows the Z-direction force on a mirror when both mirror and gimbal rotate. Only half of the mirror is plotted in Fig. 4-12 in order to show a clear view of the force distribution. Fig. 4-13 shows that the simulation solution quickly converges as the discretization is refined. Fig. 4-14 shows the CPU time. The simulation was finished in less than an hour when 42,340 panels were used.

6. SUMMARY

In this paper we summarized the algorithms in FastStokes, and in particular described several specializations of the precorrected-FFT accelerated BEM algorithm to the Stokes flow problem. In addition, we gave timing results on several examples to demonstrate that FastStokes scales almost linearly with problem complexity, can easily analyze structures as complicated as an entire comb drive in under an hour, and can produce results that accurately match measured data.

The techniques in FastStokes have been extended to include slip boundary conditions, as these conditions are used to model non-continuum microfluidic effects[14,15,16]. As devices are scaled, slip effects will become important, but geometries in current common designs are still so large that the impact of slip effects on net drag is limited. Future work is on developing more efficient methods for handling substrate ground planes, and extending these fast fluid solver techniques to unsteady problems, convection-diffusion problems, cells-in-flow problems, and non-continuum problems.

ACKNOWLEDGEMENTS

The authors would like to thank Joel Phillips and Bjarne Buchmann for providing their software for the PFFT algorithm, and D. Freeman and W. Hemmert for supplying the comb drive example and measured data.

REFERENCES

1. J. Voldman, M. L. Gray, M. A. Schmidt, "Microfabrication in Biology and Medicine," *Annu. Rev. Biomed. Engr.*, vol. 1, pp. 421-425 (1999).
2. A. F. Mills, *Heat Transfer*, 2nd edition (Prentice-Hall Inc., Upper Saddle River, New Jersey, 1999).
3. C. Pozrikidis, *Boundary integral and singularity methods for linearized viscous flow* (Cambridge University Press, Cambridge, U.K., 1992).
4. J.N. Newman, "Distribution of sources and normal dipoles over a quadrilateral panel", *J. of Eng. Math.*, 20, pp. 113-126 (1986).
5. V. Rokhlin, "Rapid solution of integral equation of classical potential theory," *J. Comput. Phys.*, vol. 60, pp. 187-207 (1985).
6. K. Nabors and J. White, "FastCap: A Multipole-Accelerated 3-D Capacitance Extraction Program," *IEEE Transactions on Computer-Aided Design*, vol. 10, no. 10, pp. 1447-1459 (Nov. 1991).
7. Y. Saad and M. Schultz, "GMRES: A generalized minimal residual algorithm for solving nonsymmetric linear systems," SIAM J. Sci. Statist. Comput., vol. 7, no. 3, pp. 856-869 (July 1986).
8. L. Greengard, The Rapid Evolution of Potential Fields in Particle Systems (MIT Press, Cambridge, MA, 1988).
9. J. R. Phillips and J. K. White, "A Precorrected-FFT method for Electrostatic Analysis of Complicated 3-D Structures", *IEEE Trans. on Computer-Aided Design*, Vol. 16, No. 10, pp. 1059-1072 (October 1997).
10. W. Ye, J. Kanapka, X. Wang, J. White, "Efficiency and Accuracy Improvements for FastStokes, A Precorrected-FFT Accelerated 3-D Stokes Solver", *Proc. of Modeling and Simulation of Microsystems (MSM)*, San Juan, PR, pp. 502-505 (1999).
11. J. Tausch. "Rapid Solution of Stokes Flow using Multiscale Galerkin BEM" PAMM, Proc. Appl. Math. Mech. 1, pp. 8-11 (2002).
12. X. Wang. "FastStokes: A Fast 3-D Fluid Simulation Program for Micro-Electro-Mechanical Systems," Ph.D. thesis (Massachusetts Institute of Technology, Massachusetts, June 2002).
13. W. Ye, X. Wang, W. Hemmert, D. Freeman and J. White, "Air Damping in Lateral Oscillating Micro Resonators: a Numerical and Experimental Study," *IEEE/ASME Journal of Microelectromechanical Systems*, Vol. 12, No. 5, pp. 557 - 566 (2003).
14. A. B. Basset, *A Treatise on Hydrodynamics* (Cambridge University Press, 1888).
15. R.W. Barber and D.R. Emerson, *Advances in Fluid Mechanics IV*, pp. 207-216 (2002).
16. J. Ding and W. Ye, "A Fast Integral Approach for Drag force Calculation Due to Oscillatory Slip Stokes Flows," *International Journal for Numerical Methods in Engineering*, Vol. 60, No. 9, pp. 1535 – 1567 (2004).

APPENDIX: ANALYTICAL FLAT PANEL INTEGRATION ALGORITHM

Accurate calculation of the elements of ***G*** in Eq. (7) associated with nearby interactions is crucial to ensuring the accuracy of the Stokes flow calculation. Although these nearby terms are few in number, they are large and very sensitive to spatial variation. The most reliable approach to computing nearby interactions is to develop analytical formulas for panel integrals of Stokes kernels. For the FastStokes program, a fast analytical kernel integration algorithm was developed based on an extension of the method presented by Newman[4]. This precise algorithm is give below.

Local coordinate system

To simplify the calculations, a local Cartesian coordinate system (ξ,η,ζ) is set up so that the panel is in the $\xi-\eta$ coordinate plane with the centroid at the origin. Major computations of the kernel integration are done in the local coordinate system and the solutions are then transferred back to the global coordinate system. Transition between the local coordinate system (ξ,η,ζ) and the global coordinate system (x, y, z) can be expressed as

$$\begin{bmatrix}\xi\\ \eta\\ \zeta\end{bmatrix}_{3\times1}=\begin{bmatrix}Coordinate\\ Transformation\\ Matrix\end{bmatrix}_{3\times3}\left\{\begin{bmatrix}x\\ y\\ z\end{bmatrix}_{3\times1}-Centroid\right\};$$
$$\begin{bmatrix}X\\ Y\\ Z\end{bmatrix}_{3\times1}=\begin{bmatrix}Coordinate\\ Transformation\\ Matrix\end{bmatrix}_{3\times3}\left\{\begin{bmatrix}X'\\ Y'\\ Z'\end{bmatrix}_{3\times1}-Centroid\right\} \tag{A.1}$$

where (X,Y,Z) is the local coordinate of the evaluation point, and (X',Y',Z') is the corresponding global coordinate.

Integration of the Stokes kernels

According to Newman[4], the Gauss-Bonnet theorem can be used to calculate the potential due to constant -4π normal dipole distributed over the flat panel. The result is:

$$\Phi=Z\iint\frac{1}{r^3}d\xi d\eta=$$
$$\sum_{i=1}^{ns}\left\{\tan^{-1}\left(\frac{\delta\eta_i\left[(X-\xi_i)^2+Z^2\right]-\delta\xi_i(X-\xi_i)(Y-\eta_i)}{R_iZ\delta\xi_i}\right)\right. \tag{A.2}$$
$$\left.-\tan^{-1}\left(\frac{\delta\eta_i\left[(X-\xi_{i+1})^2+Z^2\right]-\delta\xi_i(X-\xi_{i+1})(Y-\eta_{i+1})}{R_{i+1}Z\delta\xi_i}\right)\right\}$$

where r is the distance between the evaluation point and a point on the panel; R_i is the distance between the evaluation point and the ith panel corner; ξ_i and η_i are the local coordinate of the ith panel corner; $\delta\xi_i = \xi_{i+1} - \xi_i$, $\delta\eta_i = \eta_{i+1} - \eta_i$; and ns is the number of corners. Integrating Φ in the direction of the panel normal yields Ψ, which is the potential due to -4π monopole distribution over a flat panel. The resulting formula is

$$\Psi = \iint \frac{1}{r} d\xi d\eta$$
$$= \sum_{i=1}^{ns} \left[(X - \xi_i)\sin\theta_i - (Y - \eta_i)\cos\theta_i \right] \cdot Q_i - Z\Phi \tag{A.3}$$

and

$$Q_i = \log \frac{R_i + R_{i+1} + s_i}{R_i + R_{i+1} - s_i}$$

Where θ_i is the polar angle of the ith edge; s_i is the length of the ith edge. Furthermore, the potentials due to linear, bilinear and higher-order dipoles distributions can be obtained in a similar way:

$$\begin{pmatrix} \Phi_x \\ \Phi_y \end{pmatrix} = Z \iint \begin{pmatrix} \xi \\ \eta \end{pmatrix} \frac{d\xi d\eta}{r^3} = \begin{pmatrix} X \\ Y \end{pmatrix} \Phi \pm Z \sum_{i=1}^{ns} Q_i \begin{pmatrix} \sin\theta_i \\ \cos\theta_i \end{pmatrix}$$

$$\Phi_{xy} = X\Phi_y + Y\Phi_x - XY\Phi$$
$$+ Z \sum_{i=1}^{ns} \cos\theta_i \left[v_i Q_i \sin\theta_i - (R_{i+1} - R_i)\cos\theta_i \right]$$

$$\Phi_{xx} = \Psi + \sum_{i=1}^{ns} \left\{ (R_{i+1} - R_i)\cos\theta_i \sin\theta_i \right.$$
$$\left. + (\xi_i + u_i\cos\theta_i - X)\sin\theta_i \ln\left(\frac{R_{i+1} - U_i}{R_i - u_i} \right) \right\} \tag{A.4}$$
$$\Phi_{yy} = \Psi - \sum_{i=1}^{ns} \left\{ (R_{i+1} - R_i)\cos\theta_i \sin\theta_i \right.$$
$$\left. - (\eta_i + u_i\sin\theta_i - Y)\cos\theta_i \ln\left(\frac{R_{i+1} - U_i}{R_i - u_i} \right) \right\}$$

where $(u_i, -v_i)$ and $(U_i, -V_i)$ are real and imaginary parts of two 2-D vectors starting from the ith corner and the $(i+1)^{th}$ corner individually; both vectors end at the projection of the evaluation point on the ith edge.

Transferring local solutions back to the global coordinate system

The above solutions are local solutions that must be transferred back to the global coordinate system. Here we offer a simple approach to the Stokes kernels. Assume that the solutions in the local coordinate system and in the global coordinate system are defined as:

$$\Phi_{m,n,k}^{local} = \iint \frac{1}{r^3} (X - \xi)^m (Y - \eta)^n Z^k \, ds$$

$$\Phi_{m,n,k}^{global} = \iint \frac{1}{r^3} (X' - x)^m (Y' - y)^n (Z' - z)^k \, ds$$

and

$$[\Phi]_1^{local} = \begin{bmatrix} \Phi_{1,0,0}^{local} \\ \Phi_{0,1,0}^{local} \\ \Phi_{0,0,1}^{local} \end{bmatrix}$$

$$[\Phi]_2^{local} = \begin{bmatrix} \Phi_{2,0,0}^{local} & \Phi_{1,1,0}^{local} & \Phi_{1,0,1}^{local} \\ \Phi_{1,1,0}^{local} & \Phi_{0,2,0}^{local} & \Phi_{0,1,1}^{local} \\ \Phi_{1,0,1}^{local} & \Phi_{0,1,1}^{local} & \Phi_{0,0,2}^{local} \end{bmatrix} \qquad \text{(A.5)}$$

$$[C] = \begin{bmatrix} Coordinate \\ Transformation \\ Matrix \end{bmatrix}_{3\times 3}$$

Then applying coordinate transition equations in Eq. (A.1) yields:

$$\Phi_{0,0,0}^{global} = \Phi_{0,0,0}^{local}$$

$$\Psi_{0,0,0}^{global} = \Psi_{0,0,0}^{local} \qquad \text{(A.6)}$$

$$[\Phi]_1^{global} = [C]^T [\Phi]_1^{local}$$

$$[\Phi]_2^{global} = [C]^T [\Phi]_2^{local} [C]$$

Chapter 5

COMPOSABLE BEHAVIORAL MODELS AND SCHEMATIC-BASED SIMULATION OF ELECTROKINETIC LAB-ON-A-CHIP SYSTEMS

Yi Wang[1], Qiao Lin[1], and Tamal Mukherjee[2]

[1]*Department of Mechanical Engineering, Carnegie Mellon University, 5000 Forbes Ave., Pittsburgh, PA 15213;* [2]*Department of Electrical and Computer Engineering, Carnegie Mellon University, 5000 Forbes Ave., Pittsburgh, PA 15213*

Abstract: This paper presents composable behavioral models and a schematic-based simulation methodology to enable top-down design of electrokinetic Lab-on-a-Chip (LoC) systems. Complex electrokinetic LoCs are shown to be decomposable into a system of elements with simple geometries and specific functions. Parameterized and analytical models are developed to describe the electrical and biofluidic behavior within each element. Electrical and biofluidic pins at element terminals support the communication between adjacent elements in a simulation schematic. An analog hardware description language implementation of the models is used to simulate LoC subsystems for micromixing and electrophoretic separation. Both direct current (DC) and transient analysis can be performed to capture the influence of system topologies, element sizes, material properties, and operational parameters on LoC system performance. Accuracy (relative error generally less than 5%) and speedup (>100×) of the schematic-based simulation methodology is demonstrated by comparison to experimental measurements and continuum numerical simulation.

Key words: Lab-on-a-Chip (LoC); electrokinetic; behavioral model; schematic-based simulation; electrophoresis; micromixer

K. Chakrabarty and J. Zeng (eds.),
Design Automation Methods and Tools for Microfluidics-Based Biochips, 109–142.

1. INTRODUCTION

Lab-on-a-chip (LoC) systems hold great promise for a wide spectrum of applications in biology, medicine, and chemistry[1, 2] due to their ability to integrate chemical analysis with other bio-processing functionalities[1]. Biofluidic LoCs have demonstrated tremendous advantages over conventional analysis methods, such as orders-of-magnitude analysis speedup, extremely low bio-sample consumption, parallel processing capability, high levels of integration, and ease of automation. Integrated biofluidic LoCs based on electrokinetic (EK) transport of charged biomolecules and biofluids are of particular interest as they are amenable to integration with EK injection, electrophoresis-based analysis, direct and accurate flow control[3], and electronics.

However, efficient modeling and simulation to assist design of such biofluidic LoCs at the system-level continue to be a challenge. This is due to the lack of an efficient design methodology to tackle the growing system complexity arising from two sources, namely: 1) more and more components are being integrated[4]; and 2) components with diverse functionalities are being integrated[5]. An additional source of the complexity is the nature of the microscale multi-physics phenomena within LoCs (e.g., the turn geometry induced skew and broadening of the species band[6] and the slow molecular diffusion-based mixing[7]), which requires accurate models and simulation for iterative design studies. Presently, detailed numerical simulation[8, 9] is the only available way to obtain desired modeling accuracy. However, their central processing unit (CPU) time and memory requirements are prohibitive for system-level design of complex LoCs. For example, a finite element based simulation of a simple microchip consisting of a pair of complementary turns for the electrophoretic separation application can cost several hours to days[10]. Reduced-order macromodels have to be built from each numerical simulation and stitched together for an overall system evaluation[11, 12]. The resulting macromodels in this bottom-up approach to design are specific to the geometry that was simulated numerically. Thus the macromodels have to be regenerated whenever the geometry is perturbed for design optimization. This leads to unacceptably long design iterations and hinders the industrial application of this approach to LoC design. To address these issues, efficient parameterized modeling as well as system (circuit) level simulation has recently attracted a lot of attention. Qiao *et al.*[13] proposed a compact model to evaluate the flow rate and pressure distribution of both EK and pressure driven flow within the network and capture the effect of the non-uniform zeta potential at the channel wall. Xuan *et al.*[14] later presented a fully analytical model to capture the effects of channel sizes and surface EK properties on microfluidic characteristics using phenomenological coefficients. Both papers focus on bulk fluid flow in microchannels and ignore the details of sample transport that often become the limiting issues in biochip design. Coventor's circuit level MEMS and microfluidics modeling and simulation environment, ARCHITECT[15], includes an EK library with simple models for injectors, straight channels, and turns that can model sample transport in electrophoretic separation. But it still requires users to extract parameters from full numerical simulation and, hence, does not allow design of electrophoresis channels of general shapes where the interaction of dispersion effects between turns can be very strong. Thus, its practical usefulness is significantly limited[10]. Zhang *et al.*[16] developed an integrated modeling and simulation environment for microfluidic systems in SystemC, which was used to evaluate and compare performance of continuous-flow and droplet based microfluidic systems on a polymerase chain reaction (PCR). Like the Coventor solution, the focus is at the system level, with an assumption that reduced-order models from detailed numerical simulation or experimental data are available. Most recently, Chatterjee *et al.*[17] combined circuit/device models to analyze fluidic transport, chemical reaction, reagent mixing as well as separation in integrated microfluidic systems. These models exploit an analogy between fluid and sample transport, effectively reducing the problem-governing partial differential equations into single ordinary differential equations or algebraic equations, leading to fast simulation speed. However, this speedup is at the cost of ignoring local geometry induced non-idealities.

This paper presents a top-down methodology that is both accurate and efficient in handling complex biofluidic LoC design. Based on the system hierarchy, we geometrically decompose a complex LoC into a collection of commonly used microfluidic channel elements. The design topology is captured by interconnecting these elements. Electrical and biofluidic information is exchanged between adjacent interconnected elements. Parameterized behavioral models are analytically derived to efficiently and accurately capture the multi-physics behavior of the elements. As a result, a complex LoC can be represented by a system-level schematic model that can be iteratively simulated to investigate the impact of changes in design topologies, element sizes, and material properties on the overall LoC performance. Examples will focus on EK passive micromixers and electrophoretic separation chips that can work as independent biofluidic devices or serve as subsystems of an integrated LoC system.

Unlike the bottom-up design methodology, where reduced-order models are obtained from numerical simulation, use of parameterized models enables a top-down design methodology similar to SoC[18] and MEMS design[19]. The top-down methodology (shown in Fig. 5-1) begins with a conceptual schematic representation of the system, gradually and hierarchically specifying components and elements of LoCs using the behavioral models stored in the library. Then the schematic representation is used for iterative performance evaluation and design optimization. The design process ends with numerical simulation to verify that the design goals have been reached, and the design is finally sent to layout and fabrication.

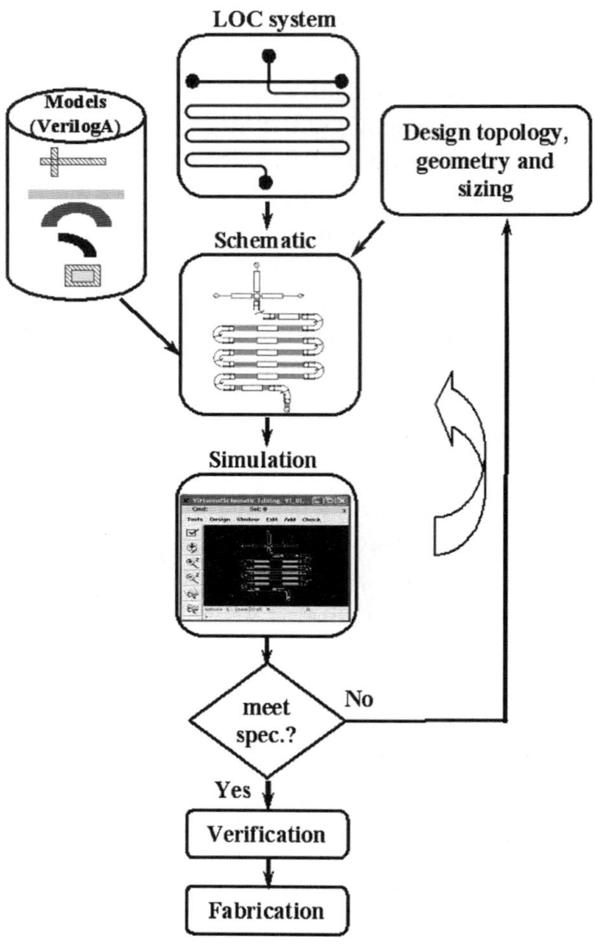

Figure 5-1. Flow chart of the modeling and simulation of biofluidic LoCs based on the top-down design methodology.

In this paper, we will consider hierarchical schematic representation of LoCs (in particular microchips used for electrophoretic separation and micromixing) in Section 2, including the system composition, operation, hierarchy of the LoCs, and definitions of pins and wires that enable communication between elements. A description of behavioral models of the elements appears in Sections 3. The behavioral models are stored in model libraries and used for schematic simulation with examples shown in Section 4. Physical design of lab-on-a-Chip systems are possible using the modeling and simulation described in this paper and are discussed elsewhere[20-22].

2. SCHEMATIC REPRESENTATION

In this section, we will first introduce the system composition and operation of a canonical LoC, as well as the functionalities that can be achieved. Then we will illustrate the process of decomposing a complex LoC into commonly used biofluidic elements based on the geometrical and functional hierarchy of the LoC. Then electrical and biofluidic pins, and analog wiring buses will be defined to link these elements and obtain a complete simulation schematic.

2.1 LoC Introduction

A variety of LoCs with diverse chemical and biological applications have been demonstrated to date. A canonical LoC system integrating functions of micromixing, reaction, injection, and separation is shown in Fig. 5-2. Its operation involves typical functions from a biochemical laboratory: synthesis and analysis. In the first phase, electrical voltages are applied between reservoirs 1, 2, 3, 4, and 5. The electrical field arising from these voltages moves the sample by EK flow[23], leading to dilution by the buffer solvent, or mixing with the reagent within the micromixer. The mixture then flows into a bio-chemical reactor. Reaction products are generated, often with the aid of external activations such as heat, light, or catalyst. Usually the sample and reagent are continuously supplied by the reservoirs; therefore concentrations of all the sample, reagent, and product in the mixer and reactor at this phase are in steady state. This completes the synthesis operation.

In the second phase (analysis), the voltage is switched on reservoirs 6 and 7 with the others left floating. Thus a band of the analyte (from the reaction products) is injected into the separation channel for further analysis (in addition to the cross injection shown in Fig. 5-2, other injection schemes are also available[24]). Because the analyte is comprised of biological species/molecules [e.g., deoxyribonucleic acid (DNA) or amino acids] with different charges and sizes, they move at different speeds and eventually can be separated by electrophoresis[25]. In this phase, the species bands broaden due to molecular diffusion and other dispersion sources; therefore the transient evolution of the band concentration is of primary importance.

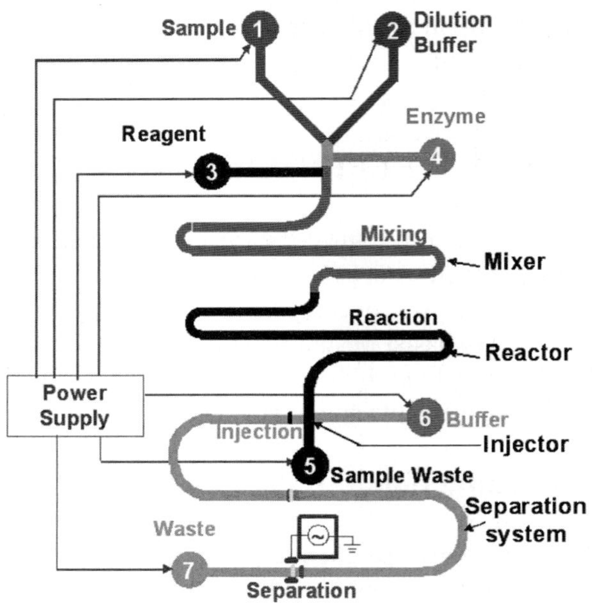

Figure 5-2. Sketch of a canonical EK biofluidic LoC.

In this paper, we will discuss behavioral models and schematic based simulation of micromixers and electrophoretic separators separately, because of their distinctly different biofluidic behavior. Injector[26, 27] and reactor[28] models are presented elsewhere. The top-down design methodology based on hierarchical decomposition that is valid to the entire integrated system[29], will be described jointly (Sections 2.2 and 2.3).

2.2 System Hierarchy

The schematic representation of the biofluidic LoC is based on its geometrical and functional hierarchy. A complex system can be decomposed into a set of commonly used elements of simple geometries, each with an associated function (e.g. mixing or separation), such as the straight mixing channel or semi-circular turn separation channel. This decomposition enables derivation of a closed-form parameterized model. The elements and their models can be reused in a top-down manner to represent various chip designs using different topologies, element sizes, and material properties.

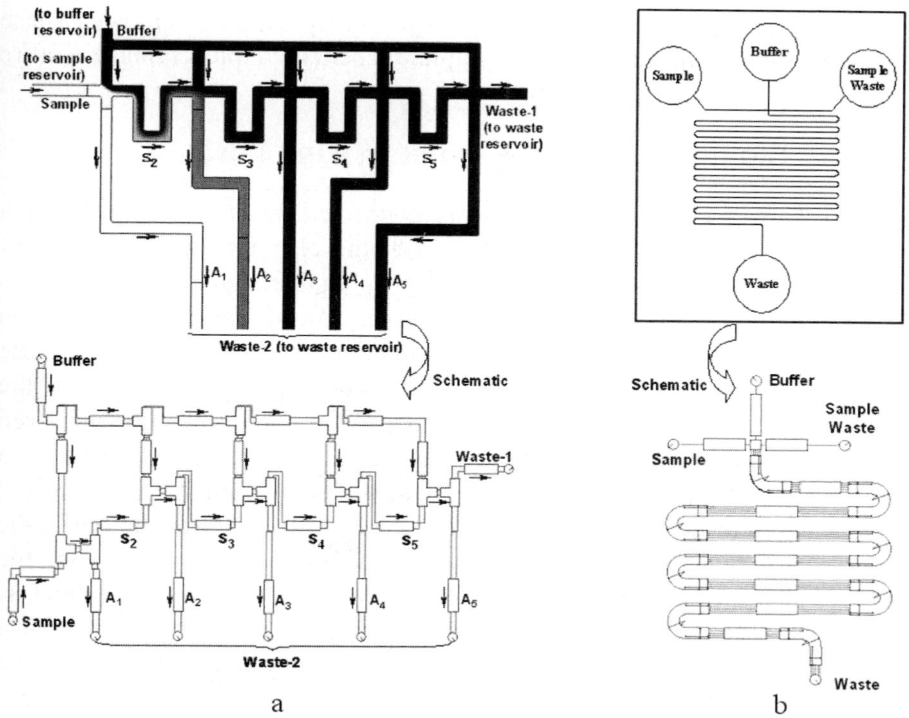

Figure 5-3. (a) An EK serial mixing network[3] and its hierarchical schematic representation. (b) A serpentine electrophoretic separation microchip[25] and its hierarchical schematic representation.

Figure 5-3a illustrates a complex EK serial mixing network[3] consisting of reservoirs, mixing channels, and T- and cross-intersections. The sample is released and collected by reservoirs at the extreme ends of the mixer. Within the cross intersection, a portion of the input sample is diverted to analysis channels A_1–A_5 and the rest continues along dilution channels S_2–S_5 for further dilution. Repeating this functional unit in series leads to an array of continuously diluted sample concentrations in channels A_1–A_5 that can be used for parallel biochemical analysis and titration tests. Variations of sample concentrations are indicated by grey levels in numerical simulation shown in Fig. 5-3a. In our approach, we represent the serial mixing network as a collection of interconnected mixing elements composed of microchannels, converging intersections, and diverging intersections (note that the double-input and double-output cross-intersection at both ends of channels S_2–S_5 is modeled as a combination of the converging and diverging intersections). Fig. 5-3b shows a serpentine electrophoretic separation microchip, which is similarly decomposed into a set of elements including reservoirs, injector, straight channels, 180° turns, and 90° elbows. These elements are then wired to form a complete schematic representing the entire LoC according to its chip topology.

2.3 Biofluidic Pins and Analog Wiring Buses

In the schematic, element terminals are connected by groups of pins. Each pin defines the state of biofluidic signal at the element terminals. Pins of adjacent elements are then linked by wires to enable signal transmission in the hierarchical LoC schematic. Therefore, pin definition affects both schematic composition and behavioral modeling of the elements. There are two types of pins defined in the network. One is the electrical pin at the element terminals. This type of pin is independent of the function achieved by the LoC and is present in all elements. It is used to construct a Kirchhoffian network, with both the voltage at the pin and the current flowing through the element. The second type of pin captures the biofluidic state, which is calculated in terms of a directional signal flow from the upstream to the downstream. That is, its value at an element outlet is determined from the corresponding value at the inlet and the element's own contribution. Pin values at the outlet are assigned to those at the inlet of the next downstream element. Schematic simulation can then serially process each element, starting from the most upstream element. The details of the information that need to be captured for complete definition of the biofluidic state depend on the functionality of the network. In this paper, we focus on two types of biofluidic networks, micromixing and electrophoretic separation.

Within the micromixer, different samples or reagents carried by EK flow mix with each other and their concentrations stay steady state provided there is a continuous supply from the inlet reservoirs. The sample concentration profile c (as a function of the widthwise position of the channel) describes the biofluidic state at the element terminals in this network, as shown in Fig. 5-4a, where $\eta = y/w$ $(0 \le \eta \le 1)$ is the normalized widthwise coordinate of mixing elements. Therefore, this pin uses a vector of concentration coefficients $\{d_n\}$, the Fourier cosine series coefficients of the widthwise concentration profile. The reason for such a choice is attributed to the fact that Fourier cosine series is the eigen-function of the convection-diffusion equation governing the sample concentration in the network, given the insulation condition at channel walls and normalized widthwise position from 0 to 1.

For the electrophoretic separation microchip, the injected species bands move through the microchannel accompanied by the band-spreading effect that is caused by dispersion (e.g., molecular diffusion and turn-geometry induced dispersion[10]). This band spreading adversely affects separation performance by reducing the detectability and separation resolution of the bands. Therefore, the state associated with the species band shape, such as the width of the band, skew, and amplitude as shown in Fig. 5-4b, is needed, as well as the time at which the band reaches the element terminal. Specifically, the concentration profile c of a skewed band is first cross-sectionally averaged, yielding a distribution of the average concentration c_m in the EK flow direction. Thus, pins are defined in terms of the variance σ^2, the square of the standard deviation of c_m distribution in the flow direction, representing the width of the band; the Fourier cosine series coefficients $\{S_n\}$ used to reconstruct the skew c_1 (the centroid positions of the axial filaments of the species band[10], see APPENDIX) caused by the non-uniform electrical field and migration distance in turns (the Fourier cosine series is used again for the same reason as the above); the separation time (t), the moment the band's centroid reaches the element terminal; and amplitude (A), the maximum average concentration.

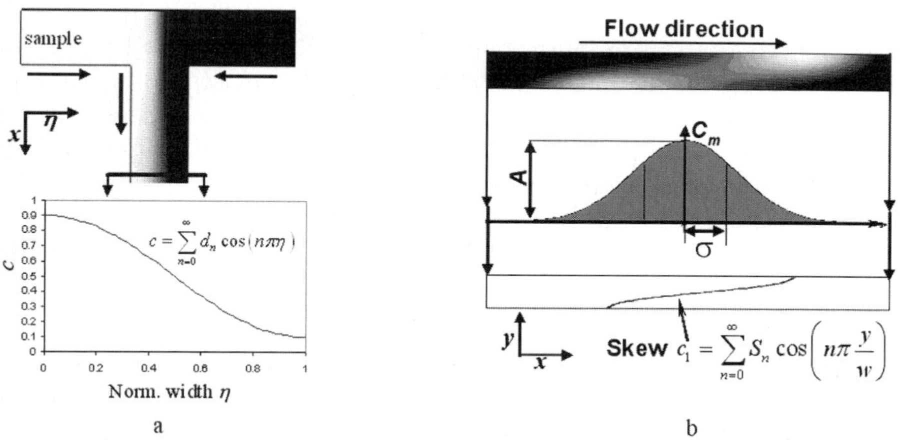

Figure 5-4. (a) Biofluidic pin definitions for (a) a micromixer and (b) an electrophoretic separation microchip.

The concentration profile of samples in the micromixer and concentration skew of species bands in the separation channel are defined in terms of a vector of Fourier cosine coefficients. For most biofluidic applications, ten terms ($n = 1,3,...19$ for separation and $n = 0...9$ for mixing) for each species/sample are found to yield sufficient computational accuracy due to quick convergence of the Fourier cosine series. These behavioral models allow for a virtually arbitrary number of different species/samples coexisting in the buffer. Each species requires its own set of pins for the biofluidic state (electrical pins can be shared among species). To reduce the wiring effort between elements, analog wiring buses are employed and the wires connecting the pins of the same discipline are grouped, resulting in only one bus (concentration coefficients) and four buses (separation time, variance, skew, and amplitude) at the terminals of the mixing and separation elements respectively, as shown in Fig. 5-5 and Fig. 5-6. Table 5-1 summarizes the numbering and disciplines of the buses used in both mixing and separation behavioral models with the implementation of three samples/species. Fig. 5-3 also illustrates the schematics with the symbol view of the behavioral models interconnected by wiring buses.

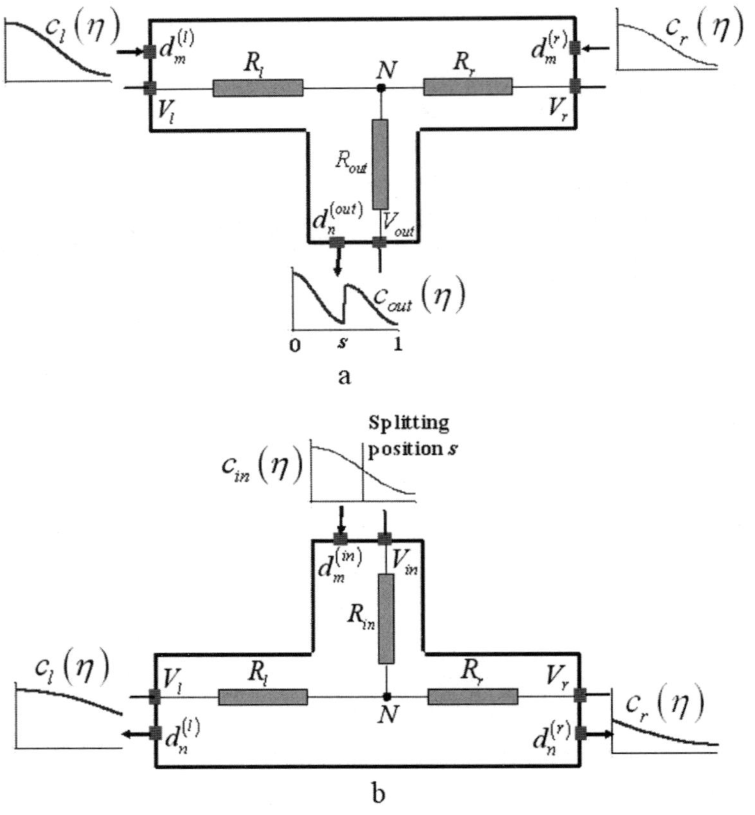

Figure 5-5. (a) Behavioral model structure for the converging intersection in the micromixer. (b) Behavioral model structure for the diverging intersection in the micromixer.

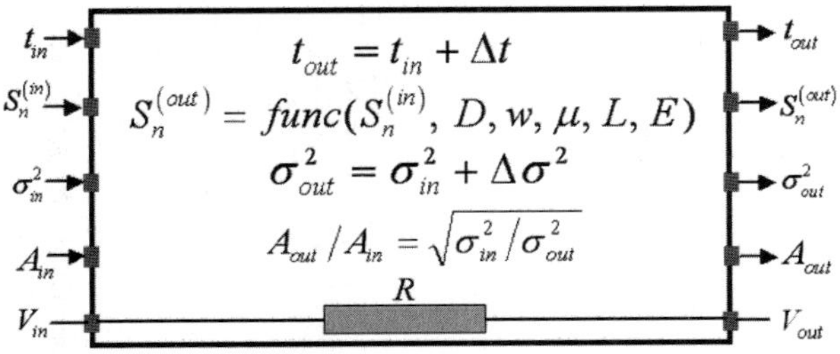

Figure 5-6. Behavioral model structure for separation channels in electrophoretic separation microchips.

Table 5-1. Definition of Biofluidic Pins.

Micromixing		
Bus	Pins connected	Description
d [0:29]	Concentration coefficients	*d* [0:9]: the 1st sample, *d* [10:19]: the 2nd, *d* [20:29]: the 3rd
Electrophoretic Separation		
Bus	Pins connected	Description
t [0:2]	Separation time	*t*[0] for the 1st species, *t*[1] the 2nd, *t*[2] the 3rd
σ^2 [0:2]	Variance	σ^2 [0] for the 1st species, σ^2 [1] the 2nd, σ^2 [2] the 3rd
A [0:2]	Amplitude	*A* [0] for the 1st species, *A* [1] the 2nd, *A* [2] the 3rd
S [0: 30]	Skew coefficients	*S* [0]: the direction of the skew caused by the 1st turn *S* [1:10]: the 1st species *S* [11:20]: the 2nd, *S* [21:30]: the 3rd

3. BEHAVIORAL MODELS

The goal of each behavioral model is to capture the input-output signal flow relationship of the pin value that defines the biofluidic state at the inlet and outlet of each element. This captures the physical phenomena being modeled in that element. In addition, an electrical resistance is associated with each element to relate the EK current flow through the element to the inlet and outlet voltages. In contrast to the bottom-up reduced-order model approaches, our behavioral models possess several important attributes to enable accurate and efficient system-level simulation of complex LoCs. Our analytical models effectively account for the same multi-physics (e.g., electrostatics, fluidics, and mass transfer) as numerical simulation tools. They do not require any parameters from user-conducted experiments or numerical simulations to capture interactions between the elements, and, hence, provide seamless model interconnectivity. Most importantly, they are in closed-form and are all parameterized by element dimensions and material properties;

therefore, they are reusable, fast to evaluate, and well suited for an iterative simulation-based design methodology.

As discussed above, depending on the physical phenomena of individual devices, contents of the behavioral model libraries will be different. Hence, models for the micromixer and electrophoretic separation system will be developed separately, and are available in separate model libraries for schematic-based simulation.

3.1 EK Passive Micromixers

The EK passive micromixer library consists of models for nine elements, which includes reservoirs (sample and waste), slightly tapered straight mixing channel, turns (90° or 180°, clockwise or counterclockwise), as well as converging and diverging intersections. In this section, we will present behavioral models for basic elements such as the slightly tapered mixing channel, converging and diverging intersections. Other elements can be modeled in a similar fashion.

3.1.1 Slightly Tapered Straight Mixing Channels

The tapered straight mixing channel, in which different samples *a* and *b* mix with each other, has one inlet and one outlet, with different cross-sectional area. It is critical in designing a geometrical focusing micromixer[30]. Electrically, it is modeled as a resistor and the resistance is given by

$$R = \int_0^L \frac{dz}{w(z)h(z)C_e} \tag{5.1}$$

where w and h are the channel width and depth (both are functions of the axial coordinate z), C_e is the electrical conductivity of the buffer solution in the channel. As a special case, in a straight channel with the uniform cross-section, Eq. (5.1) can be reduced to

$$R = \frac{L}{whC_e} \tag{5.2}$$

To obtain the sample concentration profile at the outlet, we partition the slightly tapered straight channel into a series of segments (segment number tends to infinity), each with uniform cross section. In each segment, the convection-diffusion equation is solved to establish the input-output relationship of concentration coefficients between the segment terminals.

Then all the segmental solutions are multiplied and the concentration coefficients $d_n^{(out)}$ ($n = 0,1,2...$) at the channel outlet are attained as[29, 31]

$$d_n^{(out)} = d_n^{(in)} e^{-\gamma n^2 \pi^2 \frac{LD}{E_{in} \mu w_{in}^2}} \tag{5.3}$$

where $d_n^{(in)}$, w_{in}, and E_{in} are the concentration coefficients, channel width, and electrical field at the inlet, respectively, γ is a factor capturing the effect of the cross-sectional shape on mixing[31], D and μ are the diffusivity and EK (including both electroosmotic and electrophoretic) mobility of the sample, and L is the channel length. The special case of a straight channel with the uniform cross-section yields $\gamma = 1$.

3.1.2 Converging Intersections

Figure 5-5 shows the behavioral model structure of converging and diverging intersections in micromixers[3]. Arrows at pins indicate the signal flow direction of computing biofluidic pin values. The converging intersection acts as a combiner to merge and compress upstream sample flows and their concentration profiles side by side at its outlet (Fig. 5-5a). As its flow path lengths are negligibly small compared with those of mixing channels, such an element can be assumed to have zero physical size, and electrically represented as three resistors with zero resistance between each terminal and the internal node N

$$R_l = R_r = R_{out} = 0 \tag{5.4}$$

Here, N is the intersection of flow paths and subscripts l, r, and out represent the left and right inlets, and the outlet, respectively. Denote $d_m^{(l)}$ and $d_m^{(r)}$ ($m = 0,1,2...$) the Fourier coefficients of the sample concentration profiles at the left and right inlets, respectively. Then the coefficients $d_n^{(out)}$ ($n = 0,1,2...$) of the profile at the outlet ($c_{out}(\eta)$) are given by

$$c_{out}(\eta) = \sum_{n=0}^{\infty} d_n^{(out)} \cos(n\pi\eta) = \begin{cases} \sum_{m=0}^{\infty} d_m^{(l)} \cos\left(\dfrac{m\pi\eta}{s}\right), & 0 \le \eta < s \\ \sum_{m=0}^{\infty} d_m^{(r)} \cos\left(\dfrac{m\pi(\eta - s)}{1-s}\right), & s \le \eta < 1 \end{cases} \tag{5.5}$$

Equation (5.5) shows that the concentration profile at the outlet can be treated as a superposition of the scaled-down profiles from both inlets, where $s = q_l/(q_l + q_r) = I_l/(I_l + I_r)$ denotes the interface position [or flow ratio, the ratio of the left flow rate q_l to the total flow rate $(q_l + q_r)$] between incoming streams in the normalized coordinate at the outlet (note that flow rates q_l and q_r are, respectively, linear with the electrical currents I_l and I_r).

Solving Eq. (5.5) yields $d_n^{(out)}$ as

$$\begin{cases} d_0^{(out)} = d_0^{(l)} s + d_0^{(r)}(1-s) \\ d_{n>0}^{(out)} = s \sum_{m=0}^{\infty,\ \text{if } m \neq ns} d_m^{(l)} \frac{f_1 \sin(f_2) + f_2 \sin(f_1)}{f_1 f_2} \\ \quad + s \sum_{m=0}^{\infty,\ \text{if } m=ns} d_m^{(l)} + (1-s) \sum_{m=0}^{\infty,\ \text{if } m=n(1-s)} (-1)^{n-m} d_m^{(r)} \\ \quad + 2(-1)^n (1-s) \sum_{m=0}^{\infty,\ \text{if } m \neq n(1-s)} d_m^{(r)} \left(\frac{\cos(F_2/2)\sin(F_1/2)}{F_1} \right. \\ \quad \left. + \frac{\cos(F_1/2)\sin(F_2/2)}{F_2} \right) \end{cases} \tag{5.6}$$

where $f_1 = (m - ns)\pi$, $f_2 = (m + ns)\pi$, $F_1 = (m + n - ns)\pi$ and $F_2 = (m - n + ns)\pi$. Since the sample concentration profiles at the inlets are scaled down, the Fourier series modes at the inlets are not orthogonal to those at the outlet. Therefore, the calculation of the coefficient for a certain Fourier mode at the outlet depends on all the modes at the inlets.

3.1.3 Diverging Intersections

The diverging intersection has one inlet and two outlets and is the dual of the converging intersection. It splits the incoming flow and electrical current into two streams that exit out of the outlets. It can also be represented by three zero-resistance resistors,

$$R_{in} = R_l = R_r = 0 \tag{5.7}$$

where subscripts *in*, *l*, and *r* represent quantities at the inlet, the left and right outlets, respectively.

Defining $d_m^{(in)}$ (m = 0, 1, 2...) as the Fourier coefficients of the sample concentration profile at the inlet. Denote $d_n^{(l)}$ and $d_n^{(r)}$ the coefficients at the left and right outlets. Then sample concentration profiles of the left and right outgoing streams are given by

$$c_l(\eta) = \sum_{n=0}^{\infty} d_n^{(l)} \cos(n\pi\eta) = \sum_{m=0}^{\infty} d_m^{(in)} \cos(m\pi s\eta) \tag{5.8}$$

and

$$c_r(\eta) = \sum_{n=0}^{\infty} d_n^{(r)} \cos(n\pi\eta) = \sum_{m=0}^{\infty} d_m^{(in)} \cos\left(m\pi(1-s)\left(\eta + \frac{s}{1-s}\right)\right) \tag{5.9}$$

Solving Eqs. (5.8) and (5.9) yields

$$\begin{cases} d_0^{(l)} = d_0^{(in)} + \sum_{m=1}^{\infty} d_m^{(in)} \sin(\phi_1)/\phi_1 \\ d_{n>0}^{(l)} = 2 \sum_{m=0}^{\infty,\ \text{if}\ m\neq n/s} d_m^{(in)} (-1)^{n+1} \phi_1 \sin(\phi_1)/f_1 f_2 + \sum_{m=0}^{\infty,\ \text{if}\ m=n/s} d_m^{(in)} \end{cases} \tag{5.10}$$

and

$$\begin{cases} d_0^{(r)} = d_0^{(in)} - \sum_{m=1}^{\infty} d_m^{(in)} \sin(\phi_1)/\phi_2 \\ d_{n>0}^{(r)} = 2 \sum_{m=0}^{\infty,\ \text{if}\ m\neq n/(1-s)} d_m^{(in)} \phi_2 \sin(\phi_1)/F_1 F_2 + \sum_{m=0}^{\infty,\ \text{if}\ m=n/(1-s)} (-1)^{m-n} d_m^{(in)} \end{cases} \tag{5.11}$$

where $f_1 = (n - ms)\pi$, $f_2 = (n + ms)\pi$, $F_1 = (n + m - ms)\pi$, $F_2 = (n - m + ms)\pi$, $\phi_1 = ms\pi$, and $\phi_2 = m(1-s)\pi$. Similar to the converging intersection, s is the normalized splitting position (or ratio).

It should be pointed out that in contrast to the resistor-based mixing models[3, 17] that exploit the analogy between fluidic and sample transport and only convey average concentration values through the entire network, our models [Eqs. (5.3), (5.6), (5.10) and (5.11)] propagate sample concentration profiles characterized by the Fourier series coefficients. This removes the constraint of complete mixing (along the channel width) at the end of each channel[3] in the network imposed by the resistor-based models and allows for optimal design of both effective and efficient micromixers.

3.2 Electrophoretic Separation Chips

The electrophoretic separation library includes models for ten basic elements: turns (90° or 180°, clockwise or counterclockwise), straight channel, detector, injector, injection channel, and reservoirs (sample and waste). In this section, behavioral models for basic elements such as separation channels (straight and turn) will be developed to analyze the band-spreading effect caused by molecular diffusion and turn dispersion. Additionally, a detector model applicable for both direct current (DC) and transient analysis will be presented. Models of the other elements can be derived using the same principles.

Figure 5-6 shows the behavioral model structure of electrophoretic separation channels (straight or turn). Arrows indicate the direction of signal flow for calculating biofluidic pin values and state. Electrically, separation channels are modeled as resistors in the same way as the uniform straight mixing channels (for a constant-radius turn, L in Eq. (5.2) is replaced by $L = r_c\theta$, where r_c and θ are the mean radius and angle included by the turn, see[10, 32] for the detailed geometrical interpretation). Additionally, symbols and characters used in this section are defined the same as those for the mixer, unless otherwise noted. The residence time Δt of a species band within a separation channel (the time for the band's centroid to move from the channel inlet to outlet) is given by

$$\Delta t = \frac{L}{\mu E} \tag{5.12}$$

The calculation of changes in the skew coefficients and variance depends on the specific element[10] and the inherent variable is the residence time Δt obtained by Eq. (5.12). For a straight separation channel

$$S_n^{(out)} = S_n^{(in)} \cdot e^{-(n\pi)^2 \Delta t D / w^2} \tag{5.13}$$

$$\sigma_{out}^2 - \sigma_{in}^2 = \Delta\sigma^2 = 2D \cdot \Delta t \tag{5.14}$$

For a separation turn,

$$S_n^{(out)} = \pm \frac{8\theta w^2 \left(1 - e^{-(n\pi)^2 \Delta t D / w^2}\right)}{(n\pi)^4 \Delta t D} + S_n^{(in)} e^{-(n\pi)^2 \Delta t D / w^2} \quad , \quad n = 1,3,5... \quad (5.15)$$

$$\begin{aligned} \sigma_{out}^2 - \sigma_{in}^2 = \Delta\sigma^2 = 2D\Delta t \pm \frac{8w^4\theta}{D\Delta t} \sum_{n=1,3,5...}^{\infty} \left(\frac{S_n^{(in)} \left(1 - e^{-(n\pi)^2 D\Delta t / w^2}\right)}{(n\pi)^4} \right) \\ + \frac{64 w^6 \theta^2}{(D\Delta t)^2} \sum_{n=1,3,5...}^{\infty} \frac{\left(-1 + e^{-(n\pi)^2 D\Delta t / w^2} + (n\pi)^2 D\Delta t / w^2\right)}{(n\pi)^8} \end{aligned} \quad (5.16)$$

where subscripts/superscripts *in* and *out* represent quantities at the inlet and outlet of the channel, respectively. In Eqs. (5.15) and (5.16), the "+" sign is assigned to the first turn and any turn strengthening the skew caused by the first; the "-" sign is assigned to any turn undoing the skew from the first. For example, in Fig. 5-3b, the first 90° elbow and the three 180° turns on the left are all given a "+" sign, in which the species band flows counterclockwise. On the contrary, the three 180° turns on the right use a "-" sign, in which the band migrates clockwise.

Assuming a Gaussian distribution of the average concentration c_m of the species band at element terminals, we can obtain the amplitude of the species band by

$$A_{out} / A_{in} = \sqrt{\sigma_{in}^2 / \sigma_{out}^2} \quad (5.17)$$

For the detector model, the variance change associated with the detector path length L_{det} is given by[25]

$$\Delta\sigma^2 = L_{\det}^2 / 12 \quad (5.18)$$

3.3 Model Implementation

To demonstrate use of the above parameterized models for top-down design, we have implemented the models in the Verilog-A analog hardware description language. Symbol view for each of the elements is used to compose a schematic within Cadence's[33] integrated circuit design framework (e.g., Fig. 5-3). The Cadence design framework is used to automatically netlist the complex topologies in the biofluidic LoC schematics, and Spectre is used as the simulator. Similar tools from other vendors, or custom schematic entry tools and solvers that can handle both signal flow and Kirchhoffian networks could also used.

```
`define NUM 10                          // The number of terms in the Fourier series
...

module Uturn1(...);                     // Module declaration
parameter real sp_D_0 = 500;            // Diffusivity of the zero-th species; unit µm²/s
parameter real R2 = 110 ;               // Outer radius of the U shape turn; unit um
parameter real R1 = 100 ;               // Inner radius of the U shape turn; unit um
...

integer Fs;                             // System flag
integer Fi;                             // Intrinsic flag
real dummy1;                            // Intermediate variable
real W;                                 // Channel width
real delta_t_0;                         // The residence time of the zero-th species within the turn; unit s
real pi;                                // π
...

analog begin                            // Behavioral description begins

Fi = 1;                                 // The present turn is clockwise
W = abs (R2-R1);
Fs = SKA(sk_in[0]);                     // sk_in is the skew coefficient; sk_in[0] stores the system flag
pi = 3.1415926;
...

if (Fs == 0) begin                      // systemflag is not set yet; no turns or elbows upstream
        SKA(sk_out[0])<+(Fi);           // set the systemflag as the intrinsic flag
        generate i(1,`NUM, 1) begin     // this loop is to calculate skew coefficient
                dummy1 = 1- exp(-(2*i-1)*(2*i-1)*pi*pi*delta_t_0*sp_D_0/(W*W));
                Part1_Sk = 8*pi*W*W*dummy1/(pow((2*i-1), 4)*pow(pi, 4)*delta_t_0*sp_D_0); // the 1st part in (5-15), use "+"
                ...
        end
...

end else if (Fs == 1) begin             // Species flow in the first turn is clockwise
        SKA(sk_out[0])<+SKA(sk_in[0]);  // Convey the system flag
        generate i(1,`NUM, 1) begin
                dummy1 = 1- exp(-(2*i-1)*(2*i-1)*pi*pi*delta_t_0*sp_D_0/(W*W));
                Part1_Sk = 8*pi*W*W*dummy1/(pow((2*i-1), 4)*pow(pi, 4)*delta_t_0*sp_D_0); // use "+" sign
                ...
        end
...


end else begin                          // Species flow in the first turn is counter-clockwise
        SKA(sk_out[0])<+SKA(sk_in[0]);  // Convey the system flag
        generate i(1,`NUM, 1) begin
                dummy1 = 1- exp(-(2*i-1)*(2*i-1)*pi*pi*delta_t_0*sp_D_0/(W*W));
                Part1_Sk = - 8*pi*W*W*dummy1/(pow((2*i-1), 4)*pow(pi, 4)*delta_t_0*sp_D_0); // use "-" sign
                ...      ↖ use "-" sign
        end
...

end
...
end                                     // End of behavioral description
endmodule                               // End of module
```

Figure 5-7. Verilog-A description for a 180° turn involving clockwise flow of the species band. It determines the signs used in Eq. (5.15), as well as the canceling and strengthening effects on the skew.

An important issue of implementing separation channel models of turn geometry [Eqs. (5.15) and (5.16)] is the real-time determination of the turn "sign". Providing this flexibility allows a single turn model to be reused for constructing arbitrary topologies such as a serpentine, spiral, or their combination thereof, as will be shown later. To address this, two sets of flags are used in the models. One is the system flag F_s, stored as the zeroth component of the skew coefficients ($S[0]$ in Table 5-1) to record the direction of the skew caused by the first turn or elbow. The other is the intrinsic flag F_i of individual elements. For example, $F_i = 1$ is for turns or elbows involving clockwise flow of species bands; $F_i = 2$ is for counterclockwise turns or elbows. Since straight channels do not incur any skew, no flag is needed. During simulations, $F_s = 0$ (i.e., $S[0] = 0$) is first generated by the injector, which is the most upstream element of a separation channel and, hence, initiates the computation of the separation state. Then as the species band migrates to the first turn or elbow, F_s is irreversibly set to their intrinsic flag F_i. Afterwards, the written F_s is compared with F_i of each downstream element as the band moves on. If they are identical, a "+" sign is used for the element, otherwise a "-" sign. Fig. 5-7 shows the codes for a 180° turn involving clockwise flow of species bands to implement this logic and determine the sign.

4. SCHEMATIC-BASED SIMULATIONS

In this section, we will first describe the simulation procedure, in which the Kirchhoffian resistor network to predict electrical current and field as well as the signal flow network to evaluate biofluidic state values (e.g., steady-state mixing concentrations and transient electrophoretic species band shapes) are solved sequentially. Then, the results of schematic simulation exploring various micromixers and separation microchips will be discussed and validated against numerical and experimental data.

4.1 Simulation Description

Schematic simulation for mixers and separation chips involves both electrical and biofluidic calculations. For DC analysis, given the applied potential at reservoirs, system topologies, and element dimensions, nodal voltages at element terminals within the entire system are first computed by Ohm's and Kirchhoff's laws using the resistor models presented in the last section. The resulting nodal voltages and branch currents are in turn used to calculate the electrical field strength (E) and its direction within each element, as well as flow and splitting ratios at intersections (for mixers). With these results and user-provided sample properties (D and μ), the sample speed is then given by

$u = \mu E$. Next, values of biofluidic pins at the outlet(s) of each element (e.g., concentration coefficients for micromixers; arrival time, variance, skew, and amplitude for separation microchips) are determined. The process starts from the most upstream element, typically the sample reservoirs in mixers and the injectors in separation chips in terms of the directional signal flow as described in Section 3. As such, both electrical and fluidic information in the entire system is obtained.

As described in Section 2.1, the mixer operates in steady state, while transient evolution is critical in separation channels. Transient analysis can be also conducted for separation chips that involve the species band's motion and broadening. An electropherogram (average concentration c_m vs. time) can be obtained at the detector, yielding an intuitive picture of separation resolution between species bands. The transient analysis first calculates for the DC operating points of the amplitude A_{det}, separation time t_{det}, and variance σ_{det}^2 of the species band at the detector as described above. Based on these points, the actual read-out time is scanned and the average concentration output c_m is calculated. Assuming the species band does not appreciably spread out as it passes by the detector, c_m is given by

$$c_m = A_{\text{det}} \cdot e^{\frac{-(E\mu)^2 (t - t_{\text{det}})^2}{2\left(\sigma_{\text{det}}^2 + \Delta\sigma^2\right)}} \tag{5.19}$$

where t is the actual read-out time and $\Delta\sigma^2$ is the variance growth associated with detection and given in Eq. (5.18).

4.2 Results and Discussion

In this section, simulation examples of complex EK passive mixers and electrophoretic separation microchips will be presented to verify the behavioral models for biofluidic elements and validate the modeling and simulation methodology. Schematic simulation results for micromixers are shown in Figs. 5-8–5-10 and Table 5-2, and those for electrophoretic separation systems are given in Figs. 5-11–5-16.

4.2.1 EK micromixers and mixing networks

EK focusing[34], which first appeared as an important sample manipulation technique in EK LoC systems, also can be applied to accelerate mixing, especially in reaction kinetics studies[7]. Fig. 5-8 illustrates an EK focusing mixer and its simulation schematic. In the discussion below, subscripts *i*, *s*, and *o*, respectively, denote the middle-input, side, and output mixing channels. Unlike the serial mixing network in Fig. 5-3a, the cross intersection where sample *a* (white) from the input channel is focused by buffer or sample *b* (black) from both side channels, is modeled as two serially concatenated converging intersections. The flow ratio (the ratio of the flow rate of the middle-input stream to the total flow rate) of sample *a* is $s = I_i/(2I_s+I_i)$.

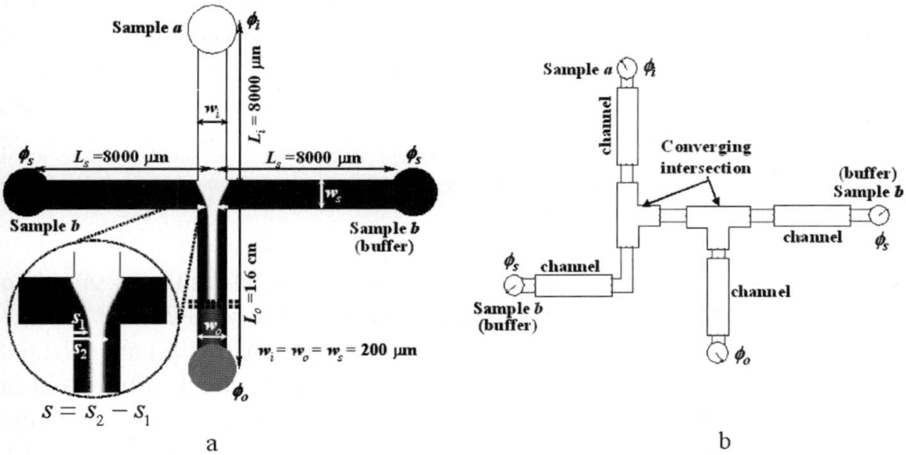

Figure 5-8. (a) An EK focusing micromixer and the contour plot of sample *a* concentration (from numerical simulation). (b) Its hierarchical schematic representation.

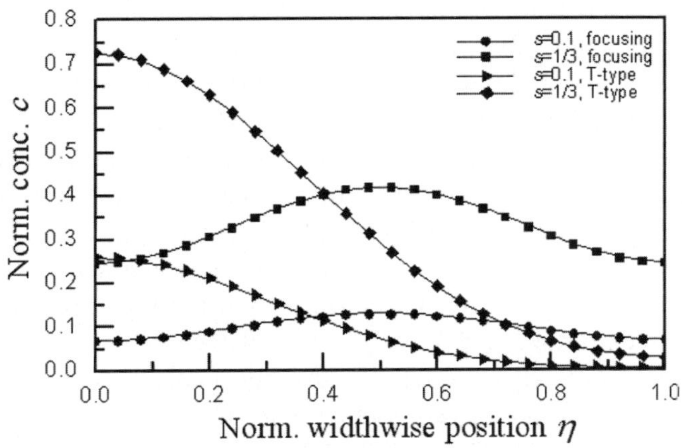

Figure 5-9. Schematic simulation results (lines) compared with numerical data (symbols) on widthwise concentration profiles *c* (sample *a*) for the EK focusing and T-type mixers.

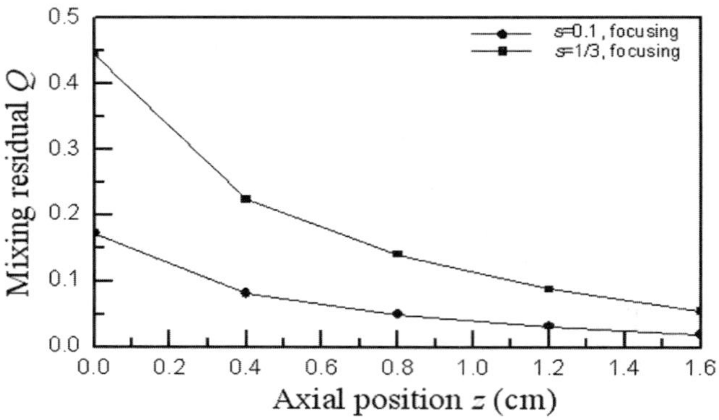

Figure 5-10. Schematic simulation results on variation of mixing residual Q along axial channel length (data points are connected by lines to guide the eye) for the EK focusing mixer involving different stream width s.

Figure 5-9 shows numerical and schematic simulation results of sample a concentration profile at the mixing channel outlet for two flow ratios $s = 0.1$ and $s = 1/3$. In both simulations, reservoir potentials (ϕ_i and ϕ_s) are selected to vary s while holding E (143 V/cm) and the sample residence time fixed in the mixing channel. Excellent agreement between numerical and schematic simulation results is found with the worst-case relative error of 3% at $s = 0.1$. The results are also compared with those from a T-type mixer that has the same electrical field in the mixing channel, channel length, and width as the focusing mixer. The focusing mixer considerably speedups sample mixing and improves sample homogeneity, which can be attributed to the reduced diffusion distance between samples (or between the sample and buffer). That is, the axial centerline of the mixing channel in the focusing-mixer is essentially an impermeable wall due to the geometrical symmetry; hence, the inter-diffusion distance between different samples is only one-half of that of the T-type mixer. Another interesting observation is that a smaller stream width (e.g., $s = 0.1$) yields a more uniform concentration profile at the end of the mixing channel. To investigate the influence of the stream width on mixing performance, an index of mixing residual, $Q = \int_0^1 |c(\eta) - c_{avg}| d\eta$, is introduced in Fig. 5-10 to characterize the non-uniformity of concentration profiles, where $c(\eta)$ and c_{avg} are the normalized concentration profile and width-averaged concentration, respectively, at the detection spot[31]. At the channel inlet ($z = 0$), mixing residual Q strongly depends on s. Asymmetric incoming streams yield a lower Q value (e.g., $Q = 0.18$ at $s = 0.1$ in contrast to $Q = 0.44$ at $s = 1/3$) and a more uniform initial profile. Along the channel, Q initially drops rapidly and then becomes saturated because the improved sample mixing reduces the concentration gradient and the driving force for further mixing. Given sufficiently long mixing channels, uniform sample concentrations can be obtained in both mixers, which, however from the design perspective, is not efficient. Thus, a tradeoff between *Q* and mixer size and complexity can be captured by our behavioral models to achieve designs of both effective and efficient micromixers.

These parameterized behavioral models are well suited to study complex mixing networks[3], in which an array of sample concentrations can be attained at multiple analysis channels by geometrically duplicating functional units with a single constant voltage applied at all reservoirs.

Table 5-2. Comparison of schematic simulation results (sch) with numerical (num) and experimental (exp) data on sample concentrations in analysis channels of serial and parallel mixing networks[3].

Serial Mixing					
Widthwise Complete				Widthwise Incomplete	
channel	*c* (sch)	*c* (exp)	*c* (num)	*c* (sch)	*c* (num)
A_1	1	1	1	1	1
A_2	0.37	0.36	0.378	0.48	0.496
A_3	0.22	0.21	0.224	0.187	0.187
A_4	0.125	0.13	0.133	0.081	0.0815
A_5	0.052	0.059	0.0628	0.029	0.0315

Parallel Mixing			
Widthwise Complete			
channel	*c* (sch)	*c*(exp)	*c* (num)
A_1	0	0	0
A_2	0.83	0.84	0.832
A_3	0.68	0.67	0.674
A_4	0.52	0.51	0.523
A_5	0.35	0.36	0.354
A_6	0.17	0.19	0.168
A_7	1	1	1

Table 5-2 shows the comparison of schematic simulation results with experimental and numerical data on sample (rhodamine B) concentrations in analysis channels A_1–A_5 in the serial mixing network (Fig. 5-3a). Both complete and partial mixing cases are investigated. When a voltage of 0.4 kV is applied at the sample and buffer reservoirs with the waste reservoirs grounded, sample mixing in channels S_2–S_5 is widthwise complete. Excellent agreement between the schematic simulation and numerical analysis and experimental data (with an average error smaller than 6%) is obtained. In contrast to the electrical resistor-based models[3, 17, 35] that take advantage of the analogy between the electrical current, EK flow, and sample transport and, hence, require post-calculations of concentration values from electrical currents in the network, our behavioral models directly deliver the concentration value in each analysis channel. In addition to complete mixing, partial mixing case is also schematically simulated. A voltage of 1.6 kV, as used in the experiments in the literature[3], is applied at the sample and buffer reservoirs with the waste grounded, which increases the EK velocity and then decreases the residence time of the sample in channels S_2–S_5. Thereby, mixing in channels S_2–S_5 is incomplete along their width, and the sample amount diverted to channels A_1–A_5 depends on not only the electrical currents in the network but also the sample concentration profiles at the exits of channels S_2–S_5, which violates the assumption for the analogy between EK flow and sample transport, and, hence, the resistor-based modeling becomes invalid. However, it can be readily simulated by our behavioral models. In the schematic, the cross-intersection is modeled as a serial concatenation of the converging and diverging intersections, in which the sample concentration profiles of the incoming and outgoing streams are accurately captured. Results from the schematic simulation are compared with numerical data in Table 5-2 (a comparison to experimental data is not available due to a lack of knowledge on sample properties. Hence a diffusivity of D=3×10^{-10} m^2/s and an EK mobility of μ =2.0×10^{-8} m^2/Vs are assumed in numerical simulation). Very good agreement is attained with an average error of 4%. At the cross-intersection following channel S_2, the sample amount diverted to channel A_2 is more than the complete-mixing case due to the non-uniform sample profiles at the intersection's inlet. Consequently, concentrations in channels A_3–A_5 show lower values, which is consistent with the experimental observations[3]. Netlisting and schematic simulation of this example take 20 s on a multi-user, 2-CPU 1-GHz Sun Fire 280 processors with 4 GB RAM for the first-time simulation, and less than a second for subsequent iterations, leading to a 1,000–20,000× speedup.

In additional to the serial mixing network, the parallel mixing network[3] can be hierarchically represented and simulated in a similar fashion and excellent agreement among schematic simulation results, numerical analysis, and experimental data (with an average error of 3.6% relative to experiments) is also found.

4.2.2 Electrophoretic separation microchips

Schematic simulation results for electrophoretic separation microchips are shown in Figs. 5-11–5-16. In Figs. 5-11 and 5-12, a serpentine electrophoresis column including two complementary turns is used to separate an analyte band comprised of two species *a* (D = 3.12×10^{-10} m^2/s, μ = 1.2×10^{-8} m^2/sV) and *b* (D = 2.72×10^{-10} m^2/s, μ = 1.1×10^{-8} m^2/sV) with E = 600 V/cm. Experimental data[6] on variance vs. time of species *a* are compared with DC schematic simulation results in Fig. 5-11, showing excellent agreement with the worst case relative error of only 5%. Again, netlisting and DC simulation for this example take 20 s for the first-time simulation and less than a second for subsequent iterations, leading to a 500–10,000× speedup (higher speedup can be obtained for a more complex chip topology or a less diffusive species as shown in Fig. 5-15). The first turn skews the species band and accordingly incurs an abrupt increase in variance. During the species band's migration in the long inter-turn straight channel, transverse diffusion smears out most of the skew, leading to a nearly uniform band before the second turn. The second turn then distorts the band again in the opposite direction, leading to another turn-induced variance increase that is equal to the one from the first turn. Fig. 5-12 shows separation electropherograms of both species from three detectors. The spacing between concentration peaks of species *a* and *b* increases as they migrate through channels, but due to the band-broadening effect, the amplitude decreases consecutively.

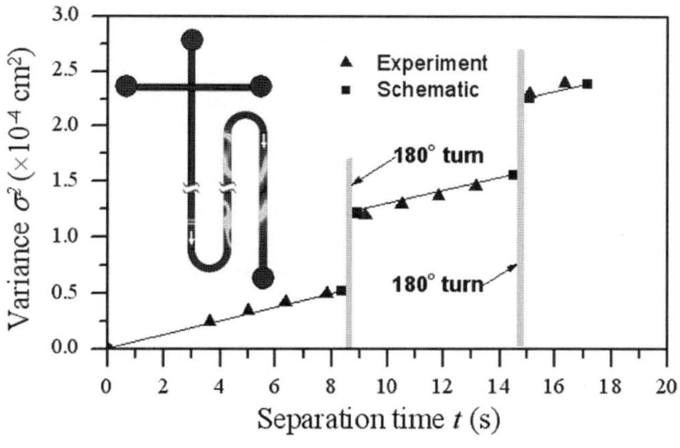

Figure 5-11. Comparison of experimental data[6] with DC schematic simulation on variance σ^2 vs. separation time t of species a in a serpentine electrophoretic separation microchip consisting of two complementary turns. The grey bars represent the residence time of the sample within the turns.

In Fig. 5-13, dispersion of Dichlorofluoroscein in a complex spiral separation microchip of five turns is simulated and compared with experimental results[36]. Spiral channels differ from the serpentine in that the species band is distorted in the same direction; therefore, its skew and variance always increase with the turn number. A scalar index of plate number N_s to characterize the resolving power of the electrophoresis chip is defined $N_s = L_{tot}^2/\sigma^2$, where L_{tot} is the total separation length from the injector to the detector. The higher the plate number, the better separation capacity achieved by the chip. The linear growth of the plate number with electrical field implies that molecular diffusion is the major dispersion source in such a system (Fig. 5-14), as molecular diffusion decreases as electrical field increases (if Joule heating is negligible[32]). The worst case relative error of 12% is considered acceptably small considering the uncertainties of the measurement of species diffusivity[36].

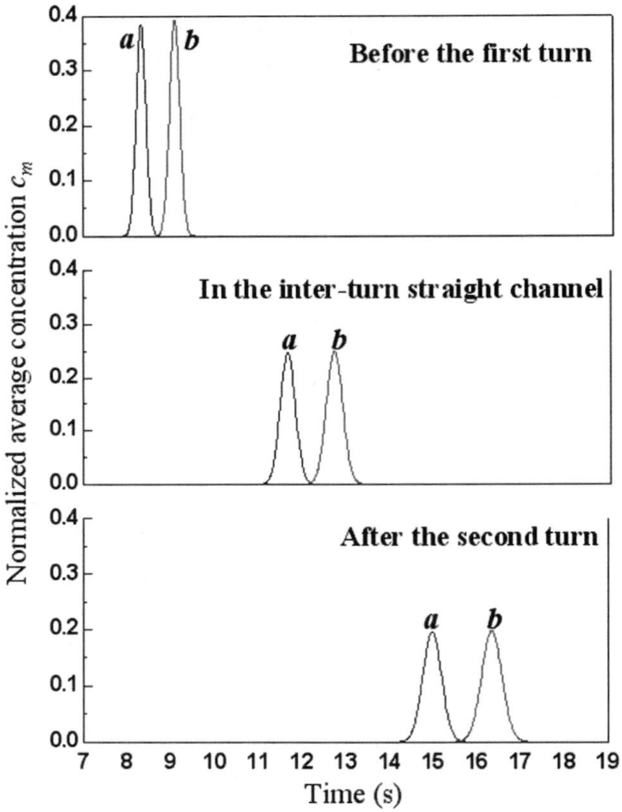

Figure 5-12. Transient analysis simulates the electropherogram outputs from three detectors, which are respectively arranged before the first turn (top trace), in the middle of the inter-turn straight channel (middle trace), and after the second turn (bottom trace).

Figure 5-15 illustrates a hybrid electrophoretic separation microchip[37] and its schematic representation including both spiral and serpentine channels. Due to the difficulty of accounting for the coexisting skew canceling and strengthening effects in such a topology, it has not been effectively investigated since it was proposed[37]. Fig. 5-16 shows schematic simulation result on the variance of a species band vs. time in such a chip, as well as its comparison with numerical data. A low species diffusivity of $D = 1\times10^{-11}$ m^2/s is chosen to analyze highly convective dispersion that has not been considered by the previous example in Fig. 5-11 (other properties and parameters are the same as those of species *a* in Fig. 5-11). Highly convective dispersion is practically important for microchip electrophoresis of the species with low diffusivity, such as the separation of DNA in a gel or sieving matrix[6, 38]. It is shown in Fig. 5-15 that since species flows in the clockwise direction in both turns T_1 and T_2 (spiral topology), T_2 strengthens the sharp skew generated by T_1, leading to a more skewed band and a higher variance. Due to the small species diffusivity, the skew almost persists through the inter-turn straight channel between T_2 and T_3, and is significantly cancelled out by T_3, which, as a result, yields a drastic variance drop in T_3 (serpentine topology). However, the skewed band after T_3 is overly corrected by T_4 and a counter-skew is shown afterward. Excellent agreement between the schematic and numerical simulation results with 1% relative error and tremendous speedup up to 400,000× have been achieved in Fig. 5-16. This is the first time that the highly convective dispersion in the hybrid electrophoresis microchip at this complexity level has been accurately and efficiently simulated by analytical models.

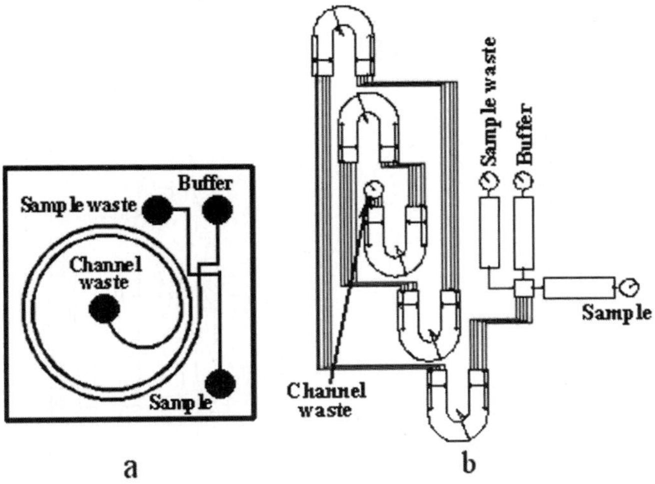

Figure 5-13. (a) A spiral electrophoretic separation microchip[36] consisting of five turns with continuously decreased radius (1.9, 1.8, 1.7, 1.6 and 0.8 cm). (b) Its hierarchical schematic representation.

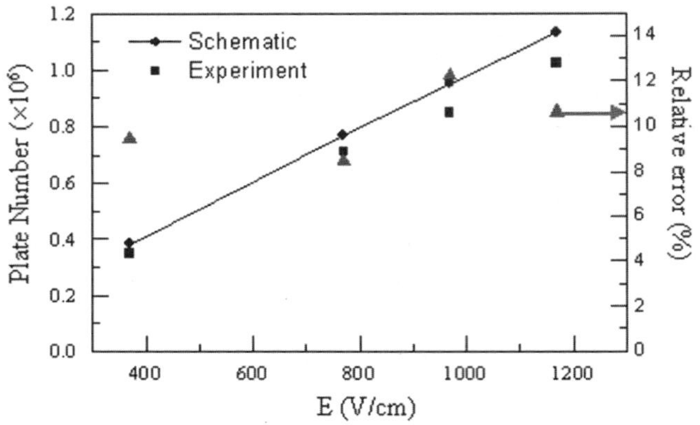

Figure 5-14. Comparison of schematic simulation results with the experimental data on plate number N_s vs. electrical field E. Right axis shows the relative error between the schematic simulation results and experimental data.

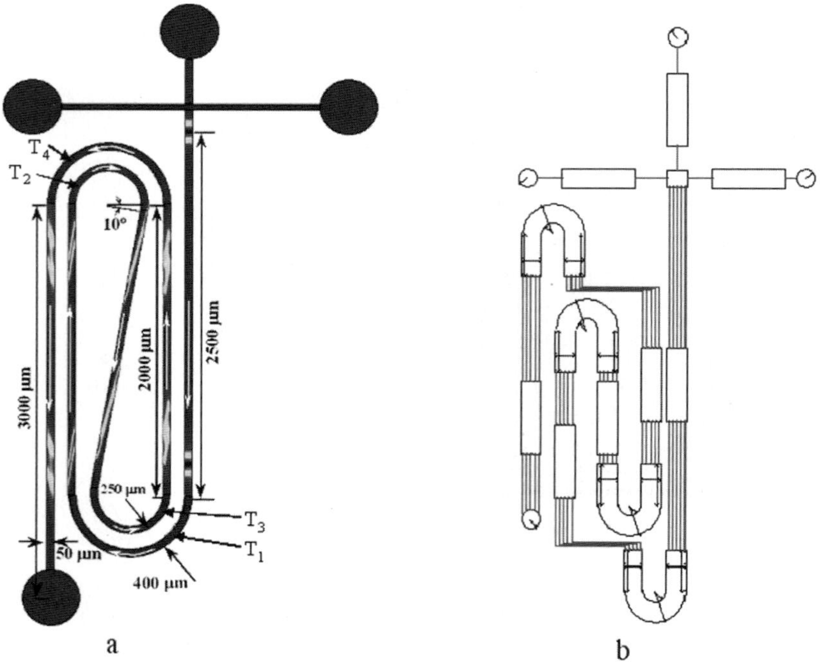

Figure 5-15. (a) A hybrid electrophoretic separation microchip consisting of both spiral and serpentine channels. (b) Its hierarchical schematic representation.

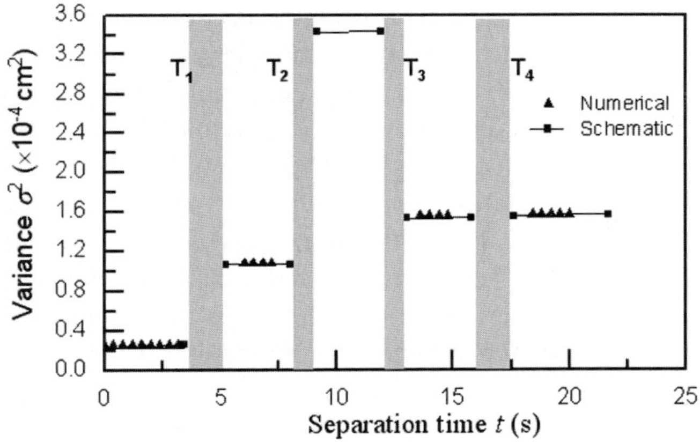

Figure 5-16. Comparison of numerical data with DC schematic simulation on variance vs. separation time in a hybrid electrophoretic separation microchip.

5. FUTURE WORK

At present, the separation and mixing models can only be used independently of each other as the mixing occurs in continuous flow of samples, while the separation exploits the transient behavior. To combine them for practical integrated LoC simulation requires the use of an injector. Additionally, to achieve the canonical assay described in Fig. 5-2, a reactor model is needed. A simple reactor model has been assembled with the separation and mixing models described above and injector models[26, 27] to simulate an integrated immunoassay microchip[29], showing the path to our envisioned LoC design methodology.

6. CONCLUSION

Modeling and simulation of EK biofluidic LoC systems (especially complex EK passive micromixers and electrophoretic separation systems) based on the top-down design methodology has been presented. Complex biofluidic LoCs have been geometrically and functionally decomposed into commonly used elements of simple geometry and specific function. Electrical and biofluidic pins have been proposed to support the communication between adjacent elements. Parameterized models that can accurately capture the element behavior have been implemented in an analog hardware description language (Verilog-A). Thus, a system-level schematic model can be developed for LoC design for iterative simulation to evaluate the impact of chip topologies, element sizes, and material properties on system performance. The simulation employs the Kirchhoff's law and directional signal flow to solve electrical and microfluidic networks.

The schematic simulation results of EK passive micromixers and electrophoretic separation microchips have been verified by numerical and experimental data. It has been shown that the proposed behavioral models are able to accurately describe the overall effects of chip topology, material properties, and operational parameters on mixing and separation performance, as well as interactions among elements. Tremendous speedup (up to 20,000× for mixers and 400,000× for electrophoretic separation chips) over full numerical simulation has been achieved by schematic simulation using behavioral models, while still maintaining high accuracy (relative error generally less than 5%). Therefore, our modeling and simulation efforts represent a significant contribution to addressing the need for efficient and accurate modeling and simulation tools to enable optimal design of integrated biofluidic LoCs.

ACKNOWLEDGMENT

This research is sponsored by the DARPA and the Air Force Research Laboratory, Air Force Material Command, USAF, under grant number F30602-01-2-0587, and the NSF ITR program under award number CCR-0325344. We also thank the other members of the SYNBIOSYS project, Xiang He, Ryan Magargle, and Anton Pfeiffer for their insightful discussion.

APPENDIX

The species band concentration $c(y,z,t)$ within a separation channel is governed by the convection-diffusion equation[10]

$$\frac{\partial c}{\partial t}+u\frac{\partial c}{\partial z}=D\left(\frac{\partial^2 c}{\partial z^2}+\frac{\partial^2 c}{\partial y^2}\right) \quad \text{(5-A1)}$$

where y and z are the widthwise and axial coordinates, respectively, and t is the separation time counted from the channel entrance. The width of the species band can be characterized by variance, the square of the standard deviation of the cross-sectional average concentration profile c_m, which is defined as

$$\sigma^2 = \frac{\int_{-\infty}^{\infty} \left(z - \bar{z}\right)^2 c_m \cdot dz}{\int_{-\infty}^{\infty} c_m \cdot dz} \tag{5-A2}$$

where $\bar{z}$ is the axial position of the species band's centroid in the channel.

Equation (5-A1) can be reformulated into a more concise and reduced-dimension form in terms of spatial moments of the species concentration. Such moments are capable of describing the species band's main characteristics such as mass distribution, skew, and variance without solving for detailed concentration distributions. We introduce a new coordinate frame, moving at the species band's average velocity U, and normalize the equation to reduce all variables into dimensionless forms. Define a dimensionless axial coordinate ξ, widthwise coordinate η, and time τ by

$$\xi = \left(z - Ut\right)/w, \quad \eta = y/w, \quad \tau = Dt/w^2 \tag{5-A3}$$

In terms of these dimensionless variables, Eq. (5-A1) is rewritten as

$$\frac{\partial c}{\partial \tau} = \frac{\partial^2 c}{\partial \xi^2} + \frac{\partial^2 c}{\partial \eta^2} - \mathrm{Pe}\,\chi \frac{\partial c}{\partial \xi} \tag{5-A4}$$

where $\mathrm{Pe} = Uw/D$ is the Peclet number representing the ratio of the convective transport rate to the diffusive transport rate, and χ is the normalized species velocity relative to the average, given as

$$\chi(\eta) = \left(u - U\right)/U \tag{5-A5}$$

We now recast Eq. (5-A4) in terms of spatial moments of the species concentration. If the species band is entirely contained in the channel, Eq. (5-A4) holds valid over the axial domain $-\infty < \xi < \infty$ (the widthwise domain is $0 < \eta < 1$), such that $c \to 0$ as $\xi \to \pm\infty$. Therefore, we can define spatial moments of the species concentration by

$$c_p\left(\eta,\tau\right) = \int_{-\infty}^{\infty} \xi^p c\left(\eta,\xi,\tau\right) d\xi, \quad m_p\left(\tau\right) = \int_0^1 c_p d\eta \tag{5-A6}$$

Here, c_p is the pth moment of the species concentration in the axial filament at η, and m_p is the pth moment of the average concentration of the band. As a result of the coordinate transformation (5-A3), all moments are defined with respect to the moving frame (ξ, η). For purposes of simulating species dispersion, it is sufficient to obtain the moments up to the second order. Specifically, c_0 provides the transverse distribution of the species mass in each axial filament within the channel and m_0 is the total species mass and can be chosen as $m_0 = 1$ without losing generality. Next, c_1 gives the axial location of the centroid of the axial filament and, hence, measures the skew of the band. Then, m_1, the widthwise average of c_1, is the axial location of the centroid of the entire species band in the frame (ξ,η) and is always zero for this study[10]. Finally, m_2 can be used to determine the variance σ^2 of the species band by $\sigma^2 = w^2\left(m_2/m_0 - m_1^2/m_0^2\right)$.

REFERENCES

1 D. R. Reyes, D. Lossifidis, P.-A. Auroux, and A. Manz, "Micro total analysis systems. 1. Introduction, theory, and technology," *Analytical Chemistry*, 74(12), 2623-2636 (2002).

2 P. A. Aurouz, D. Lossifidis, D. R. Reyes, and A. Manz, "Micro total analysis systems. 2. Analytical standard operations and applications," *Analytical Chemistry*, 74(12), 2637-2652 (2002).

3 S. C. Jacobson, T. E. McKnight, and J. M. Ramsey, "Microfluidic devices for electrokinetically driven parallel and serial mixing," *Analytical Chemistry*, 71(20), 4455-4459 (1999).

4 C. A. Emrich, H. J. Tian, I. L. Medintz, and R. A. Mathies, "Microfabricated 384-lane capillary array electrophoresis bioanalyzer for ultrahigh-throughput genetic analysis," *Analytical Chemistry*, 74(19), 5076-5083 (2002).

5 D. J. Harrison, C. Skinner, S. B. Cheng, G. Ocvirk, S. Attiya, N. Bings, C. Wang, J. Li, P. Thibault, and W. Lee, "From micro-motors to micro-fluidics: The blossoming of micromachining technologies in chemistry, biochemistry and biology," *Proceedings of 10th International Conference on Solid-State Sensors and Actuators*, Sendai, Japan, 12-15 (1999).

6 C. T. Culbertson, S. C. Jacobson, and J. M. Ramsey, "Dispersion sources for compact geometries on microchips," *Analytical Chemistry*, 70(18), 3781-3789 (1998).

7 J. B. Knight, A. Vishwanath, J. P. Brody, and R. H. Austin, "Hydrodynamic focusing on a silicon chip: Mixing nanoliters in microseconds," *Physical Review Letters*, 80(17), 3863-3866 (1998).

8 S. Krishnamoorthy and M. G. Giridharan, "Analysis of sample injection and band-broadening in capillary electrophoresis microchips," *Technical Proceedings of the 2000 International Conference on Modeling and Simulation of Microsystems*, San Diego, CA, U.S.A, 528-531 (2000).

9 P. M. S. John, T. Woudenberg, C. Connell, M. Deshpande, J. R. Gilbert, M. Garguilo, P. Paul, J. Molho, A. E. Herr, T. W. Kenny, and M. G. Mungal, "Metrology and simulation of chemical transport in microchannels," *Proceedings of the 8th IEEE Solid-State Sensor and Actuator Workshop*, Hilton Head Island, SC, U.S.A, 106-111 (1998).

10 Y. Wang, Q. Lin, and T. Mukherjee, "System-oriented dispersion models of general-shaped electrophoresis microchannels," *Lab on a Chip*, 4(5), 453-463 (2004).

11 J. C. Harley, R. F. Day, J. R. Gilbert, M. Deshpande, J. M. Ramsey, and S. C. Jacobson, "System design of two dimensional microchip separation devices," *Technical Proceedings of the Fifth International Conference on Micro Total Analysis Systems (MicroTAS 2001)*, Monterey, CA, U.S.A, 63-65 (2001).

12 M. Turowski, Z. J. Chen, and A. Przekwas, "Automated generation of compact models for fluidic microsystems," *Analog Integrated Circuits and Signal Processing*, 29(1-2), 27-36 (2001).

13 R. Qiao and N. R. Aluru, "A compact model for electroosmotic flows in microfluidic devices," *Journal of Micromechanics and Microengineering*, 12(5), 625-635 (2002).

14 X. C. Xuan and D. Q. Li, "Analysis of electrokinetic flow in microfluidic networks," *Journal of Micromechanics and Microengineering*, 14(2), 290-298 (2004).

15 http://www.coventor.com.

16 T. H. Zhang, K. Chakrabarty, and R. B. Fair, "Behavioral modeling and performance evaluation of microelectrofluidics-based PCR systems using SystemC," *IEEE Transactions on Computer-Aided Design of Integrated Circuits and Systems*, 23(6), 843-858 (2004).

17 A. N. Chatterjee and N. R. Aluru, "Combined circuit/device modeling and simulation of integrated microfluidic systems," *Journal of Microelectromechanical Systems*, 14(1), 81-95 (2005).

18 H. Chang, *A Top-down Constraint-driven Design Methodology for Analog Integrated Circuits* (Kluwer Academic, Boston, 1997).

19 T. Mukherjee, G. K. Fedder, and R. D. S. Blanton, "Hierarchical design and test of integrated microsystems," *Ieee Design & Test of Computers*, 16(4), 18-27 (1999).

20 A. J. Pfeiffer, T. Mukherjee, and S. Hauan, "Design and optimization of compact microscale electrophoretic separation systems," *Industrial & Engineering Chemistry Research*, 43(14), 3539-3553 (2004).

21 A. J. Pfeiffer, T. Mukherjee, and S. Hauan, "Synthesis of Multiplexed Biofluidic Microchips," *IEEE Transactions on Computer-Aided Design of Integrated Circuits and Systems*, DOI: 10.1109/TCAD.2006.855931 (2006).

22 A. J. Pfeiffer, T. Mukherjee, and S. Hauan, "Synthesis of Multiplexed Biofluidic Microchips," in *Design Automation Methods and Tools for Microfluidics-Based Biochips*, K. Chakrabarty and J. Zeng, Eds. Norwell, MA: Springer, 2006.

23 R. F. Probstein, *Physicochemical Hydrodynamics : An Introduction* (John Wiley & Sons, New York, 2003).

24 S. J. Haswell, "Development and operating characteristics of micro flow injection analysis systems based on electroosmotic flow - A review," *Analyst*, 122(1), R1-R10 (1997).

25 S. C. Jacobson, R. Hergenroder, L. B. Koutny, R. J. Warmack, and J. M. Ramsey, "Effects of injection schemes and column geometry on the performance of microchip electrophoresis devices," *Analytical Chemistry*, 66(7), 1107-1113 (1994).

26 R. M. Magargle, J. F. Hoburg, and T. Mukherjee, "Microfluidic injector models based on neural networks," *IEEE Transactions on Computer-Aided Design of Integrated Circuits and Systems*, DOI: 10.1109/TCAD.2006.855936 (2006).

27 R. M. Magargle, J. F. Hoburg, and T. Mukherjee, "Microfluidic Injector Models Based On Artificial Neural Networks," in *Design Automation Methods and Tools for Microfluidics-Based Biochips*, K. Chakrabarty and J. Zeng, Eds. Norwell, MA: Springer, 2006.

28 X. He, "Electrokinetic control of microfluidic reaction-mixing systems," *AICHE Annual Meeting*, Austin TX, 430f (2004).

29 Y. Wang, R. M. Magargle, Q. Lin, J. F. Hoburg, and T. Mukherjee, "System-oriented modeling and simulation of biofluidic lab-on-a-Chip," *Proceedings of the 13th International Conference on Solid-State Sensors and Actuators*, Seoul, Korea, 1280-1283 (2005).

30 P. Lob, K. S. Drese, V. Hessel, S. Hardt, C. Hofmann, H. Lowe, R. Schenk, F. Schonfeld, and B. Werner, "Steering of liquid mixing speed in interdigital micro mixers - from very fast to deliberately slow mixing," *Chemical Engineering & Technology*, 27(3), 340-345 (2004).

31 Y. Wang, Q. Lin, and T. Mukherjee, "A model for laminar diffusion-based complex electrokinetic passive micromixers," *Lab on a chip*, 5(8), 877-887 (2005).

32 Y. Wang, Q. Lin, and T. Mukherjee, "A model for Joule heating-induced dispersion in microchip electrophoresis," *Lab on a Chip*, 4(6), 625-631 (2004).

33 http://www.cadence.com,

34 S. C. Jacobson and J. M. Ramsey, "Electrokinetic focusing in microfabricated channel structures," *Analytical Chemistry*, 69(16), 3212-3217 (1997).

35 N. H. Chiem and D. J. Harrison, "Microchip systems for immunoassay: an integrated immunoreactor with electrophoretic separation for serum theophylline determination," *Clinical Chemistry*, 44(3), 591-598 (1998).

36 C. T. Culbertson, S. C. Jacobson, and J. M. Ramsey, "Microchip devices for high-efficiency separations," *Analytical Chemistry*, 72(23), 5814-5819 (2000).

37 S. K. Griffiths and R. H. Nilson, "Design and analysis of folded channels for chip-based separations," *Analytical Chemistry*, 74(13), 2960-2967 (2002).

38 B. M. Paegel, L. D. Hutt, P. C. Simpson, and R. A. Mathies, "Turn geometry for minimizing band broadening in microfabricated capillary electrophoresis channels," *Analytical Chemistry*, 72(14), 3030-3037 (2000).

Chapter 6

FFTSVD: A FAST MULTISCALE BOUNDARY ELEMENT METHOD SOLVER SUITABLE FOR BIO-MEMS AND BIOMOLECULE SIMULATION

Michael D. Altman,[1,+] Jaydeep P. Bardhan,[2,+] Bruce Tidor,[2,3] and Jacob K. White[2]
[1]*Department of Chemistry,*
[2]*Department of Electrical Engineering and Computer Science,*
[3]*Biological Engineering Division,*
Massachusetts Institute of Technology, Cambridge MA, USA
{*maltman, jbardhan, tidor, white*}*@mit.edu*
[+]*These authors contributed equally to this work.*

Abstract: We present a fast boundary element method (BEM) algorithm that is well-suited for solving electrostatics problems that arise in traditional and Bio-MEMS design. The algorithm, FFTSVD, is Green's function independent for low-frequency kernels and efficient for inhomogeneous problems. FFTSVD is a multiscale algorithm that decomposes the problem domain using an octree and uses sampling to calculate low-rank approximations to dominant source distributions and responses. Long-range interactions at each length scale are computed using the FFT. Computational results illustrate that the FFTSVD algorithm performs better than precorrected-FFT style algorithms or the multipole style algorithms in FastCap.

Keywords: Bio-MEMS, biomolecule, boundary element, electrostatic, fast solver, FFTSVD.

1. INTRODUCTION

Microelectromechanical systems (MEMS) have recently become a popular platform for biological experiments because they offer new avenues for investigating the structure and function of biological systems. Their chief advantages over traditional *in vitro* methods are reduced sample requirements, potentially improved detection sensitivity, and structures of approximately the

K. Chakrabarty and J. Zeng (eds.),
Design Automation Methods and Tools for Microfluidics-Based Biochips, 143–168.

same dimensions as the systems under investigation [Voldman99]. Devices have been presented for sorting cells [Gray04], separating and sequencing DNA [Lee01], and biomolecule detection [Burg03]. Furthermore, because arrays of sensors can be batch fabricated on a single device, parallel experiments and high-throughput analysis are readily performed. However, since microfabrication is relatively slow and expensive, numerical simulation of MEMS devices is an essential component of the design process [Korsmeyer04, White04]. Design tools for integrated circuits cannot address multiphysics problems, and this has motivated the development of several computer-aided MEMS design software packages, most of which are based on the finite element method (FEM) and the boundary element method (BEM) [Senturia92].

BioMEMS, when applied to such problems as biomolecule detection, are often functionalized with receptor molecules that bind targets of interest [Savran04]. Molecular labels can also be used to aid in the detection process [Potyrailo98]. However, the interactions between these molecules, the MEMS device, and the solvent environment are often neglected during computational prototyping. In other fields, such as computational chemistry and chemical engineering, continuum models of solvation are often used to study the electrostatic component of these interactions [Honig95]. These mean-field models permit the efficient calculation of many useful properties, including solvation energies and electrostatic fields [Klapper86, Gilson88], and have been shown to correlate well with more expensive calculations that include explicit solvent [Jean–Charles91]. However, continuum models are unable to resolve specific molecular interactions between solvent molecules and the solute. A variety of numerical techniques can be used to simulate the continuum models, including the finite difference method (FDM), the finite element method (FEM), and the boundary element method (BEM) [Gilson87, Holst00, Yoon90].

The boundary element method has a number of advantages relative to FDM and FEM, such as requiring only surface discretizations and exactly treating boundary conditions at infinity. However, discretizing boundary integral equations produces dense linear systems whose memory costs scale as $O(n^2)$ and solution costs scale with $O(n^3)$, where n is the number of discretization unknowns. This rapid rise in cost with increasing problem complexity has motivated the development of accelerated BEM solvers. Preconditioned Krylov subspace techniques, combined with fast algorithms for computing matrix–vector (MV) products, can require as little as $O(n)$ memory and time to solve BEM problems [Nabors94]. Many such algorithms have been presented, including the fast multipole method (FMM) [Greengard87, Greengard88], $\mathcal{H}$-matrices [Hackbusch99, Hackbusch00, Borm03], the precorrected-FFT method [Phillips97], wavelet techniques [Shi98, Tausch99], FFT on multipoles [Ong03a, Ong04], kernel-independent multipole methods [Biros04, Ying04], the hierarchical SVD method [Kapur97, Kapur98], plane-wave expansion based approaches [Greengard98], and the pre-determined interaction list oct-tree (PILOT) algorithm [Gope03]. Some algorithms, such as the original FMM, exploit the decay of the integral equation kernel; the precorrected-FFT method makes use of kernel shift-invariance. This paper introduces an algorithm that combines the benefits of both of these approaches, leading to a method that has excellent memory and time efficiency even on highly inhomogeneous problems.

Fast BEM algorithms whose structures depend on kernel decay suffer from a common, well-known problem: computing medium- and long-range interactions is still expensive, even when their numerical low rank is exploited. For instance, in the fast multipole method, computing the M2L (multipole to local) products dominates the matrix–vector product time, since each cube can have as many as 124 or 189 interacting cubes, depending on the interaction list definition, and the work per M2L multiplication scales as $O(p^4)$, where p is the expansion order and is related to accuracy [Greengard87, Greengard88, Nabors91]. Much work has focused on reducing this cost; for the FMM, plane-wave expansions [Greengard98] diagonalize the M2L translation, but are typically only efficient for large p. The precorrected-FFT (pFFT) algorithm [Phillips97] relies on not the kernel's decay but rather its translation invariance to achieve high efficiency. The pFFT method is Green's function independent, even for highly oscillatory kernels. Consequently, the method has been applied in a number of different fields, including wide-band impedance extraction [Zhu03], microfluidics [Aluru98, Ye00, Wang02] and biomolecule electrostatics [Kuo02]. One weakness of the precorrected-FFT method is that its efficiency decreases as the problem domain becomes increasingly inhomogeneous [Phillips97].

In this paper, we introduce a fast BEM algorithm called FFTSVD. The method is well-suited to MEMS device simulation because it is Green's function independent and maintains high efficiency when solving inhomogeneous problems. The FFTSVD algorithm is similar to the PILOT algorithm introduced by Gope and Jandhyala [Gope03], in that our algorithm is multiscale and based on an octree decomposition of the problem domain. Similar to PILOT and IES^3, our algorithm uses sampling and QR decomposition to calculate reduced representations for long-range interactions. The FFT is used to efficiently compute the interactions, as in the kernel-independent multipole method [Ying04]. Numerical results from capacitance extraction problems demonstrate that FFTSVD is more memory efficient than FastCap or pFFT and that the algorithm does not have the homogeneity problem. In addition, we illustrate electrostatic force analysis by simulating a MEMS comb drive [Wang02]. Finally, we demonstrate the method's kernel-independence by calculating the electrostatic free energy of transferring a small fluorescent molecule from the gas phase to aqueous solution, using an integral formulation of a popular continuum electrostatics model [Yoon90, Kuo02].

The following section briefly describes a representative MEMS electrostatics problem, a boundary element method used to solve the problem, and a more complicated surface formulation for calculating the electrostatic component of the solvation energy of a biomolecule. Section 3 presents the FFTSVD algorithm. Computational results and performance comparisons appear in Section 4. Section 5 describes several algorithm variants and summarizes the paper.

2. BACKGROUND EXAMPLES

In this section we describe two electrostatics problems that arise in BioMEMS design and describe how they can be addressed using BEM.

2.1 MEMS Electrostatic Force Calculation

Consider the electrostatically actuated MEMS comb drive illustrated in Figure 6-1. Two interdigitated polysilicon combs form the drive; one comb is fixed to the substrate and the other is attached to a flexible tether. Applying a voltage difference to the two combs results in an electrostatic force between the two structures, and the tethered comb moves in response [Wang02]. The electrostatic response of the system to an applied voltage difference can be calculated by solving the first kind integral equation

$$\int_S \sigma(r')G(r;r')\,dr' = V(r), \tag{6.1}$$

where S is the union of the comb surfaces, $V(r)$ is the applied potential on the comb surfaces, $G(r;r') = 1/\|r-r'\|$ is the free-space Green's function, and $\sigma(r)$ is the charge density on the comb surfaces. Note that this is a standard capacitance extraction problem.

We can compute the axial electrostatic force between the combs by the relation

$$F(s) \quad = \quad -\frac{d}{ds}E = -\frac{d}{ds}\frac{1}{2}V^T C(s)V, \tag{6.2}$$

where $F(s)$ is the force in the axial direction, s is the separation between the combs, E is the electrostatic energy of the system, V is the vector of conductor potentials, and $C(s)$ is the capacitance matrix, written as a function of the comb separation.

To solve (6.1) numerically,we discretize the surfaces into n_p panels and represent $\sigma(r)$, the charge density on the surface as a weighted combination of compactly supported basis functions defined on the panels:

$$\sigma(r) = \sum_{i=1}^{n_p} x_i f_i(r). \tag{6.3}$$

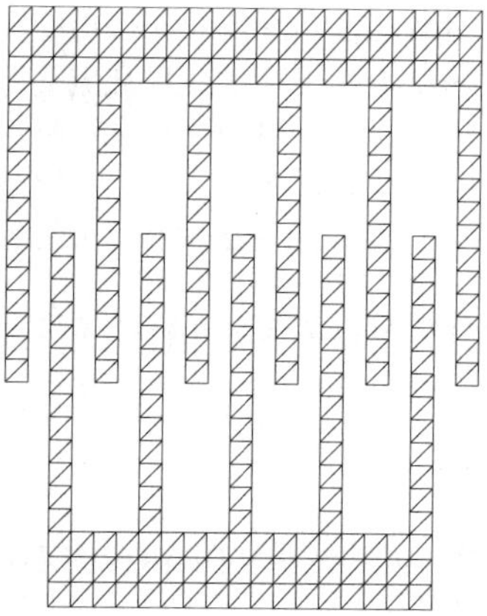

Figure 6-1. An electrostatically actuated MEMS comb drive.

Here, $f_i(r)$ is the ith basis function and x_i the corresponding weight. Forcing the integral over the discretized surface to match the known potential at a set of collocation points, we form the dense linear system

$$Gx = b. \tag{6.4}$$

The Green's function matrix G is defined by

$$G_{ij} = \int f_j(r')G(r_i, r')\,da', \tag{6.5}$$

where r_i is the ith collocation point and $b_i = V(r_i)$. Alternatively, one can use a Galerkin method, in which case

$$G_{ij} = \int\int f_i(r) f_j(r') G(r; r')\,dr\,dr' \tag{6.6}$$

and

$$b_i = \int f_i(r)\psi(r)\,dr. \tag{6.7}$$

The linear system of equations (6.4) is solved using preconditioned GMRES [Saad86].

2.2 BEM Simulation of Biomolecule Electrostatics

Electrostatic solvation energy, the cost of transferring a molecule from a non-polar low dielectric medium to an aqueous solution with mobile ions, plays an important role in understanding molecular interactions and properties. To calculate solvation energy, continuum electrostatic models are commonly employed. Figure 6-2 illustrates one such model. The Richards molecular surface [Richards77] is taken to define the boundary a that separates the biomolecule interior and the solvent exterior. The interior is modeled as a homogeneous region of low permittivity ϵ_I, where the potential $\varphi(r)$ is governed by the Poisson equation, and partial atomic charges on the biomolecule atoms are modeled as discrete point charges at the atom centers:

$$\nabla^2\varphi(r) = -\sum_{i=1}^{n_c}\frac{q_i}{\epsilon_I}\delta(r-r_i), \tag{6.8}$$

where n_c is the number of discrete point charges and q_i and r_i are the ith charge's value and location, respectively. In the solvent region, the linearized Poisson–Boltzmann equation

$$\nabla^2\varphi(r) = \kappa^2\varphi(r) \tag{6.9}$$

governs the potential, where κ, the inverse Debye screening length, depends on the concentration of ions in the solution and a higher permittivity ϵ_{II}. We write Green's theorem in the interior and exterior regions and then enforce continuity conditions at the boundary to produce a pair of coupled integral equations,

$$\frac{1}{2}\varphi(r_a) + \fint_a dr'\varphi(r')\frac{\partial G_1}{\partial n}(r_a;r') - \fint_a dr'\frac{\partial \varphi}{\partial n}(r')G_1(r_a;r') = \sum_{i=1}^{n_c}\frac{q_i}{\epsilon_I}G_1(r_a;r_i) \tag{6.10}$$

$$\frac{1}{2}\varphi(r_a) - \fint_a dr'\varphi(r')\frac{\partial G_2}{\partial n}(r_a;r') + \frac{\epsilon_I}{\epsilon_{II}}\fint_a dr'\frac{\partial \varphi}{\partial n}(r')G_2(r_a;r') = 0, \tag{6.11}$$

where r_a is a point on the surface, $\fint$ denotes the Cauchy principal value integral, G_1 is the Laplace Green's function, G_2 is the real Helmholtz Green's function, $\frac{\partial G_i}{\partial n}$ denotes the appropriate double layer Green's function, $\varphi(r)$ is the potential on the surface, and $\frac{\partial\varphi}{\partial n}(r)$ is the normal derivative of the potential on the surface. Readers are referred to [Yoon90, Kuo02] for detailed derivations of the formulation. To solve (6.10, 6.11) numerically we define a set of basis functions on the discretized surface and represent the surface potential and its normal derivative as weighted combinations of these basis functions:

$$\varphi(r) \approx \sum_i x_i f_i(r) \tag{6.12}$$

$$\frac{\partial\varphi}{\partial n}(r) \approx \sum_i y_i f_i(r). \tag{6.13}$$

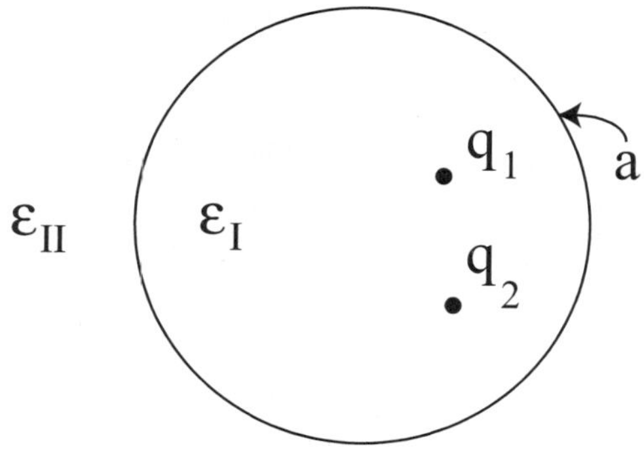

Figure 6-2. Continuum model for calculating biomolecule solvation.

We force the discretized integrals to exactly match the known surface conditions at the panel centroids; this produces the dense linear system

$$\begin{bmatrix} \frac{1}{2}I + \frac{\partial G_1}{\partial n} & -G_1 \\ \frac{1}{2}I - \frac{\partial G_2}{\partial n} & +\frac{\epsilon_I}{\epsilon_{II}}G_2 \end{bmatrix} \begin{bmatrix} x \\ y \end{bmatrix} = \begin{bmatrix} \sum_k \frac{q_k}{\epsilon_I} G_1(r; r_k) \\ 0 \end{bmatrix}, \tag{6.14}$$

where, denoting the ith panel centroid as r_i, the block matrix entries are

$$G_{1,ij} = \int f_j(r')G_1(r_i;r')dr' \tag{6.15}$$

$$\left(\frac{\partial G_1}{\partial n}\right)_{ij} = \int f_j(r')\frac{\partial G_1}{\partial n(r')}(r_i;r')dr' \tag{6.16}$$

and the block matrices G_2 and $\frac{\partial G_2}{\partial n}$ are similarly defined. Note that boundary element method solution of this problem requires a Green's function independent fast algorithm.

3. THE FFTSVD ALGORITHM

The FFTSVD is a multiscale algorithm like most fast algorithms for low frequency applications: to compute the total action of the integral operator on a vector, we separate its actions at different length scales and compute them separately, combining them only at the end. In describing the FFTSVD algorithm, it is helpful to think of the basis functions as sources, $\int f_i(r')G(r;r')dr'$ as the potential produced by source i, and the collocation points r_i as destinations. Multiplying x by G in Equation (6.4) is then computing potentials at all the destinations due to all sources. Figure 6-3 illustrates the multiscale approach to fast matrix multiplication: the square S denotes a source, and the squares denoted I represent destinations.

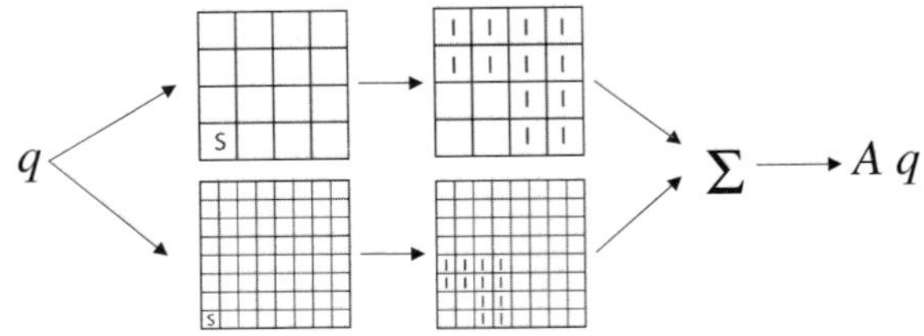

Figure 6-3. The multiscale approach to fast matrix multiplication.

3.1 Notation

Let d and s denote two sets of panels: then $G_{d,s}$ is the submatrix of G that maps sources in s to responses in d. The number of panels in set i is denoted by n_i.

3.2 Octree Decomposition

We first define the problem domain to be the union of all the sets of panels that comprise the discretized surfaces. We then place a bounding cube around the domain and recursively decompose the cube using octrees. Given a cube s at level i, the *nearest neighbors* N_s are those cubes at level i that share a face, edge, or vertex with s. The *interaction list* for s is denoted as I_s and defined to be the set of cubes at level i that are not nearest neighbors to s and not descended from any cube in an interaction list of an ancestor of s [Kanapka00]. Figure 6-4 illustrates the exclusion process for a 2-D domain. At every level, each panel is assigned to the cube that contains its centroid. Where ambiguity will not result, s denotes either the cube itself or the set of panels assigned to it. This assignment rule ensures that each panel–panel interaction is treated exactly once.

The coarsest decomposition is termed level 0 and has 4^3 cubes; coarser decompositions have null interaction lists. We continue decomposing the domain until we reach a level l at which no cube is assigned more than $n_{p,max}$ destinations. At each level i, every cube s has a set of interacting cubes I_s that are well-separated from s with respect to the current cube size. Note that the definition of an interaction list is symmetric: $d \in I_s \rightarrow s \in I_d$.

3.3 Sampling Dominant Sources and Responses

One can compute the potential response φ_{I_s} in I_s due to a source q_s in s by the dense matrix-vector product

$$\begin{aligned} \varphi_{I_s} &= G_{I_s,s} q_s \\ G_{I_s,s} &\in \Re^{n_{I_s} \times n_s}. \end{aligned} \tag{6.17}$$

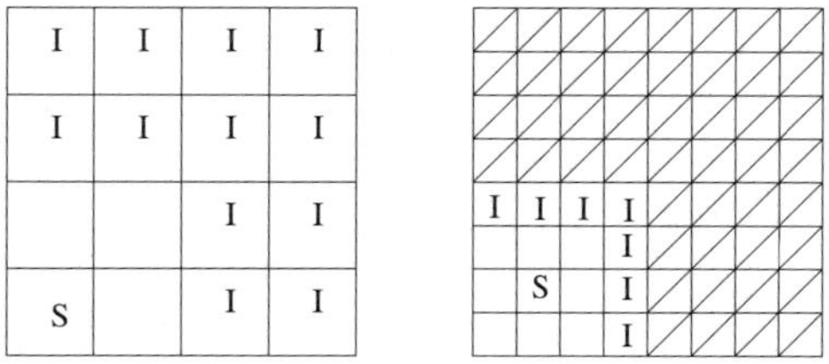

Figure 6-4. Interacting squares at two levels of decomposition.

However, the separation between s and I_s motivates the approximation

$$\begin{aligned} G_{I_s,s} &\approx U_{I_s} V_{s,\mathrm{src}}^T \\ U_{I_s} &\in \Re^{n_{I_s} \times k} \\ V_{s,\mathrm{src}}^T &\in \Re^{k \times n_s} \\ k &\ll n_{I_s} \end{aligned} \quad (6.18)$$

where $V_{s,\mathrm{src}}$ has orthogonal columns [Kapur97]. The matrix $V_{s,\mathrm{src}}$ is small and represents the k source distributions in s that produce dominant effects in I_s. It is a reduced row basis for $G_{I_s,s}$. The projection of q_s onto $V_{s,\mathrm{src}}$ loosely parallels the fast multipole method's calculation of multipoles from sources, in the sense that both the multipole expansion and the product $V_{s,\mathrm{src}}^T q_s$ capture the important pieces of q_s when calculating far-field interactions. We call $V_{s,\mathrm{src}}$ the source compression matrix.

A similar low-rank approximation can be made to find the response in a cube d given a source distribution in I_d:

$$\begin{aligned} \varphi_d &= G_{d,I_d} q_{I_d} \\ &\approx U_{d,\mathrm{dest}} V_{I_d}^T q_{I_d} \\ U_{d,\mathrm{dest}} &\in \Re^{n_d \times k} \\ V_{I_d}^T &\in \Re^{k \times n_{I_d}} \\ k &\ll n_{I_d}. \end{aligned} \quad (6.19)$$

Here, $U_{d,\mathrm{dest}}$ is small and represents the k dominant potential responses in d, the destination cube, due to source distributions in I_d. We call $U_{d,\mathrm{dest}}$ the destination compression matrix; $U_{d,\mathrm{dest}}$ is a reduced column basis for G_{d,I_d}.

Since it is impractical to compute $G_{I_s,s}$ and G_{s,I_s} for each cube s, we use a sampling procedure inspired by the Kapur and Long hierarchical SVD method [Kapur97]. Figures 6-5 and 6-6 illustrate the process of finding a reduced row basis $V_{s,\mathrm{src}}$. To determine the row basis, we begin by selecting one destination per interacting cube, computing the corresponding rows of $G_{I_s,s}$, and performing rank-revealing QR factorization with reorthogonalization on the transpose

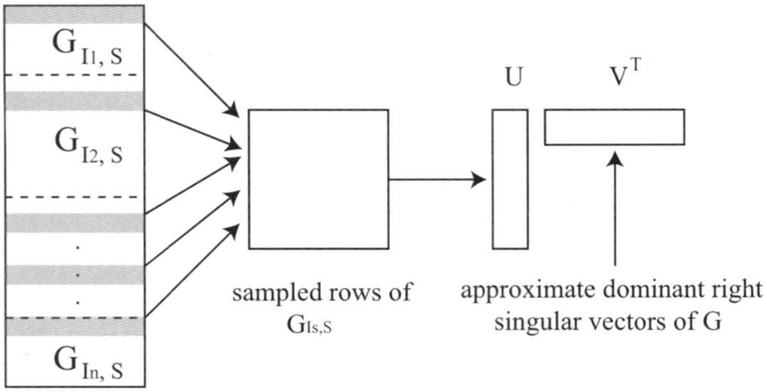

Figure 6-5. Computing dominant row basis for $G_{I_s,s}$ using sampling.

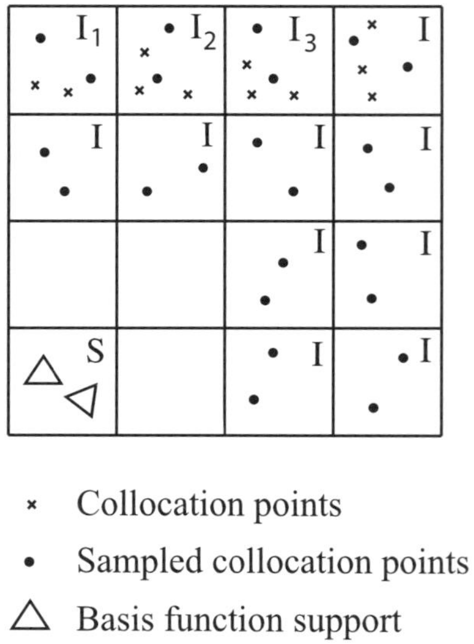

Figure 6-6. Sampling a small set of long-range interactions.

of the submatrix. If the submatrix rank is less than half the number of sampled destinations, the QR-determined row basis is considered to be adequate. Otherwise, an additional destination is sampled for each interacting cube; the extra destination is chosen to be well-separated from the originally chosen destination. The transpose of the new submatrix is factorized and again required to have rank less than half the total number of samples. The process of resampling is continued until the required rank threshold is met.

To compute the reduced column basis $U_{d,\text{dest}}$ for the matrix G_{d,I_d}, we select a set of well-separated panels in I_d, compute the corresponding columns of G_{d,I_d}, and QR factorize the submatrix.

3.4 Computing Long-range Interactions

Consider two well separated cubes s and d. Because the cubes are well separated, we could find a low-rank approximation to $G_{d,s}$ by truncating its SVD:

$$\varphi_d = G_{d,s} q_s \tag{6.20}$$
$$= U_{d,s} \Sigma_{d,s} V_{d,s}^T q_s \tag{6.21}$$
$$\approx \hat{U}_{d,s} \hat{\Sigma}_{d,s} \hat{V}_{d,s}^T q_s \tag{6.22}$$

where the hat denotes trunctation to k columns, $k < n_s$. Since the source compression matrix $V_{s,\text{src}}$ finds an approximation to the dominant row space of $G_{I_s,s}$, we expect that it also approximates the dominant row space of $G_{d,s}$, which is a submatrix of $G_{I_s,s}$. Similarly, we expect that $U_{d,\text{dest}}$ approximates the dominant column space of $G_{d,s}$. A small matrix $K_{d,s}$ maps source distributions in the reduced basis $V_{s,\text{src}}$ to responses in the reduced basis $U_{d,\text{dest}}$:

$$\varphi_d \approx U_{d,\text{dest}} K_{d,s} V_{s,\text{src}}^T q_s, \tag{6.23}$$

and it is easy to see that

$$K_{d,s} = U_{d,\text{dest}}^T G_{d,s} V_{s,\text{src}}. \tag{6.24}$$

Note that $K_{d,s}$ is not diagonal because $U_{d,\text{dest}}$ and $V_{s,\text{src}}$ only approximate the singular vectors of $G_{d,s}$. If $V_{s,\text{src}} \in \Re^{n_s \times k_s}$ and $U_{d,\text{dest}} \in \Re^{n_d \times k_d}$, then $K_{d,s} \in \Re^{k_d \times k_s}$.

The action of the K matrices can be computed in a number of different ways: they can be computed explicitly, via multipoles, or via an FFT. Explicit storage is memory intensive, and multipole representations are Green's function dependent. We have therefore chosen to implement the memory-efficient, Green's function-independent FFT translation method presented by Ying *et al.* [Ying04].

3.5 Diagonalizing Long-range Interactions with the FFT

Our method projects sources to a grid, uses an FFT convolution to accomplish translation between source and destination, and interpolates results back from the grid. Figure 6-7 illustrates the approach. We introduce two matrices: $P_{g,j}$ projects sources in cube j to the cube grid, and $I_{j,g}$ interpolates from the grid in cube j to the evaluation points in j. We use an equivalent density scheme similar to those used by Phillips and White [Phillips97] and Biros *et al.* [Biros04] to determine the projection and interpolation matrices.

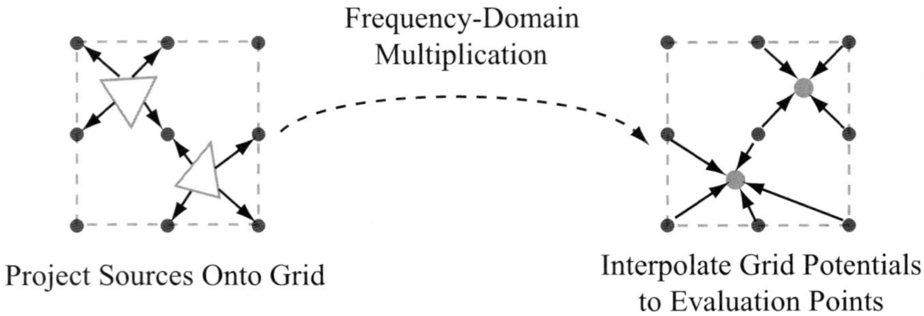

Figure 6-7. Schematic of the FFTSVD method for computing long-range interactions.

3.5.1 Projection Matrix Calculation. Given a cube s and the basis function weights q_s for panels in s, we wish to find a set of grid charges $q_{g,s}$ that reproduce the potential field far from s. We accomplish this by defining a sphere Γ bounding s and picking a set of quadrature points [Fliege99] on the sphere. Denoting quadrature point i on Γ by $r_{\Gamma,i}$, the mapping between q_s and the responses at the quadrature points can be written as $G_{\Gamma,s}$, where

$$G_{\Gamma,s,ij} = \int_{panel\ j} G(r_{\Gamma,i}; r')dr'. \tag{6.25}$$

The mapping between grid charges and responses at the quadrature points can be written as

$$G_{\Gamma,g,ij} = G(r_{\Gamma,i}, r_{g,j}) \tag{6.26}$$

where $r_{g,j}$ is the position of the jth grid point. If more quadrature points than grid points are used for the matching, solving a least squares problem gives the desired projection $P_{g,s}$:

$$P_{g,s} = G_{\Gamma,g}^{-1} G_{\Gamma,s}. \tag{6.27}$$

In practice, one uses the singular value decomposition to solve for $P_{g,s}$.

3.5.2 Interpolation Matrix Calculation. Given grid potentials q_d in a cube d, we find the potentials φ_d at the panel centroids in d by interpolation. For problems in which centroid collocation is used to generate a linear system of equations, the interpolation matrix is calculated as

$$I_{d,g} = (G_{\Gamma,g}^{-1} G_{\Gamma,d})^T \tag{6.28}$$

where $G_{\Gamma,d}$ denotes the Green's function matrix from the quadrature points on Γ to the panel centroids in d. If Galerkin methods are used rather than centroid collocation, the interpolation matrix is the transpose of the projection matrix.

3.5.3 Diagonal Translation. Once the grid charges in s are known, a spatial convolution with the Green's function produces the potentials at the grid points in the destination cube d. This spatial convolution is diagonalized by the Fourier transform; we write the transform matrix as $\mathcal{F}$, its inverse by $\mathcal{F}^{-1}$, and the transform of the Green's function matrix by $\tilde{G}_{d,s}$. After calculating the grid potentials in d, interpolation produces the potentials at the desired evaluation points. The matrix $G_{d,s}$ is therefore written as

$$G_{d,s} = I_{d,g}\mathcal{F}^{-1}\tilde{G}_{d,s}\mathcal{F}P_{g,s}. \tag{6.29}$$

The products $I_{d,g}\mathcal{F}^{-1}$ and $\mathcal{F}P_{g,s}$ could be stored, but in our experience this precomputation only marginally improves the matrix–vector product time while increasing memory use since $\mathcal{F}$ and $\mathcal{F}^{-1}$ are padded and complex.

In addition to diagonalizing the translation operation between cubes, the FFT significantly decreases memory requirements. Using explicit K matrices requires storing a small dense matrix for each pair of cubes; using FFT translation eliminates the expensive per-pair matrix cost. Instead, each cube has its own P_g and I_g matrices, which are used for all long-range interactions. In addition, because the Green's function is translationally invariant, we only need to store a small number of $\tilde{G}$ matrices for each octree level; each one represents a particular relative translation between source and destination cubes. Because these matrices are diagonal, storage requirements are minimal.

Since translation is the dominant cost in the FFTSVD matrix–vector product, efficient implementation of the translation procedure is essential to maximizing performance. The translation operation is simply an element-wise multiplication of two complex vectors, therefore, for g_p grid points per cube side, each translation vector is $(2g_p - 1)^2[(2g_p - 1)/2 + 1]$ complex numbers long when using the FFTW library [Frigo98]. This number takes into account padding and symmetry. For example, with $g_p = 3$, 75 complex numbers are required, resulting in 250 individual multiplies during the translation operation. This number has been reduced by taking advantage of vectorization. Many modern CPUs include instructions that can assist in multiplying complex numbers within a register, effectively halving the number of required multiplies. For comparison, standard fast multipole method translations require more multiplications since they are not diagonal, and cannot be vectorized as easily since they involve matrix–vector products. In addition, we have yet to exploit additional ways to accelerate the FFTSVD translation operation. These include using symmetries between related translation vectors ($\tilde{G}$), such as those that translate in opposite directions, and exploiting the fact that for axial translations, many $\tilde{G}$ elements are purely real.

3.6 Local Interactions

At the finest level of the decomposition, interactions between nearest neighbor cubes are computed directly by calculating the corresponding dense submatrices of G. These submatrices are denoted by $D_{i,j}$ where j is the source cube and i the destination. We bound the complexity of the local interaction computation by continuing the octree decomposition until each cube has fewer than $n_{p,\max}$ panels.

3.7 Algorithm Detail

The mapping from source cube s to destination cube d can thus be written as

$$\varphi_d = U_d\left(U_d^T I_{d,g}\right)\mathcal{F}^{-1}\tilde{G}\mathcal{F}\left(P_{g,s}V_s\right)V_s^T q_s \tag{6.30}$$

The computations are grouped to eliminate redundant multiplications; the matrix products $U_d^T I_{d,g}$ and $P_{g,s}V_s$ are stored for each cube rather than recomputed at every iteration. Below, we introduce the restriction operator $M_j^{(i)}$ that restricts a global vector to a local vector associated with cube j at level i; let the inverse operator map a local vector to the global by inserting appropriate zeros. Let L^i denote the set of cubes at level i. Given a charge vector q, the matrix–vector product is computed by the following procedure:

1 DOWNWARD PASS FOR LONG-RANGE INTERACTIONS: For levels $i = 0, 1, \ldots, l$:

(a) PROJECT INTO DOMINANT SOURCE SPACE: For each cube $j \in L^i$, compute

$$\zeta_j = \mathcal{F}(P_{g,j}V_{j,\text{src}})V_{j,\text{src}}^T M_j^{(i)} q. \tag{6.31}$$

(b) COMPUTE LONG-RANGE INTERACTIONS: For each cube $j \in L^i$, compute

$$\nu_j = \sum_{s \in I_j} \tilde{G}\zeta_s. \tag{6.32}$$

(c) DETERMINE TOTAL DOMINANT RESPONSE: For each cube $j \in L^i$, compute

$$\varphi = \varphi + M_j^{(i),-1} U_{j,\text{dest}}(U_{j,\text{dest}}^T I_{j,g})\mathcal{F}^{-1}\nu_j. \tag{6.33}$$

2 SUM DIRECT INTERACTIONS: For each cube d at level l, add the contributions from neighboring cubes N_d:

$$\varphi = \varphi + M_d^{(l),-1} \sum_{s \in N_d} D_{d,s} M_s^{(l)} q. \tag{6.34}$$

4. COMPUTATIONAL RESULTS

To demonstrate the accuracy, speed, and memory efficiency of the FFTSVD algorithm, we have used FFTSVD to solve for self and mutual capacitances in various geometries. A MEMS comb drive example [Wang02] illustrates electrostatic force calculation using FFTSVD. In addition, to show Green's function independence and use of double layer kernels, we have used FFTSVD to solve for the electrostatics of solvation for the highly charged dye molecule fluorescein. Fluorescein is often used as a fluorescent label in BioMEMS applications [Mosier02, Cho05], and its electrostatic properties in aqueous solution modulate its interaction with other molecules and surfaces.

The FFTSVD algorithm has several adjustable parameters: ϵ_{QR} is the reduced basis tolerance; g_p is the number of FFT grid points on each side of a finest-level cube; $n_{p,\max}$ is the maximum number of panels in a finest-level cube; n_{quad} is the number of quadrature points used on the equivalent density sphere, tol_{GMRES} is the tolerance on the relative residual that the resulting linear equations are solved to. At the two finest levels, g_p FFT grid points per cube edge are used, and the number of grid points per edge increases by one for each successively coarser level; experience has shown that using different numbers of grid points per edge provides significant accuracy improvements for marginal memory and time costs. The parameters used for the following results are 10^{-4} for ϵ_{QR}, 3 for g_p, 32 for $n_{p,\max}$, 25 for n_{quad}, and 10^{-4} for tol_{GMRES} unless otherwise specified.

For capacitance calculations, we compare performance to FastCap, based on the fast multipole method [Nabors91], and fftcap++, based on the pFFT++ implementation of the precorrected-FFT method [Zhu02]. All programs were compiled with full optimizations using the Intel C++ compiler version 8.1 and benchmarked on an Intel Pentium 4 3.0-GHz desktop computer with 2 GB of RAM. All parameter settings in FastCap and fftcap++ were left at their defaults, except for the tolerance on solving the resulting linear equations, which was set to 10^{-4} unless otherwise specified.

4.1 Self-Capacitance of a Sphere

In order to test the accuracy of the FFTSVD method, we have applied it to solving for the self-capacitance of a unit 1-m radius sphere, a quantity known analytically. Figure 6-8 shows the improvement in accuracy with increasing sphere discretization for FFTSVD with values of 3 and 5 for g_p, 2nd and 4th order multipoles in FastCap, and default settings for fftcap++. A tolerance of 10^{-6} for the relative residual when solving the BEM equations was used in all programs. The analytical value for the self-capacitance of a 1-m radius sphere is 0.111265 nF as computed by Gauss' law. The results show that FFTSVD with a value of 3 for g_p tends to be more accurate than 2nd order multipoles in FastCap. In addition, FFTSVD with low values of g_p tends to overshoot the analytical solution while FastCap tends to undershoot with truncation of multipole order. These findings are consistent across many geometries when examining convergence behavior.

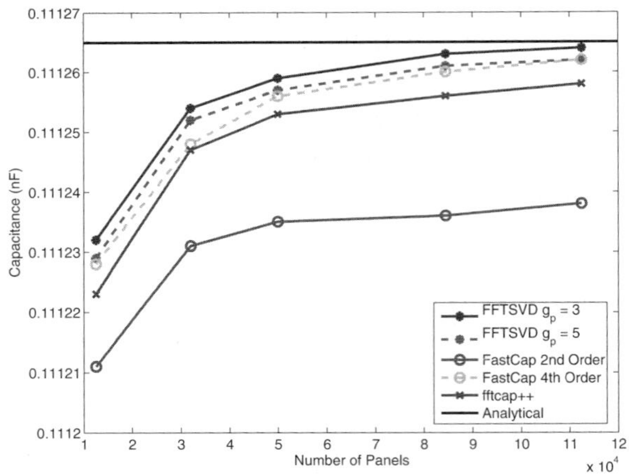

Figure 6-8. Accuracy versus number of panels for FFTSVD, FastCap and fftcap++ solving the unit sphere self-capacitance problem.

4.2 Woven Bus Example (Homogeneous Problem)

As stated previously, one of the advantages of the FFTSVD method is its use of diagonal translation operators. This advantage becomes apparent in cases of homogeneous geometry, since a large number of translation operations are required. To examine performance in a problem with homogeneous geometry, we have applied FFTSVD to solving for the mutual capacitances between woven bus conductors as in Figure 6-9. Table 6-1 summarizes the results for several woven bus capacitance problems. FFTSVD can achieve slightly better speed and memory performance than precorrected-FFT, which is expected to excel at problems with uniform distribution, and significantly better performance as compared to FastCap.

4.3 Inhomogeneous Capacitance Problem

One of the disadvantages of the precorrected-FFT method is that it lays down a uniform grid over the entire problem domain, and the simulation time grows

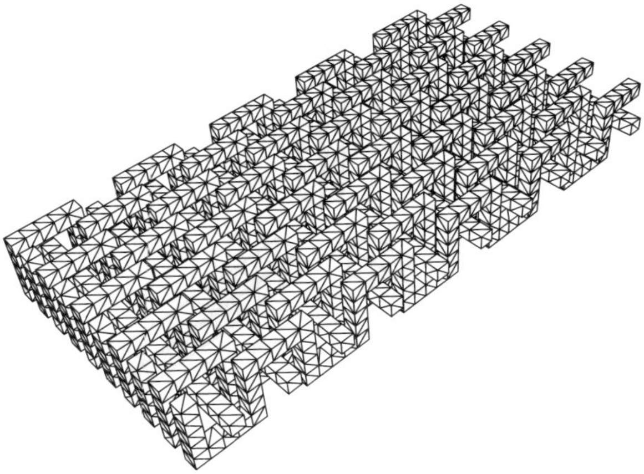

Figure 6-9. Homogeneous woven bus capacitance problem (woven10n01).

roughly in proportion to the number of grid points. For simulations in which most of the domain is empty, therefore, the precorrected-FFT algorithm is inefficient. We have demonstrated this inefficiency, and FFTSVD's relative advantage, by configuring a set of conductors as shown in Figure 6-10. Almost

Table 6-1. Comparison of FastCap (FC), fftcap++ (FFT++) and FFTSVD (FS) performance in terms of matrix–vector product time in seconds (MV) and memory usage in megabytes (MEM) on homogeneous woven bus capacitance problems with 2, 5, and 10 crossings (woven02n03, woven05n03, woven10n03) and 10 crossings with lower discretization (woven10n01).

Problem	Panels	FC MV	FC MEM	FFT++ MV	FFT++ MEM	FS MV	FS MEM
woven02n03	3168	0.03	30	0.02	23	0.01	11
woven05n03	18720	0.17	205	0.22	411	0.09	110
woven10n01	8160	0.08	89	0.04	69	0.04	41
woven10n03	73440	0.73	901	0.51	818	0.41	466

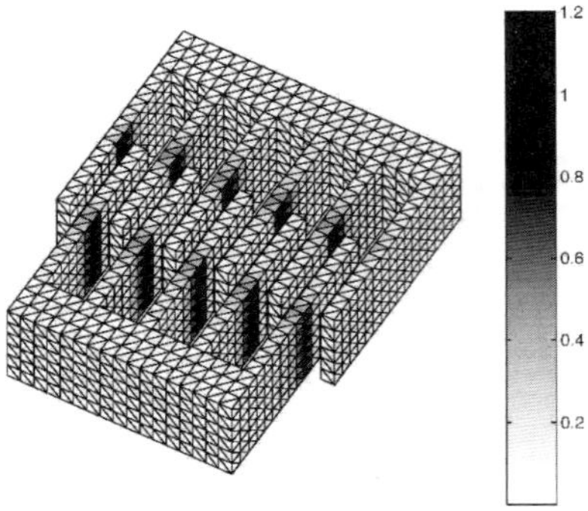

Figure 6-13. Magnitudes of panel contributions to the axial electrostatic force. Units are pN.

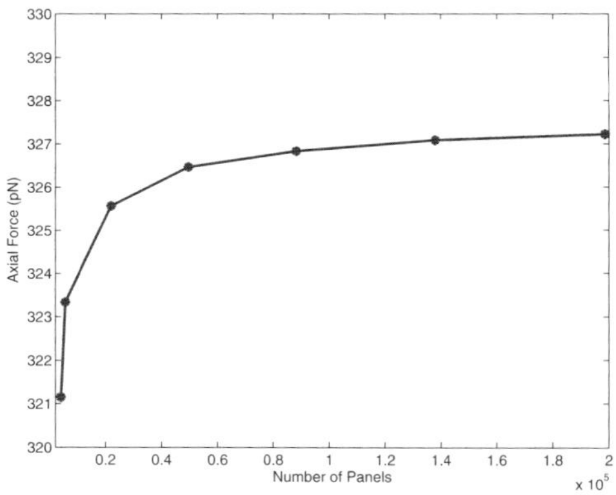

Figure 6-14. Calculated total axial electrostatic force on one comb.

general triangles and note that the discretization scheme is poorly tuned for the calculation of electrostatic forces; nonuniform meshes achieve superior accuracy at reduced panel counts [Ruehli73]. The force can also be calculated by integrating the squared charge density over the conductor surface, but this approach requires specialized treatment because the charge density becomes infinite at the edges and corners of the conductors [Ong03b, Ong05].

4.5 Solvation of Fluorescein

We have used the integral formulation in Equations (6.10) and (6.11) to calculate the solvation energy of fluorescein. To prepare a model for solvation calculations, its structure and partial atomic charges were determined from quantum mechanical calculations. Radii were assigned to each atom and used

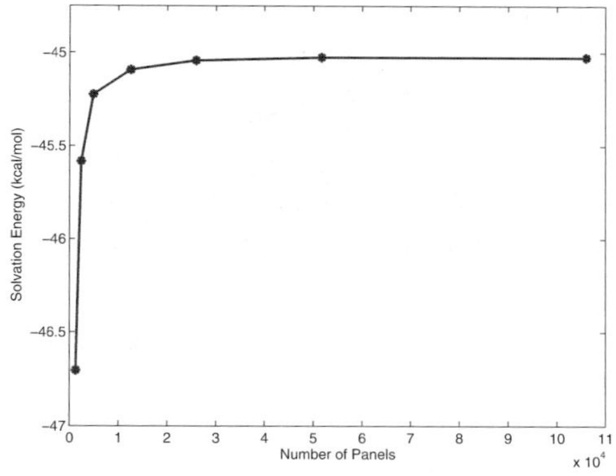

Figure 6-15. Computed electrostatic solvation energy of fluorescein with increasing problem discretization.

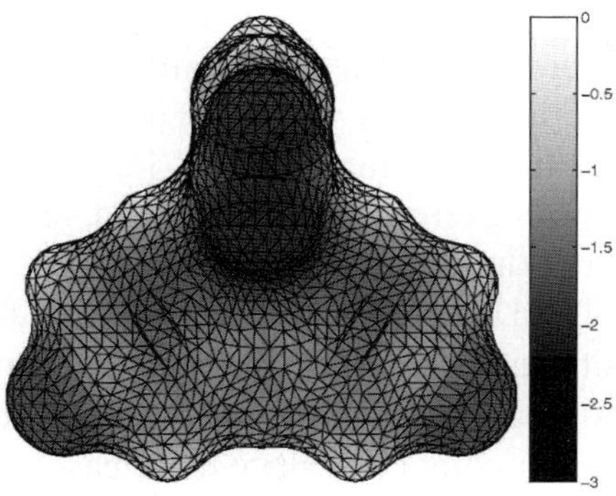

Figure 6-16. Electrostatic solvation potentials on the molecular surface of fluorescein. Units are kcal mol^{-1} e^{-1}.

to generate a triangulation of the molecular surface. The interior of the fluorescein molecule was assigned a dielectric constant of 4, and the exterior was assigned a dielectric constant of 80 (for water) with an ionic strength of 0.145 M ($\kappa = 0.124$ Å^{-1}). FFTSVD was used to solve for both the electrostatic solvation energy (Figure 6-15), as well as the total electrostatic potential on the surface of the fluorescein molecule (Figure 6-16). We note that the long-range single and double layer integrals can be computed using only one set of translation operations. Different projection operators are used to find the corresponding grid charges due to monopole and dipole distributions, and the grid charges can then be summed for translation.

5. DISCUSSION

5.1 Algorithm Variants

For problems with a small number of integral operators, memory constraints may not be a significant consideration. In these cases, the matrices $K_{d,s}$ can be stored explicitly. These $K_{d,s}$ matrices are computed using Equation (6.24), but instead of computing $G_{d,s}$ explicitly, we project, translate, and interpolate an identity matrix using the methodology outlined in Section 3.5. Although setup time and memory use increase when explicit K-matrices are used, the matrix–vector product time is significantly reduced. We have also implemented a parameter that allows a tradeoff between speed and memory use through K-matrices. Pairs of interacting octree cubes that contain fewer panels than the parameter are handled with explicit K-matrices, while all other cubes use the FFT-based translation. In this manner, the balance between speed and memory can be fine-tuned for the given application.

It is also straightforward to create an FFTSVD variant that runs in linear time; the same method used to generate the projection and interpolation matrices can be used to create "upward pass" and "downward pass" operators such as those found in multipole algorithms. This variant algorithm is essentially equivalent to the kernel-independent method by Ying *et al.* [Ying04], except that we allow all the grid charges to be nonzero. The Ying method, in contrast, uses only grid charges on the surface of the cube.

The linear-time FFTSVD method requires a greater number of grid points per cube, due to the loss of degrees of freedom during each upward pass from child to parent cube. In addition, the SVD-based compression of dominant sources and responses is no longer computed, since these bases are now taken directly from child cubes. This method is extremely memory efficient since dominant source and response bases are no longer stored, but it trades off performance to achieve it due to the larger required grid sizes.

Finally, the multilevel structure of FFTSVD allows easy parallelization. Each processor can be assigned responsibility for a set of cubes on coarse levels, and the computation can proceed independently until the final potential responses are summed. We have implemented parallel FFTSVD using both OpenMP and MPI libraries with good results.

5.2 SUMMARY

We have developed a fast algorithm for computing the dense matrix–vector products required to solve boundary element problems using Krylov subspace iterative methods. The FFTSVD method is a multiscale algorithm; an octree decomposes the matrix action into different length scales. For each length scale, we use sampling to calculate reduced bases for the interactions between well-separated groups of panels. The FFT is used to diagonalize the translation operation that computes the long-range interactions. The method described here relies on both kernel decay and translation invariance.

Numerical results illustrate that FFTSVD is much more memory-efficient than FastCap or precorrected-FFT, and that it is generally faster than either technique on a variety of problems. In addition, FFTSVD is Green's function independent, unlike FastCap, and the method performs well even when the problem domain is sparsely populated, unlike precorrected-FFT. Our implementation is well-suited to solve problems with multiple dielectric regions. Finally, we note that the structure of the algorithm permits treatment of kernels that are not translation-invariant; for such problems,the K-matrix algorithm variant should be used rather than the FFT. Together, the algorithm's performance and flexibility make FFTSVD an excellent candidate for fast BEM solvers for microfluidic and microelectromechanical problems that appear in BioMEMS design.

ACKNOWLEDGMENTS

This work was partially supported by the National Institutes of Health (GM065418 and GM066524), the National Science Foundation, and grants from the Semiconductor Research Corporation, the MARCO Interconnect Focus Center, and the Singapore-MIT Alliance. J. Bardhan is supported by a Department of Energy Computational Science Graduate Fellowship. The authors are grateful to Z. Zhu and D. Willis for numerous helpful discussions, and to S. Kuo for discussions about integral equation formulations for biomolecule electrostatics.

REFERENCES

J. Voldman, M. L. Gray, and M. A. Schmidt. Microfabrication in biology and medicine. *Ann. Rev. Biomed. Eng.*, 1:401–425, 1999.

D. S. Gray, T. L. Tan, J. Voldman, and C. S. Chen. Dielectrophoretic registration of living cells to a microelectrode array. *Biosensors and Bioelectronics*, 19(7):771–80, 2004.

L.-P. Lee and B. Tidor. Optimization of binding electrostatics: Charge complementarity in the barnase-barstar protein complex. *Protein Science*, 10:362–377, 2001.

T. P. Burg and S. R. Manalis. Suspended microchannel resonators for biomolecular detection. *Appl. Phys. Lett.*, 83(13):2698–2700, 2003.

T. Korsmeyer. Design tools for bioMEMS. In *Design Automation Conference*, pages 622–627, 2004.

J. White. CAD challenges in BioMEMS design. In *IEEE Design Automation Conference (DAC)*, pages 629–632, 2004.

S. D. Senturia, R. M. Harris, B. P. Johnson, S. Kim, K. Nabors, M. A. Shulman, and J. K. White. A computer-aided design system for microelectromechanical systems (MEMCAD). *Journal of Microelectromechanical Systems*, 1(1):3–13, 1992.

C. A. Savran, S. M. Knudsen, A. D. Ellington, and S. R. Manalis. Micromechanical detection of proteins using aptamer-based receptor molecules. *Analytical Chemistry*, 76:3194–3198, 2004.

R. A. Potyrailo, R. C. Conrad, A. D. Ellington, and G. M. Hieftje. Adapting selected nucleic acid ligands (aptamers) to biosensors. *Analytical Chemistry*, 70:3419–3425, 1998.

B. Honig and A. Nicholls. Classical electrostatics in biology and chemistry. *Science*, 268:1144–1149, 1995.

I. Klapper, R. Hagstrom, R. Fine, K. Sharp, and B. Honig. Focusing of electric fields in the active site of Cu-Zn superoxide dismutase: Effects of ionic strength and amino-acid modification. *Proteins: Structure, Function, Genetics*, 1:47–59, 1986.

M. K. Gilson and B. Honig. Calculation of the total electrostatic energy of a macromolecular system: Solvation energies, binding energies, and conformational analysis. *Proteins: Structure, Function, Genetics*, 4:7–18, 1988.

A. Jean-Charles, A. Nicholls, K. Sharp, B. Honig, A. Tempczyk, T. F. Hendrickson, and W. C. Still. Electrostatic contributions to solvation energies: Comparison of free energy perturbation and continuum calculations. *J. Am. Chem. Soc.*, 113(4):1454–1455, 1991.

M. K. Gilson, K. A. Sharp, and B. H. Honig. Calculating the electrostatic potential of molecules in solution: Method and error assessment. *Journal of Computational Chemistry*, 9(4):327–335, 1987.

M. Holst, N. Baker, and F. Wang. Adaptive multilevel finite element solution of the Poisson-Boltzmann equation I. Algorithms and examples. *Journal of Computational Chemistry*, 21(15):1319–1342, 2000.

B. J. Yoon and A. M. Lenhoff. A boundary element method for molecular electrostatics with electrolyte effects. *Journal of Computational Chemistry*, 11(9):1080–1086, 1990.

K. Nabors, F. T. Korsmeyer, F. T. Leighton, and J. K. White. Preconditioned, adaptive, multipole-accelerated iterative methods for three-dimensional first-kind integral equations of potential theory. *SIAM Journal on Scientific Computing*, 15(3):713–735, 1994.

L. Greengard and V. Rokhlin. A fast algorithm for particle simulations. *Journal of Chemical Physics*, 73:325–348, 1987.

L. Greengard. *The Rapid Evaluation of Potential Fields in Particle Systems*. MIT Press, 1988.

W. Hackbusch. A sparse matrix arithmetic based on H-matrices. I. Introduction to H-matrices. *Computing*, 62(2):89–108, 1999.

W. Hackbusch and B. N. Khoromskij. A sparse H-matrix arithmetic. II. Application to multi-dimensional problems. *Computing*, 64(1):21–47, 2000.

S. Borm, L. Grasedyck, and W. Hackbusch. Introduction to hierarchical matrices with applications. *Eng. Anal. Bnd. Elem.*, 27(5):405–22, 2003.

J. R. Phillips and J. K. White. A precorrected-FFT method for electrostatic analysis of complicated 3-D structures. *IEEE Transactions on Computer-Aided Design of Integrated Circuits and Systems*, 16(10):1059–1072, 1997.

W. Shi, J. Liu, N. Kakani, and T. Yu. A fast hierarchical algorithm for 3-D capacitance extraction. In *Design Automation Conference*, 1998.

J. Tausch and J. K. White. A multiscale method for fast capacitance extraction. In *Design Automation Conference*, pages 537–542, 1999.

E. T. Ong, K. M. Lim, K. H. Lee, and H. P. Lee. A fast algorithm for three-dimensional potential fields calculation: fast Fourier transform on multipoles. *Journal of Computational Physics*, 192(1):244–61, 2003.

E. T. Ong, H. P. Lee, and K. M. Lim. A parallel fast Fourier transform on multipoles (FFTM) algorithm for electrostatics analysis of three-dimensional structures. *IEEE Transactions on Computer-Aided Design of Integrated Circuits and Systems*, 23(7):1063–1072, 2004.

G. Biros, L. Ying, and D. Zorin. A fast solver for the Stokes equations with distributed forces in complex geometries. *Journal of Computational Physics*, 193(1):317–348, 2004.

L. Ying, G. Biros, and D. Zorin. A kernel-independent adaptive fast multipole algorithm in two and three dimensions. *Journal of Computational Physics*, 196(2):591–626, 2004.

S. Kapur and D. E. Long. IES^3: A fast integral equation solver for efficient 3-dimensional extraction. In *IEEE/ACM ICCAD*, pages 448–55, 1997.

S. Kapur and D. E. Long. IES^3: Efficient electrostatic and electromagnetic simulation. *IEEE Computational Science and Engineering*, 5(4):60–7, 1998.

L. Greengard, J. Huang, V. Rokhlin, and S. Wandzura. Accelerating fast multipole methods for the Helmholtz equation at low frequencies. *IEEE Comp. Sci. and Eng.*, 5(3):32–38, 1998.

D. Gope and V. Jandhyala. PILOT: A fast algorithm for enhanced 3D parasitic extraction efficiency. In *IEEE Electrical Performance of Electronic Packaging*, 2003.

K. Nabors and J. White. FASTCAP: A multipole accelerated 3-D capacitance extraction program. *IEEE Transactions on Computer-Aided Design of Integrated Circuits and Systems*, 10(10):1447–1459, 1991.

Z. Zhu, B. Song, and J. White. Algorithms in FastImp: A fast and wideband impedance extraction program for complicated 3D geometries. *IEEE/ACM Design Automation Conference*, 2003.

N. R. Aluru and J. White. A fast integral equation technique for analysis of microflow sensors based on drag force calculations. In *Modeling and Simulation of Microsystems*, pages 283–286, 1998.

W. Ye, X. Wang, and J. White. A fast Stokes solver for generalized flow problems. In *Modeling and Simulation of Microsystems*, pages 524–527, 2000.

X. Wang. *FastStokes: A fast 3-D fluid simulation program for micro-electro-mechanical systems.* PhD thesis, Massachusetts Institute of Technology, Department of Electrical Engineering and Computer Science, 2002.

S. S. Kuo, M. D. Altman, J. P. Bardhan, B. Tidor, and J. K. White. Fast methods for simulation of biomolecule electrostatics. *International Conference on Computer Aided Design (ICCAD)*, 2002.

Y. Saad and M. Schultz. GMRES: A generalized minimal residual algorithm for solving nonsymmetric linear systems. *SIAM Journal of Scientific and Statistical Computing*, 7:856–869, 1986.

F. M. Richards. Areas, volumes, packing, and protein structure. *Annual Review of Biophysics and Bioengineering*, 6:151–176, 1977.

J. Kanapka, J. Phillips, and J. White. Fast methods for extraction and sparsification of substrate coupling. In *Design Automation Conference*, pages 738–743, 2000.

J. Fliege and U. Maier. The distribution of points on the sphere and corresponding cubature formulae. *IMA J. of Num. Anal.*, 19:317–334, 1999.

M. Frigo and S. G. Johnson. FFTW: An adaptive software architecture for the FFT. In *Proc. 1998 IEEE Intl. Conf. Acoustics Speech and Signal Processing*, volume 3, pages 1381–1384. IEEE, 1998.

B. P. Mosier, J. I. Malho, and J. G. Santiago. Photobleached-fluorescence imaging of microflows. *Exp. in Fluids*, 33(4):545–554, 2002.

S. I. Cho, S.-H. Lee, D. S. Chung, and Y.-K. Kim. Bias-free pneumatic sample injection in microchip electrophoresis. *J. Chromat. A*, 1063(1-2):253–256, 2005.

Z. Zhu. Efficient techniques for wideband impedance extraction of complex 3-dimensional geometries. Master's thesis, Massachusetts Institute of Technology, 2002.

A. E. Ruehli and P. A. Brennan. Efficient capacitance calculations for three-dimensional multiconductor systems. *IEEE Transactions on Microwave Theory and Techniques*, 21:76–82, 1973.

E. T. Ong, K. H. Lee, and K. M. Lim. Singular elements for electro-mechanical coupling analysis of micro-devices. *J. Micromech. Microeng.*, 13:482–490, 2003.

E. T. Ong and K. M. Lim. Three-dimensional singular boundary elements for corner and edge singularities in potential problems. *Engineering Analysis with Boundary Elements*, 29:175–189, 2005.

Chapter 7

MACROMODEL GENERATION FOR BIOMEMS COMPONENTS USING A STABILIZED BALANCED TRUNCATION PLUS TRAJECTORY PIECEWISE LINEAR APPROACH

Dmitry Vasilyev,[1] Michał Rewieński,[2] and Jacob White[1]

[1] *Research Laboratory of Electronics,*
Massachusetts Institute of Technology,
Cambridge, MA
vasilyev@mit.edu
white@mit.edu

[2] *Synopsys Inc.,*
Mountain View, CA
Michal.Rewienski@synopsys.com

Abstract: In this short paper we present a technique for automatically extracting nonlinear macromodels of bioMEMS devices from physical simulation. The technique is a modification of the recently developed Trajectory Piecewise-linear (TPWL) approach, but uses ideas from balanced truncation to produce much lower-order and more accurate models. The key result is a perturbation analysis of an instability problem with the reduction algorithm, and a simple modification that makes the algorithm more robust. Results are presented from examples to demonstrate dramatic improvements in reduced model accuracy and show the limitations of the method.

Keywords: model order reduction, nonlinear dynamical systems, piecewise linear models, biomedical microelectromechanical devices (bioMEMS), microelectromechanical devices (MEMS), perturbation methods

K. Chakrabarty and J. Zeng (eds.),
Design Automation Methods and Tools for Microfluidics-Based Biochips, 169–187.

1. INTRODUCTION

The application of micromachining to biological applications, such as labs-on-a-chip [1–3], require complicated combinations of individual bioMEMS devices which process fluids, cells and molecules (e.g. mixers, separators and pumps). In order to simulate systems of these devices, models have been developed for common components, such as mixers and separators [4–7]. The wide variety of devices currently in development, and the need to rapidly assess the impact of candiate device performance on system behavior, will accelerate the demand for techniques which more automatically extract models of these bioMEMS devices from detailed physical simulation. The required automatic techniques may include approaches similar to the robust nonlinear model order reduction strategies being developed for nonlinear circuit model reduction [8–12], though bioMEMS devices can be more challenging because they are both nonlinear and typically much less damped than circuits.

In this short paper we describe an effective model reduction algorithm for bioMEMS devices that is a modification of the Trajectory-Piecewise Linear model order reduction (TPWL MOR) algorithm [10]. In the following section, we describe TPWL MOR algorithm, and then in section 3 we present an improvement on that algorithm based on using truncated balanced realizations (TBR) [13]. In section 4, we describe several example problems and in section 5 we use those examples to demonstrate both the effectiveness of our TPWL-TBR algorithm, as well as an instability problem. Also, we demonstrate a fundamental difficulty of the TPWL approach when applied to travelling wave problems. In section 6, we describe a perturbation analysis of the instability problem, and give a second algorithm modification which resolves this problem. Conclusions and acknowledgements end the paper.

2. TPWL NONLINEAR MODEL REDUCTION

After spatial discretization of the coupled PDEs that describe a bioMEMS component, the dynamic behavior of the component can often be represented using the standard state space form:

$$\begin{cases} \dot{x}(t) = f(x(t), u(t)) \\ y(t) = Cx(t) \end{cases} \tag{7.1}$$

where $x(t) \in R^N$ is a vector of states (e.g. mechanical displacements, fluid velocities) at time t, $f : R^N \times R^M \to R^N$ is a nonlinear vector-valued function, $u : R \to R^M$ is an input signal, C is an $N \times K$ output matrix and $y : R \to R^K$ is the output signal.

We assume nonliner function f being differentiable for all values of x and u:

$$f(x,u) = f(x_0, u_0) + A(x - x_0) + B(u - u_0) + h.o.t., \tag{7.2}$$

where matrices A and B (which are dependent on the linearization point (x_0, u_0)) contain derivatives of f with respect to the components of the state and input signals respectively.

The goal of applying model order reduction to (7.1) is to construct a macromodel capable of approximately simulating the input-output behavior of the systems in (7.1), but at a significantly reduced computational cost.

In order to achieve this goal, one needs to reduce the dimensionality of the state-space vector, which is usually achieved by employing projections. However, only projecting the nonlinear system (7.1) is not a complete solution to the nonlinear model reduction problem, because direct evaluation of the projected function is still directly proportional to the size of the unreduced system, and is too computationally expensive[1].

To reduce the cost of the projected function evaluation, consider the following generalized quasi-piecewise-linear approximate representation of the nonlinear function f, which has been proposed, in a slightly simpler form, in [10]:

$$f(x,u) \approx \sum_{i=1}^{s} \tilde{w}_i(x,u)\left(f(x_i,u_i) + A_i(x-x_i) + B_i(u-u_i)\right), \tag{7.3}$$

where x_i's and u_i's ($i = 1,\ldots,s$) are selected linearization points (samples of state and input values), A_i and B_i are derivatives of f with respect to x and u, evaluated at (x_i, u_i), and finally $\tilde{w}_i(x,u)$'s are state-and-input-dependent weights which satisfy:

$$\sum_{i=1}^{s} \tilde{w}_i(x,u) = 1 \quad \forall (x,u), \qquad \tilde{w}_i(x,u) \to 1 \text{ as } (x,u) \to (x_i,u_i). \tag{7.4}$$

Equation (7.4) implies that the trajectory piecewise-linear approximation in (7.3) is simply a convex combination of samples of f and f's derivatives.

Projecting the piecewise-linear approximation in (7.3) using biorthonormal projection bases V and W yields the following reduced-order nonlinear dynamical system:

$$\begin{cases} \dot{z} = \gamma \cdot w(z,u) + (\sum_{i=1}^{s} w_i(z,u)A_{ir})z + (\sum_{i=1}^{s} w_i(z,u)B_{ir})u \\ y = C_r z \end{cases}, \tag{7.5}$$

where $z(t) \in R^q$ is the q-dimensional vector of states:

$$\gamma = \left[W^T(f(x_1,u_1) - A_1x_1 - B_1u_1) \quad \ldots \quad W^T(f(x_s) - A_sx_s - B_su_s)\right].$$

Here, $w(z,u) = [w_1(z,u)\ldots w_s(z,u)]^T$ is a vector of weights, $A_{ir} = W^TA_iV$, $B_{ir} = W^TB_i$, and $C_r = CV$. One should note that $\sum_{i=1}^{s} w_i(z,u) = 1$ for all (z,u),

[1]This issue is discussed in detail in [10]

$w_i \rightarrow 1$ as $(z,u) \rightarrow (W^T x_i, u)$, and that the evaluation of the right hand side of equation (7.5) requires at most $O(sq^2)$ operations, where s is the number of linearization points.

As proposed in [10, 12], linearization points (x_i, u_i) used in system (7.5) are usually selected from a 'training trajectory' of the initial nonlinear system, corresponding to some appropriately determined 'training input'. The choice of the training input is an important aspect of the reduction procedure, since this choice directly influences accuracy. As the general rule, the training signal should be as close as possible to the signals for which the reduced system will be used. Additionally, this input signal should be rich enough to collect all "important" states in the set of linearization points (x_i, u_i) [12].

In order to obtain a reduced system in form (7.5), biorthonormal projection bases V and W must also be determined. This issue is addressed below.

3. CHOICE OF LINEAR REDUCTION METHOD

Consider a simple linearization of (7.1) about the initial state (x_0, u_0):

$$\begin{cases} \dot{x} = f(x_0) - \hat{A}x_0 - \hat{B}u_0 + \hat{A}x + \hat{B}u \\ y = \hat{C}x. \end{cases} \tag{7.6}$$

For the system in (7.6), a projection basis can be obtained using one of the many projection-based linear MOR procedures. One common choice is to reduce using the projection basis spanning the Krylov subspace [10, 14, 15]:

$$span(V) = span\{\hat{A}^{-1}\hat{B}, \ldots, \hat{A}^{-q}\hat{B}\}.$$

Reduction of (7.6) using the Krylov subspace projection is not guaranteed to provide a stable reduced model, even in this linearlized case [16, 17]. Therefore, TPWL macromodels obtained using Krylov projection are not guaranteed to be stable even if the original system is nearly linear.

Alternatively, one can apply a balanced truncation model reduction (TBR) procedure [18–20], which is presented here as algorithm 1.

The projection bases V and W obtained using algorithm 1 can then be used to compute the reduced TPWL approximation in (7.5).

TBR reduction can be more accurate than Krylov-subspace reduction as it posesses a uniform frequency error bound [21], and TBR preserves the stability of the linearized model. This superior performance for the linear cases suggests that TPWL approximation models obtained using TBR will be stable and accurate as well. This is not necessarily the case, as will be shown below.

Algorithm 1: TBR
Input: System matrices $\hat{A}$, $\hat{B}$, and $\hat{C}$.
Output: Projection bases V and W.
(1) Find observability Gramian P:
$\hat{A}P + P\hat{A}^T = -\hat{B}\hat{B}^T$;
(2) Find controllability Gramian Q:
$\hat{A}^T Q + Q\hat{A} = -\hat{C}^T\hat{C}$;
(3) Compute q dominant eigenvectors of PQ:
$(PQ)V = V\Sigma^2$, where $\Sigma^2 = diag(\Lambda_q^{dom}(PQ))$
(4) Compute q dominant eigenvectors of QP:
$(QP)W = W\Sigma^2$ and scale columns of W such that
$W^T V = I_{q\times q}$

4. EXAMPLES OF NONLINEAR SYSTEMS

In this Section we consider two examples of nonlinear systems which arise in the modeling of bioMEMS devices that have nonlinear dynamical behaviors, which make good test cases for reduction algorithms.

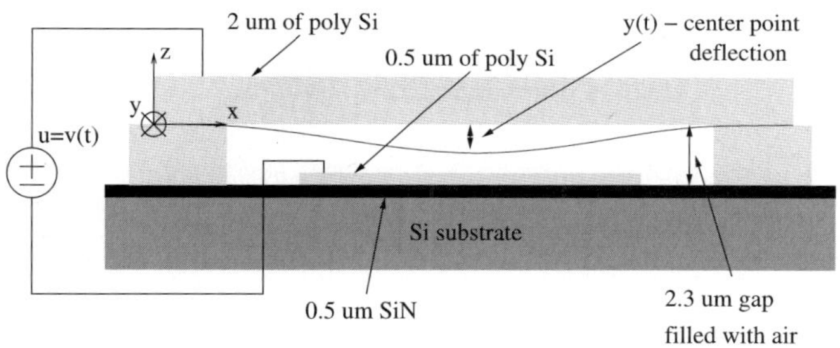

Figure 7-1. Micropump example (following Hung et al. [22]).

The first example is a fixed-fixed beam structure, which might be used as part of a micropump or valve, shown in Figure 7-1. Following Hung et al. [22], the dynamical behavior of this coupled electro-mechanical-fluid system can be modeled with a 1D Euler's beam equation and the 2D Reynolds' squeeze film damping equation [22]:

$$\begin{cases} \hat{E}I\frac{\partial^4 w}{\partial x^4} - S\frac{\partial^2 w}{\partial x^2} = F_{elec} + \int_0^d (p - p_0)dy - \rho\frac{\partial^2 w}{\partial t^2} \\ \nabla\cdot((1+6K)w^3 p\nabla p) = 12\mu\frac{\partial(pw)}{\partial t}. \end{cases} \tag{7.7}$$

Here, the axes x, y and z are as shown on figure 7-1, $\hat{E}$ is a Young's modulus, I is the moment of inertia of the beam, S is the stress coefficient, K is the Knudsen number, d is the width of the beam in the y direction, $w = w(x,t)$ is the height of the beam above the substrate, and $p(x,y,t)$ is the pressure distribution in the fluid below the beam. The electrostatic force is approximated assuming nearly parallel plates and is given by $F_{elec} = \frac{\varepsilon_0 d v^2}{2w^2}$, where v is the applied voltage.

Spatial discretization of (7.7) using a standard finite-difference scheme leads to a nonlinear dynamical system in form of (7.1), with $N = 880$ states. After discretization, the state vector, x, consists of the concatenation of: heights of the beam above the substrate w, values of $\partial(w^3)/\partial t$, and values of the pressure below the beam. For the considered example, the output $y(t)$ was selected to be the deflection of the center of the beam from the equilibrium point (cf. Figure 7-1).

The remarkable feature of this example is that the system is strongly nonlinear, and no feasible Taylor expansion made at the initial state can correctly represent the nonlinear function f, especially in the so called *pull-in* region[2] [22].

The exact actuation mechanism of the real micropumps may be quite different than the above simple structure, but this example is illustrative in that it combines electrical actuation with the structural dynamics and is coupled to fluid compression. We expect model reduction methods that are effective for this example problem to be extendable to realistic micropumps.

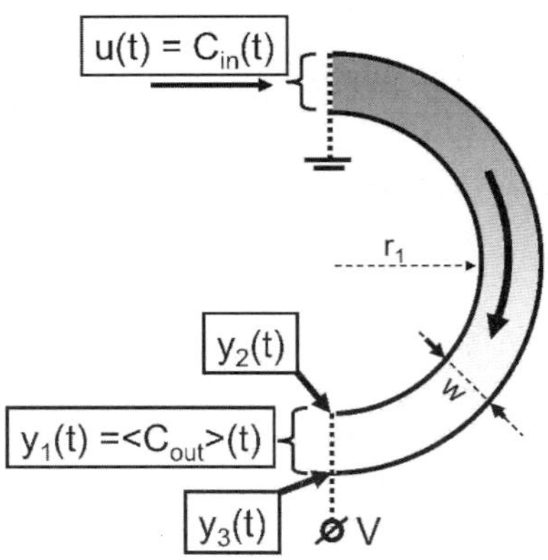

Figure 7-2. The microfluidic channel.

The second example, suggested in [23], is the injection of a (marker) fluid into a U-shaped three-dimensional microfluidic channel. The fluid is driven electrokinetically as depicted in Figure 7-2, and the channel has a rectangular cross-section of height d and width w. In this example, the electrokinetically driven flow of a buffer (carrier) fluid was considered to be steady, with the fluid

[2] If the beam is deflected by more than $\approx$1/3 of the initial gap, the beam will be pulled-in to the substrate.

velocity directly proportional to the electric field as in:

$$\vec{v}(\underbrace{x,y,z}_{\vec{r}}) = -\mu\nabla\Phi(\vec{r}),$$

where μ is an electroosmotic mobility of the fluid. The electric field can be determined from Laplace's equation

$$\nabla^2\Phi(\vec{r}) = 0,$$

with Neumann boundary conditions on the channel walls [24]. If the concentration of the marker is not small, the electroosmotic mobility can become dependent on the concentration, i.e. $\mu \equiv \mu(C(\vec{r},t))$, where $C(\vec{r},t)$ is the concentration of a marker fluid. Finally, the marker can diffuse from the areas with the high concentration to the areas with low concentration. The total flux of the marker, therefore, is:

$$\vec{J} = \vec{v}C - D\nabla C, \tag{7.8}$$

where D is the diffusion coefficient of the marker. Again, as the concentration of the marker grows, the diffusion will be governed not only by the properties of the carrying fluid, but also by the properties of a marker fluid, therefore D can depend on concentration. Conservation applied to the flux equation (7.8) yields a convection-diffusion equation [25]:

$$\frac{\partial C}{\partial t} = -\nabla\cdot\vec{J} = \nabla\Phi\cdot(C\nabla\mu(C) + \mu(C)\nabla C) + \nabla D(C)\cdot\nabla C + D(C)\nabla^2 C. \tag{7.9}$$

The standard approach is to enforce zero normal flux at the channel wall boundaries, but since $\vec{v}$ has a zero normal component at the walls, zero normal flux is equivalent to enforcing zero normal derivative in C. The concentration at the inlet was determined by the input, and the normal derivative of C was assumed zero at the outlet.

Note that equation (7.9) is nonlinear with respect to marker concentration unless both electroosmotic mobility or diffusion coefficient are concentration independent.

A state-space system was generated from (7.9) by applying a second order three-dimensional coordinate-mapped finite-difference spatial discretization to (7.9) on the half-ring domain in Figure 7-2. The states were chosen to be concentrations of the marker fluid at the spatial locations inside the channel. The concentration of the marker at the inlet of the channel is the input signal, and there are three output signals: the first being the average concentration at the outlet, the second and third signals being the concentrations at the inner and outer radii of the outlet of the channel, respectively.

Figure 7-3 illustrates the way an impulse of concentration at the inlet propagates through the channel: diffusion spreads the pulse, and due to the curvature of the channel, the front of the impulse becomes tilted with respect to the channel's cross-section. That is, the marker first reaches the points at the inner radius (point 1).

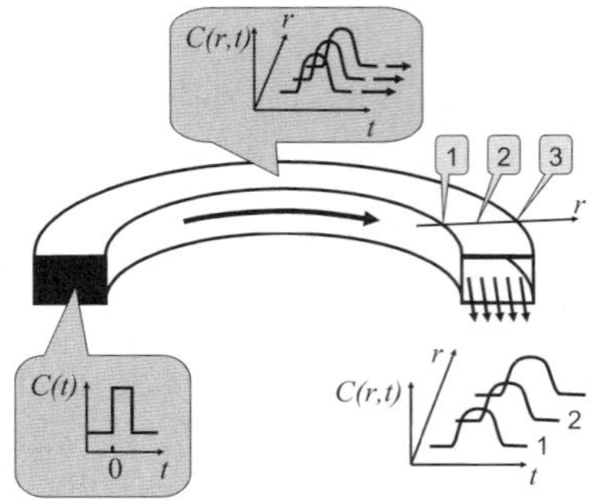

Figure 7-3. Propagation of the square inpulse of concentration of the marker. Due to the difference in lengths of the inner and outer arc, the marker reaches different points at the outlet with different delay.

5. COMPUTATIONAL RESULTS

In this section results are first presented for the linear microfluid channel model, in order to emphasize the efficiency of the TBR linear reduction. Then, results are presented for the micromachined pump model. The most challenging example was a nonlinear microfluidic channel.

5.1 Microchannel - linear Model

First, in order to demonstrate the effectiveness of TBR linear reduction, consider applying balanced-truncation algorithm to the linearized microchannel model. This corresponds to the problem of a very diluted solution of a marker in the carrier liquid (a widely used approximation in the literature). The values used for the electroosmotic mobility and diffusion coefficients are from [23]: $\mu = 2.8 \times 10^{-8} m^2 V^{-1} s^{-1}$, $D = 5.5 \times 10^{-10} m^2 s^{-1}$. Physical dimensions of the channel were chosen to be $r_1 = 500 \mu m$, $w = 300 \mu m$, $d = 300 \mu m$. Finite-difference discretization led to a linear time-invariant system (A, B, C) of order $N = 2842$ (49 discretization points by angle, 29 by radus, and 2 by height). Since Algorithm 1 requires $O(n^3)$ computation, the discretized system was too costly to reduce using original TBR algorithm. Instead, we used a fast-to-compute approximation to the TBR called modified AISIAD [26, 27].

As shown in Figure 7-4 and 7-5, applying reduction to the spatial discretization of (7.9) demonstrates the excellent efficiency of the TBR reduction algorithm. The reduction error decreases exponentially with increasing reduced model order, both in frequency-domain and in time-domain measurements. For example, in the time-domain simulations, the maximum error in the unit step response for the reduced model of order $q = 20$ (over a 100 times reduction) was lower than 10^{-6} for all three output signals.

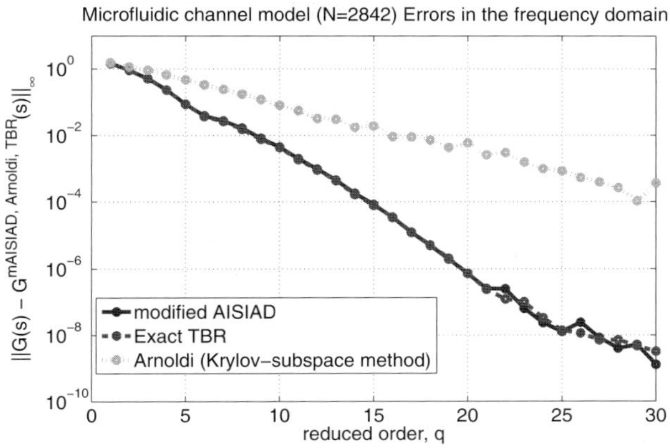

Figure 7-4. H-infinity errors (maximal discrepancy over all frequencies between transfer functions of original and reduced models) for the Krylov, TBR and modified AISIAD reduction algorithms.

The modified AISIAD method was compared with Krylov subspace-based reduction (Arnoldi method [14]) and the original TBR method in both time and frequency domains. As shown in Figure 7-4, TBR and modified AISIAD are much more accurate than the Krylov method, and are nearly indistinguishable. Though, the modified AISIAD model is much faster to compute.

To demonstrate the time-domain accuracy of the reduced model, we first redefined the outputs of the model as concentrations at the points 1, 2 and 3 on Figure 7-3, and then performed approximate TBR reduction using the modified AISIAD method.

In Figure 7-5, the output produced by a 0.1 second unit pulse is shown. The results for the 2842 state model and modified AISIAD reduced model of order 13 are compared in Figure 7-5. One can clearly see that the reduced model nearly perfectly represents different delay values and the spread of the outputs.

5.2 Micromachined Pump Example

The TBR TPWL model order reduction strategy was applied to generate macromodels for the micromachined pump example described in Section 4. The reduced basis was generated using the linearized model of system (7.1) only at the initial state, and the initial state was included in the bases V and W.

Surprisingly, unlike in several nonlinear circuit examples [13], the output error did not decrease monotonically as the order q of the reduced system grew. Instead, macromodels with odd orders behaved very differently than macromodels with even orders. Models of even orders were substantially more

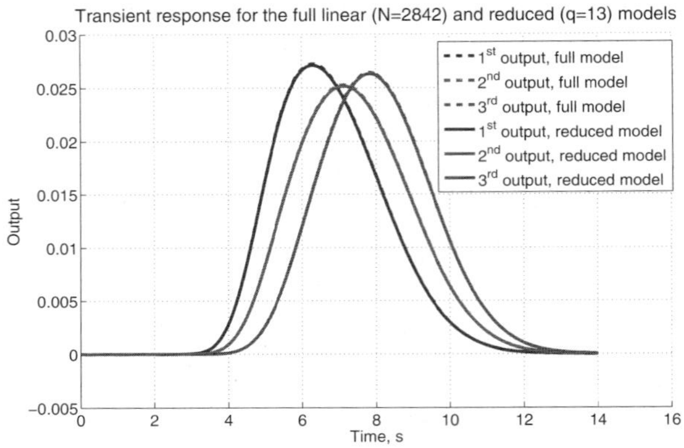

Figure 7-5. Transient responses of the orgininal linear (dashed lines) and reduced (solid lines) model (order $q = 13$). Input signal: unit pulse with duration 0.1 seconds. The maximum error between these transients is $\approx 1 \times 10^{-4}$, therefore the difference is barely visible. The different outputs correspond to the different locations along the channel's outlet (from left to right: innermost point, middle point, outermost point).

accurate than models of the same order generated by Krylov reduction – cf. Figure 7-6. However, if q was odd, inaccurate and unstable reduced order models were obtained. This phenomenon is reflected in the error plot shown in Figure 7-6. Figure 7-7 illustrates that a fourth-order (even) reduced model accurately reproduces transient behaviour.

This 'even-odd' phenomenon was observed in [28] and explained in the very general sense in [29]. The main result of [29] is described in the following section. However, there is also an insightful, but less general way of looking at this phenomenon.

The 'even-odd' phenomenon can be viewed by examining eigenvalues of the reduced order Jacobians from different linearization points. For the pump example, the initial nonlinear system is stable and Jacobians of f at all linearization points are also stable. Nevertheless, in this example, the generated reduced order basis provides a truncated balancing transformation only for the linearized system from the initial state x_0. Therefore, only the reduced Jacobian from x_0 is guaranteed to be stable. Other Jacobians, reduced with the same projection bases, may develop eigenvalues with positive real parts.

Figure 7-8 shows spectra of the reduced order Jacobians for models of order $q = 7$ and $q = 8$. One may note that, for $q = 8$, the spectra of the Jacobians from a few first linearization points are very similar. They also follow the same pattern: two of the eigenvalues are real, and the rest form complex-conjugate pairs. Increasing or decreasing the order of the model by 2 creates or eliminates a complex-conjugate pair of stable eigenvalues from the spectra of the

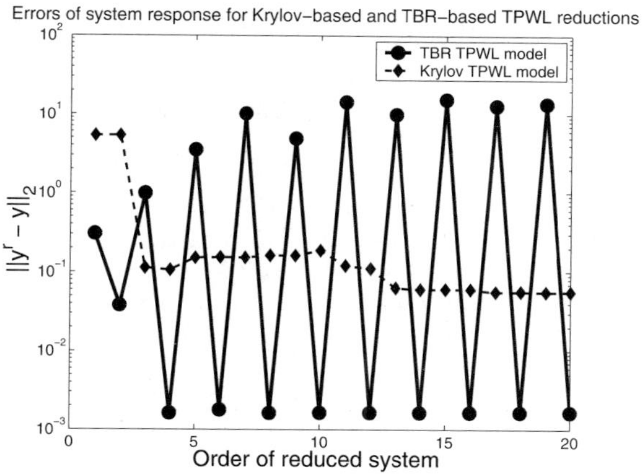

Figure 7-6. Errors in output computed by TPWL models generated with different MOR procedures (micromachined pump example); $N = 880$; 5.5-volt step testing and training input voltage.

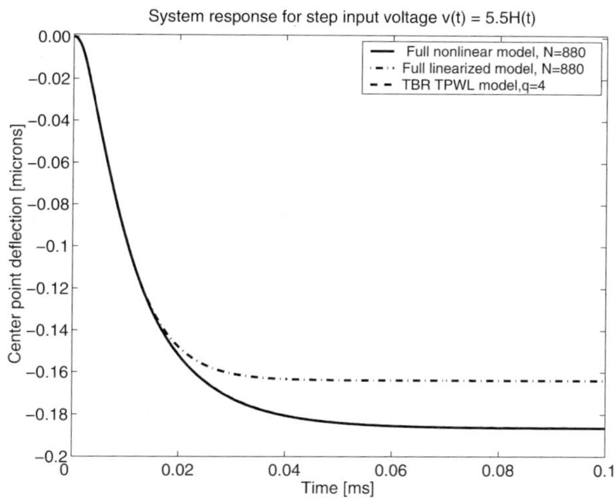

Figure 7-7. Comparison of system response (micromachined pump example) computed with both nonlinear and linear full-order models, and TBR TPWL reduced order model (7 models of order $q = 4$); 5.5-volt step testing and training input voltage. Note: solid and dashed lines almost fully overlap.

Jacobians. If the order of the model is increased or decreased by 1 (cf. Figure 7-8 (left)), the situation is very different. A complex-conjugate pair will be broken, and a real eigenvalue will form. At the first linearization point this eigenvalue is a relatively small negative number. At the next linearization point, the corresponding eigenvalue shifts significantly to the right half-plane to form an unstable mode of the system. An obvious workaround for this problem in the considered example is to generate models of even order. Nevertheless, a true

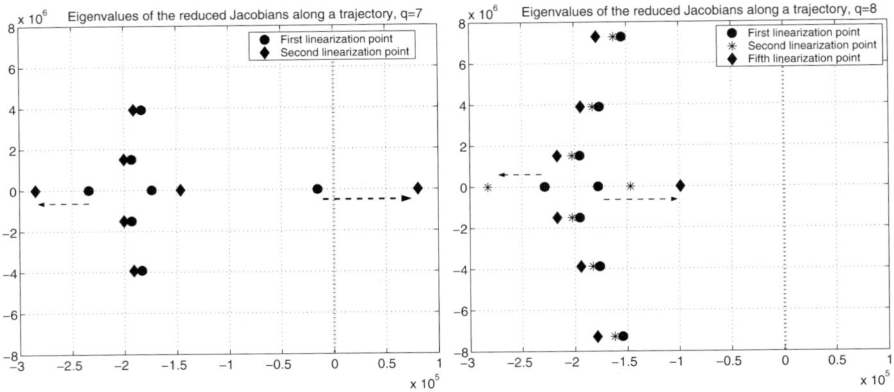

Figure 7-8. Eigenvalues of the Jacobians from the first few linearization points (micromachined pump example, Krylov-TBR TPWL reduction). Order of the reduced system $q = 7$ (left), $q = 8$ (right).

solution to this problem would involve investigating how perturbations in the model affect the balanced reduction, and this is examined in Section 6.

5.3 Nonlinear Microfluidic Example

Consider introducing a mild nonlinearity into the mobility and diffusion coefficients in (7.9):

$$\mu(C) = (28 + C \cdot 5.6) \times 10^{-9} m^2 V^{-1} s^{-1},$$
$$D(C) = (5.5 + C \cdot 1.1) \times 10^{-10} m^2 s^{-1}$$

Our experiments showed that even such a small nonlinearity creates a challenging problem for the TPWL algorithm. For this problem, the choice of training input significantly affects the set of the inputs signals for which the reduced model produces accurate outputs. For the case of a pulsed marker, this example has, in effect, a travelling wave solution. Therefore, linearizing at different timepoints implies linearizing different spatially local regions of the device, and many linearizations will be needed to cover the entire device.

Our experiments showed that a workable choice of projection matrices V and W for this example is an aggregation of the TBR basis and some of the linearization states x_i. Therefore, the projection used was a mix between TBR and snapshots-based projection [30]. For example, the reduced model whose transient step response is presented in Figure 7-9 was obtained using an aggregation of an order-15 TBR basis and 18 linearization states. The resulting system size was $q = 33$, and the number of linearization points was 23 (the initial model size was $N = 2842$). The linearization points were generated using the same step input for which the reduced simulation was performed. Although the results from the reduced model match when the input is the same as the training input, the errors become quite large if other inputs are used. For these nonlinear wave propagation problems, one needs to use a richer set of training inputs, which will result in a larger set of TPWL linearization points. In addition, instability in this simulation is still an issue, which makes the exact choice of projection basis an ad-hoc procedure.

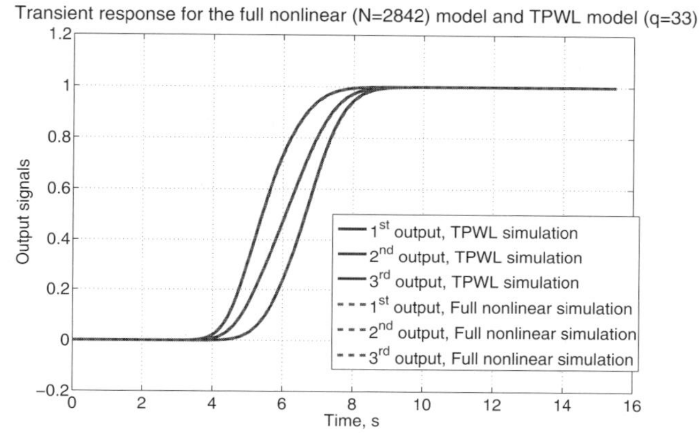

Figure 7-9. Step response of reduced and initial microfluidic model. Solid lines: order-33 TPWL reduced model obtained by using step training input. Dashed lines - full nonlinear model, N=2842. Note: solid and dashed lines almost fully overlap. The leftmost lines is the second input, which corresponds to the concentration closer to the center of the channel's curvature. The middle lines correspond to the first output signal (average concentration at the outlet). The rightmost lines correspond to the concentration at the outlet's points away from the center of curvature.

6. PERTURBATION ANALYSIS OF TBR REDUCTION ALGORITHM

For the micropump example, the even-odd behaviour of the model reduction can be analyzed using perturbation analysis. Assume the projection bases V and W are computed using TBR reduction from a single linearization point. The key issue is whether or not the TBR basis obtained at one linearization point is still suitable for reducing piecewise-linear models further along the trajectory. To understand this issue, consider two linearizations of the nonlinear system (7.1) (A_0,B,C) (initial) and (A,B,C) (perturbed). Suppose TBR reduction is performed for both of these models, resulting in projection bases V,W and $\tilde{V},\tilde{W}$ respectively. If these two bases are not significantly different, then perhaps V and W can be used to reduce the perturbed system, as is done for TPWL macromodels. This is true given some care, as will be made clear below.

6.1 Effect of Perturbation on Gramians

Consider the case for the controllability gramian P only, the results are valid for Q as well. Let $A = A_0 + \delta A$, $P = P_0 + \delta P$, where P_0 is an unperturbed gramian corresponding to unperturbed matrix A_0, and δA is relatively small so that δP is also small.

Using the perturbed values of A and P in the Lyapunov equation and neglecting $\delta P \delta A$ yields

$$A_0 \delta P + \delta P A_0^T + (\delta A P_0 + P_0 (\delta A)^T) = 0. \tag{7.10}$$

Note that (7.10) is a Lyapunov equation with the same matrix A_0 as for unperturbed system. This equation has a unique solution, assuming that the initial system is stable. The solution to (7.10) can be expressed using the following integral formula:

$$\delta P = \int_0^\infty e^{A_0^T t} (\delta A P_0 + P_0 (\delta A)^T) e^{A_0 t} dt. \tag{7.11}$$

Assuming A is diagonalizable, δP can be bounded as

$$||\delta P|| \leq 2(cond(T))^2 ||\delta A|| ||P_0|| \int_0^\infty e^{2\Re e(\lambda_{max}(A_0))t} dt, \tag{7.12}$$

where T is the matrix which diagonalizes A.

Since A is stable, the integral in (7.12) exists and yields an upper bound on infinitesimal perturbations of the gramian:

$$||\delta P|| \leq \frac{1}{|\Re e(\lambda_{max}(A_0))|} (cond(T))^2 ||P_0|| ||\delta A||. \tag{7.13}$$

Equation (7.13) shows that the bound on the norm of δP increases as the maximal eigenvalue of A_0 approaches the imaginary axis. In addition, note that perturbations in A will result in small perturbations in the gramian P as long as the system remains "stable enough", i.e. its eigenvalues are bounded away from the imaginary axis.

6.2 Effect of Perturbations on the Balancing Transformation

The balancing transformation in the algorithm 1 can be viewed essentially as a symmetric eigenvalue problem [21]:

$$RPR^T = U diag(\Sigma^2) U^T, \qquad T = \Sigma^{-1} U^T R, \tag{7.14}$$

where $R^T R = Q$ (R is a Cholesky factor of Q) and T is the coordinate transformation which diagonalizes both gramians. In the algorithm 1, matrix W consists of the first q rows of T, and matrix V consists of the first q columns of T^{-1}.

Applying the same perturbation analysis to the Cholesky factors, it can be shown that the perturbations in the Cholesky factors due to the perturbations in the original gramian are also small, provided that the system remains "observable enough", that is the eigenvalues of Q are bounded away from zero. Therefore we can state that the perturbation properties of the TBR algorithm are dictated by the symmetric eigenvalue problem $RPR^T = U\Sigma^2 U^T$.

The perturbation theory for the eigenvalue problem has been developed quite thoroughly [31], and one of the first observations is that small perturbations of a symmetric matrix can lead to large changes in the eigenvectors, if there are subsets of eigenvalues in the initial matrix which are very near to each other.

Below we summarize a perturbation theory for a symmetric eigenvalue problem with a nondegenerate spectrum.

Consider a symmetric matrix $M = M_0 + \delta M$, where M_0 is the unperturbed matrix with known eigenvalues and eigenvectors, and no repeated eigenvalues. Eigenvectors of M can be represented as a linear combination of eigenvectors of M_0:

$$x_k = \sum_{i=1}^{N} c_i^k x_i^0,$$

where x_k is the k-th eigenvector of the perturbed matrix M and x_i^0 is the i-th eigenvector of the unperturbed matrix. Coefficients c_i^k show how the eigenvectors of matrix M_0 are intermixed due to the perturbation δM, as in

$$(M_0 + \delta M)\sum_{i=1}^{N} c_i^k x_i^0 = \lambda_k \sum_{i=1}^{N} c_i^k x_i^0 \quad \Rightarrow \quad \sum_{i=1}^{N} c_i^k \delta M_{ji} = (\lambda_k - \lambda_j^0) c_j^k$$

where λ_k and λ_k^0 are the k-th eigenvalues of M and M_0 respectively and $\delta M_{ij} = (x_i^0)^T \delta M x_j^0$ is a matrix element of the perturbation in the basis of the unperturbed eigenvectors.

Now assume small perturbations and represent $\lambda_k = \lambda_k^0 + \lambda_k^{(1)} + \lambda_k^{(2)} + \ldots$ and $c_k^n = \delta_{kn} + c_k^{n(1)} + c_k^{n(2)} \ldots$ where each subsequent term represents smaller orders in magnitude. The first-order terms are:

$$\lambda_k^{(1)} - \lambda_k^0 = \delta M_{jj} \tag{7.15}$$

and

$$c_k^n = \frac{\delta M_{kn}}{\lambda_n^0 - \lambda_k^0}, k \neq n. \tag{7.16}$$

Equation (7.16) implies that the greater the separation between eigenmodes, the less they tend to intermix due to small perturbations. If a pair of modes have eigenvalues which are close, they change rapidly with perturbation. The following recipe for choosing an order of projection basis exploits this observation.

6.3 Recipe for using TBR with TPWL

Pick a reduced order to ensure that the remaining Hankel singular values are small enough and the last kept and first removed Hankel singular values are well separated.

The above recipe yields a revised TBR-based TPWL algorithm:

6.3.1 TBR-based TPWL with the Linearization at the Initial State

1 Perform the TBR linear reduction at the initial state x_0. Add x_0 to the projection matrices V and W by using biorthonormalization.

2 Choose the reduced order q such that the truncated Hankel singular values are :

- Small enough to provide sufficient accuracy
- separated enough from the Hankel singular values that are kept

3 Simulate the training trajectory and collect linearizations

4 Reduce linearizations using the projection matrices obtained in step 1.

6.4 Even-odd Behavior Explained

The perturbation analysis suggests that the sensitivity of TBR projection basis is strongly dependent on the separation of the corresponding Hankel singular values. The Hankel singular values for the linearization point of the micromachined pump example are shown in Figure 7-10.

As one can clearly see, the Hankel singular values for the micropump example are arranged in pairs of values, and evidently, even-order models violates the recipe for choice of reduction basis.

7. REMARK: SYSTEM-LEVEL AND INDIVIDUAL MODEL STABILITY

The above approach attempts to create individual reduced models of devices which are stable. However, interconnections of stable models do not, in

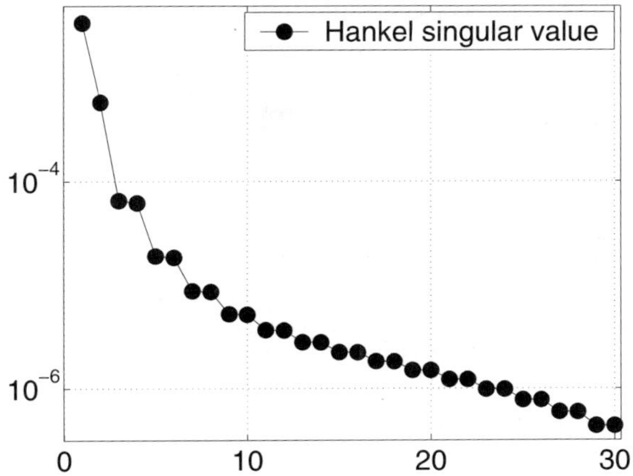

Figure 7-10. Hankel singular values of the balancing transformation at the initial state, Micromachined pump example.

general, result in a stable system (even for linear systems). In order to ensure system-level stability, it may be necessary to enforce stronger criteria on the individual models. One possibility would be to ensure an individual system's *dissipativity* (also termed *passivity*). That is, the energy in the output signal should not exceed energy in the input signal. This is a much more challenging problem than the one considered in this paper. In addition, particular constraints associated with the system's dissipativity are problem-dependent: they depend on the physical nature of the input and output signals. On the other hand, the proposed method is *generic*, that is, it does not assume any particular physical nature in the signals under consideration.

8. CONCLUSIONS

In this short paper we demonstrated that replacing Krylov-subspace methods with TBR as the linear reduction method in a TPWL algorithm dramatically improves reduced model accuracy for a given order, or substantially reduces the order needed for a given accuracy. In addition we discovered, analyzed, and resolved an instability problem with the TPWL-TBR approach. In particular, we gave a perturbation analysis which showed that when TBR is used in combination with TPWL, one should not truncate at an order that splits nearly equal Hankel singular values. Finally, we also demonstrated that the TPWL-TBR approach has much more difficulty when applied to problems with nonlinear wave propagation.

ACKNOWLEDGMENTS

The authors would like to acknowledge the support of the the National Science Foundation, the DARPA NeoCAD program, the Semiconductor Research Corporation, and the Singapore-MIT alliance.

REFERENCES

[1] P.G. Glavina, D.J. Harrison and A. Manz. Towards miniaturized electrophoresis and chemical analysis systems on silicon: an alternative to chemical sensors. *Sensors and actuators B*, 10(1):107–116, 1993.

[2] C.H. Ahn, J.W. Choi, G. Beaucage, J. Nevin, J.B. Lee, A. Puntambekar, and J.Y. Lee. Disposable smart lab on a chip for point-of-care clinical diagnostics. In *Proceedings of the IEEE, Special Issue on Biomedical Applications for MEMS and Microfluidics*, volume 92, pages 154 – 173, 2004.

[3] F. Pourahmadi, K. Lloyd, G. Kovacs, R. Chang, M. Taylor, S. Sakai, T. Schafer, W. McMillan, K. Petersen, and M.A. Northrup. Versatile, adaptable and programmable microfluidic platforms for dna diagnostics and drug discovery assays. In *Proceedings of the MicroTAS 2000 Symposium*, 2000.

[4] Y. Wang, Q. Lin, and T. Mukherjee. Applications of behavioral modeling and simulation on a lab-on-a-chip: micro-mixer and separation system. In *Proceedings of the 2004 IEEE International Behavioral Modeling and Simulation Conference, 2004. BMAS 2004.*, pages 8–13, 2004.

[5] Y. Wang, Q. Lin, and T. Mukherjee. System-oriented dispersion models of general-shaped electrophoresis microchannels. *Lab-on-a-chip*, 4(5):453–463, 2004.

[6] Y. Wang, Q. Lin, and T. Mukherjee. System simulations of complex electrokinetic passive micromixers. In *Technical Proceedings of the 2005 NSTI Nanotechnology Conference*, pages 579 – 582, 2005.

[7] T. Korsmeyer, J. Zeng, and K. Greiner. Design tools for biomems. In *Proceedings of the 41th conference on Design automation*, pages 622–627, 2004.

[8] P.Li and L.T. Pileggi. Norm: Compact model order reduction of weakly nonlinear systems. In *Proceedings of the 40th conference on Design automation*, 2003.

[9] N. Dong and J. Roychowdhury. Piecewise polynomial nonlinear model reduction. In *Proceedings of the 40th conference on Design automation*, 2003.

[10] M. Rewieński and J. White. A trajectory piecewise-linear approach to model order reduction and fast simulation of nonlinear circuits and micromachined devices. *IEEE Transactions on Computer-Aided Design of Integrated Circuits and Systems*, 22(2):155–170, 2003.

[11] A. Oliveira J.R. Phillips, J. Afonso and L.M. Silveira. Analog macromodeling using kernel methods. In *Proceedings of International Conference on Computer Aided-Design*, 2003.

[12] S.K. Tiwary and R.A. Rutenbar. Scalable trajectory methods for on-demand analog macromodel extraction. In *Proceedings of the 42nd annual conference on Design automation*, pages 403–408, New York, NY, USA, 2005. ACM Press.

[13] D. Vasilyev, M. Rewieński, and J. White. A tbr-based trajectory piecewise-linear algorithm for generating accurate low-order models for nonlinear analog circuits and mems. In *Proceedings of the 40th conference on Design automation*, pages 490–495. ACM Press, 2003.

[14] E.J. Grimme. *Krylov projection methods for model reduction.* PhD thesis, University of Illinois at Urbana Champaign, 1997.

[15] P. Feldman and R. Freund. Efficient linear circuit analysis by padé approximation via lanczos process. *IEEE Transactions on Computer-Aided Design*, 14(5):639 – 649, 1995.
[16] L. Miguel Silveira, Mattan Kamon, Ibrahim Elfadel, and Jacob White. A coordinate-transformed arnoldi algorithm for generating guaranteed stable reduced-order models of rlc circuits. In *ICCAD '96: Proceedings of the 1996 IEEE/ACM international conference on Computer-aided design*, pages 288–294, Washington, DC, USA, 1996. IEEE Computer Society.
[17] A. Odabasioglu, M. Celik, and L.T. Pileggi. Prima: passive reduced-order interconnect macromodeling algorithm. *IEEE Trans. Computer-Aided Design*, 17(8):645–654, 1998.
[18] Moore B.C. Principal component analysis in linear systems: controllability, observability, and model reduction. *IEEE Trans. Autom. Control*, AC-26(1):17–32, 1981.
[19] Glover K., Goddard P.J., and Chu Y.C. *Model Reduction for Classes of Uncertain, Multi-Dimensional, Parameter Varying and Non-Linear Systems.*, volume 240 of *Lecture Notes in Control and Information Sciences*, pages 269–280. Springer-Verlag, 1998.
[20] Jing-Rebecca Li. *Model reduction of large linear systems via low rank system gramians*. PhD thesis, MIT, 2000.
[21] K. Glover. All optimal hankel-norm approximations of linear multivariable systems and their l^{∞} -error bounds. *International Journal of Control*, 39(6):1115–1193, 1984.
[22] E. Hung, Y. Yang, and S. Senturia. Low-order models for fast dynamical simulation of mems microstructures. In *Proc. of IEEE Int. Conf. on Solid State Sensors and Actuators*, 1997.
[23] Z. Tang, S. Hong, D. Djukic, V. Modi, A.C. West, J. Yardley, and R.M. Osgood. Electrokinetic flow control for composition modulation in a microchannel. *Journal of Micromechanics and Microengineering*, 12(6):870–877, 2002.
[24] E.B. Cummings, S.K. Griffiths, and R.H. Nilson. Irrotationality of uniform electroosmosis. In *SPIE Conference on Microfluidic Devices and Systems II (Santa Clara, CA)*, volume 3877, pages 180–189, 1999.
[25] L.D. Landau and E.M. Lifshitz. *Fluid Mechanics*, volume 6. Butterworth-Heinemann, second edition, 1977.
[26] Yunkai Zhou. *Numerical Methods for Large Scale Matrix Equations with Applications in LTI System Model Reduction*. PhD thesis, Rice University, 2002.
[27] D. Vasilyev and J. White. A more reliable reduction algorithm for behavioral model extraction. In *Proceedings of the 2005 International conference on Computer-aided design (ICCAD)*, pages 813 - 820, 2005.
[28] M.I. Younis, E.M. Abdel-Rahman, and Ali Nayfeh. A reduced-order model for electrically actuated microbeam-based mems. *Journal of MEMS*, 12(5):672–680, 2003.
[29] D. Vasilyev, M. Rewieński, and J. White. Perturbation analysis of tbr model reduction in application to trajectory-piecewise linear algorithm for mems structures. In *Proceedings of the 2004 NSTI Nanotechnology Conference*, volume 2, pages 434 – 437, 2004.
[30] K. Willcox and J. Peraire. Balanced model reduction via the proper orthogonal decomposition. *AIAA Journal*, 40(11):2323–2330, November 2002.
[31] L.D. Landau and E.M. Lifshitz. *Quantum Mechanics: Non-Relativistic Theory*, volume 3, chapter 36, pages 133–137. Butterworth-Heinemann, third edition, 1977.

Chapter 8

SYSTEM-LEVEL SIMULATION OF FLOW INDUCED DISPERSION IN LAB-ON-A-CHIP SYSTEMS

Aditya S. Bedekar, Yi Wang, S. Krishnamoorthy, Sachin S. Siddhaye, and Shivshankar Sundaram
CFD Research Corporation, Huntsville, AL 35805

Abstract: Development of lab-on-a-chip systems has moved from the demonstration of individual components to a complex assembly of components. Due to the increased complexities associated with model setup, and computational time requirements, current design approaches using spatial and time resolved multiphysics modeling, though viable for component-level characterization, become unaffordable for system-level design. To overcome these limitations, we present models for the system-level simulation of fluid flow, electric field and analyte dispersion in microfluidic devices. Compact models are used to compute the flow (pressure-driven and electroosmotic) and are based on the integral formulation of the mass, momentum and current conservation equations. An analytical model based on the method of moments approach has been developed to characterize the dispersion induced by combined pressure and electrokinetic driven flow. The methodology has been validated against detailed 3D simulations and has been used to analyze hydrostatic pressure effects in electrophoretic separation chips. A 100-fold improvement in the computational time without significantly compromising the accuracy (error less than 10%) has been demonstrated.

Key words: compact models; network modeling; system design; microfluidics; dispersion

1. INTRODUCTION

Lab-on-a-chip systems are increasingly used in the areas of genomics, proteomics, biodiagnostics and drug discovery. These complex systems consist of networks of channels and reservoirs along with interfaces to the macroworld[1,2]. Increasing use of these systems calls for tools and techniques to perform the design analysis in a fast and efficient manner[3]. The

K. Chakrabarty and J. Zeng (eds.),
Design Automation Methods and Tools for Microfluidics-Based Biochips, 189–214.

functioning of a microfluidics chip involves the interplay of several physicochemical phenomena, including fluid flow, heat and mass transfer, electrokinetics, surface-tension effects, electrostatics, magnetics, particle transport, electrochemistry, biochemistry, and structural mechanics. This complexity means that design techniques based purely on experiments are prone to costly delays and failure. Computer-aided design techniques based on numerical simulations based on high-fidelity multiphysics models have been playing critical role in the design various components and subsystems of these lab-on-a-chip devices[4-8]. The numerical techniques used include finite difference methods, finite volume methods, finite element methods and boundary element methods. A number of commercially available software packages which employ these methods have been used previously in simulation of electrokinetic and pressure-driven flow as well as species transport and biochemical reactions[9]. Together with advances in techniques for rapid prototyping of microfabricated devices, these methods have improved the design process by providing accurate estimates of chip performance[10,11]. These multiphysics-based high fidelity (3D) simulation tools allow coupled analysis of these phenomena and provide detailed information of spatio-temporal variations of the field variables. Besides, they also provide experimentally inaccessible information and have the potential to reduce the need for extensive experimental testing which could be both time-consuming and expensive. This will allow exploration of more design possibilities. However, use of these tools for system-level analysis is computationally very expensive resulting in high turnaround times. The 3D modeling approach is inadequate for future lab-on-a-chip design tools where fast response and ease of use will be two major considerations. This necessitates the development of models that have the ability to perform simulations rapidly without significantly compromising accuracy.

To address these needs, compact[12], analytical[13-15] or reduced-order models[16,17] have been reported in literature. Qiao and Aluru[12] presented a compact model to compute flow rate and pressure in microfluidic devices driven by either an applied electric field or a combined electric field and pressure gradient while also considering the effects of varying zeta-potential. Wang *et al.*[13-15] have presented analytical models to study dispersion effects in electrokinetic flow induced by both turn geometry and Joule heating using a 'method of moments' approach. These models capture the effect of chip topology, separation element size, material properties and electric field on the separation performance. Recently, a behavioral model that accurately considers the analyte mixing and tradeoffs among chip size, complexity and mixer performance within laminar-diffusion based complex electrokinetic micromixers has also been presented[15]. Qiao and Aluru[18] have used mixed-domain simulation of electroosmotic flow to extract reduced order models for electroosmotic transport. They have also studied[17] the transient behavior of the electroosmotic flow using a. weighted Karhunen–Loeve decomposition method-based reduced model approach. Magargle *et al.*[19] and Mikulchenko *et al.*[20] have used neural network models for electrokinetic injection and a microflow sensor, respectively that are parameterized by the device geometry and operational parameters (e.g., electric field and flow velocity). Though these models may not capture all the details elucidated by grid-based 3D modeling techniques, they are adequate enough to quickly and accurately capture the basic physical behavior of the system[21] in a manner that is amenable to system-level simulation and design of microfluidic systems. Zhang *et al.* developed an integrated modeling and simulation framework of microelectrofluidic systems in SystemC, which enabled hierarchical analysis of composite microfluidic systems at various abstraction levels with an assumption that behavioral models or reduced-order models are available, and was used to evaluate and compare the performance of a polymerase chain reaction (PCR) system[22]. More recently, Wang *et al.*[23] have developed a behavioral modeling and schematic simulation environment based on element-level multi-physics models and system hierarchy in Verilog-A for efficiently analyzing the integrated and multi-functional (mixing, reaction, injection and separation) lab-on-a-chip systems.

We present a compact model for solving pressure-driven and electroosmotic flow in microfluidic devices. Electroosmotic flow (EOF) refers to the motion of a buffer solution past a stationary solid surface due to an externally applied electric field. Model equations for the flow field are obtained using an integral formulation of the mass (continuity), momentum, and current conservation equations. The coupling between the mass and momentum conservation equations is achieved using an implicit pressure-based scheme[24], as opposed to taking the divergence of the momentum equation and applying the continuity condition[21]. In addition, we also present an analytical model for computing analyte dispersion. The analyte is introduced in the buffer in the form of a 'plug' and transported under the action of buffer flow or by electrophoresis (migration of charged analytes under the action of electric field). The dispersion model involves the solution of the advection-diffusion equation. Specifically, we extend the method of moments[13,14] to describe dispersion effects in combined pressure-driven and electrokinetic (electroosmosis and electrophoresis) flow. The equations are solved on a network representation of the microfluidic system. Model development is discussed in Section II. Model validation and application to the analysis of hydrostatic pressure effects in electrophoretic separation chips is presented in Section III. The results are summarized with conclusions in Section IV.

2. MODELING APPROACH

2.1 Schematic Representation of a Lab-on-a-chip System

For a system-level solution, a lab-on-a-chip is represented as a network of components connected by edges. The edges can be considered as 'wires' of zero resistance. The solution variables (pressure, voltage, concentration) are computed on the components, while the flow rate and electric current from

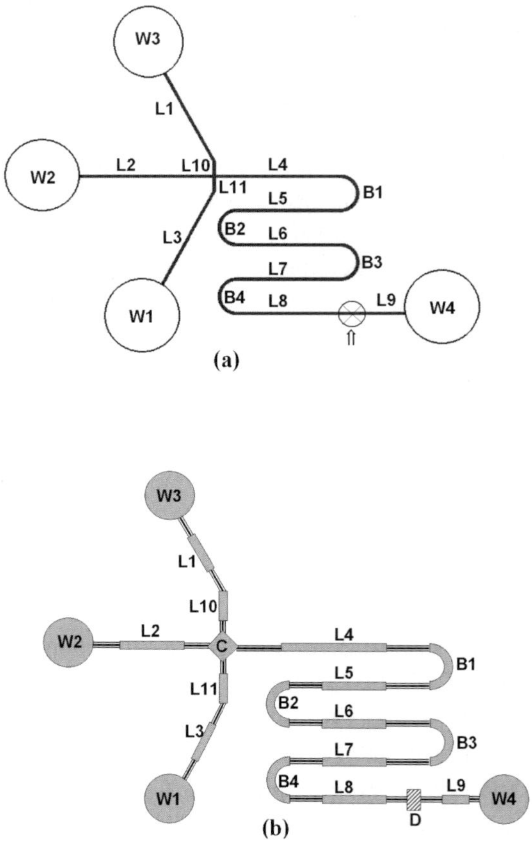

Figure 8-1. (a) Layout of a lab-on-a-chip (b) Schematic representation showing various components.

one component to another is computed on the edge. Fig. 8-1(a) shows the layout of the separation chip analyzed in this study. The system consists of components such as a cross junction (C), eleven straight channels (L1–L11), four bends (B1–B4), four reservoirs (W1–W4) and a detector (D). The separation chip can be considered a network of these components as shown schematically in Fig. 8-1(b). Our formulation does not restrict the number of edges emanating from a component. However, two components are connected uniquely by an edge.

2.2 Compact Models for Fluid Flow and Electric Field

The compact model describing the fluid flow is derived using the integral form of the continuity and momentum equations, while that of the electric current is derived from the current conservation equation. The model assumes the following:

1. Both flow and electric fields are at steady state.
2. The fluid is assumed Newtonian and incompressible.
3. A dilute electrolyte approximation is used, i.e., the hydrodynamic interaction between the analytes with the surface charges (important for electrokinetic systems) is accounted for only through modification of electroosmotic mobility.
4. The electrical conductivity of the liquid is assumed constant.
5. The buffer solution is assumed to be electrically neutral.
6. The pressure and electric fields are assumed to be completely decoupled, i.e., the presence of an electric field does not result in the generation of any internal pressure gradients[25].
7. Each component has a uniform cross section.

The conservation equations in their integral form are formulated for each component in the network. The continuity equation on component i can be written as:

$$\sum_j \dot{m}_{ij} = \dot{m}_{i\text{-}source} \tag{8.1}$$

where, $\dot{m}_{ij}$ is the mass flow from i^{th} to j^{th} component and $\dot{m}_{i-source}$ is the mass source at the component i. The momentum equation is written for each edge connecting the components i and j:

$$P_i - P_j = R_{ij}\dot{m}_{ij} \tag{8.2}$$

where P_i and P_j are the pressures at components i and j, while R_{ij} is the resistance to fluid flow (arising from viscous effects) from component i to component j. For a channel of width w, height h and length l, the resistance R_{ij} can be expressed as[26]:

$$R_{ij} = \left(\frac{12 \mu l}{\rho h w^3} \right) \Big/ \left(1 - \frac{192 w}{\pi^5 h} \sum_{i=1,3,5,\ldots}^{\infty} \frac{\tanh(i \pi h / 2w)}{i^5} \right) \quad (8.3)$$

where ρ is the buffer viscosity. The above relation is valid only in the laminar flow regime. Similar relations are available for channels of variety of cross-sections in literature[26]. Thus our compact model is fully parameterized and the effect of geometry of the component can be accounted by computing the suitable value of the resistance coefficient. For inertia-dominated flows, the resistance becomes a function of the velocity. In that scenario, our model has the ability to include such resistances by solving the governing equation in an iterative fashion. However, in microfluidic devices, the Reynolds number of the flow is very small (<<1) and the nonlinear effects can be neglected.

For the solution of electric fields, current conservation law is solved at every component (analogous to the continuity equation) along with a constitutive equation relating the current and the voltage. At a component, we write:

$$\sum_j I_{ij} = I_{i-source} \quad (8.4)$$

where I_{ij} is the current from component i to component j, while $I_{i\text{-}source}$ is the current source associated with component i. The voltage drop across each edge can be related to the current I_{ij} in the form of a constitutive equation as:

$$G_{ij} \left(V_i - V_j \right) = I_{ij} \quad (8.5)$$

where the electrical conductance G_{ij} is defined as the inverse of the electrical resistance to current flow from component i to component j, and V is the voltage.

2.3 Solution Methodology for Fluid Flow and Electric Field

Equations (8.1) and (8.2) are solved in a coupled manner using the SIMPLE (Semi-Implicit Method for Pressure-Linked Equations) algorithm proposed by Patankar[24]. This method has been widely used in the Computational Fluid Dynamics (CFD) community, and is well validated in literature. In this methodology, we write the mass flow rate in terms of a pressure correction term (derived from a Taylor series expansion of the mass flow rate in terms of pressure):

$$\dot{m}_{ij} = \dot{m}_{ij}^{*} + \frac{\partial \dot{m}_{ij}}{\partial P_i} p_i' + \frac{\partial \dot{m}_{ij}}{\partial P_j} p_j' \tag{8.6}$$

where the pressure correction $p_i' = P_i - P_i^*$. The superscript * represents the values at the previous iteration. Substituting above equation in the continuity equation and rearranging the terms, we get:

$$[K]\{p'\} = \{f\} \tag{8.7}$$

where

$$K_{ij} = \sum_j \frac{\partial \dot{m}_{ij}}{\partial P_j} \tag{8.8}$$

and

$$f_i = -\sum_j \dot{m}_{ij} + \dot{m}_{i-source} \tag{8.9}$$

In the above equations, the square brackets represent matrix quantities, while the curly brackets denote vector quantities. By taking the derivative of the mass flux with respect to the pressure in (8.2), we obtain

$$K_{ij} = \frac{\pm 1}{R_{ij}} \tag{8.10}$$

The coefficient matrix K_{ij} is sparse, but diagonally dominant. The fluidic resistance term R_{ij} in (8.10), represents the momentum losses when the fluid flows from component *i* to component *j*. An upwind approach is used in assigning proper values to R_{ij}. For example, if the fluid flows from component *i* to component *j*, the resistance of component *i* is used. The solution is obtained using the following methodology.

1. Equation (8.2) is solved to obtain the mass flow rates between each component.
2. The coefficient matrix (K_{ij}) is assembled following the upwind approach.

3. The system of linear equations given by Eq. (8.7) is solved to obtain pressure correction.
4. Pressure is updated.
5. Steps 1 through 4 are repeated until convergence. At convergence, the mass imbalance at each component (f_i) is less than the specified tolerance.

The electric potential is also computed in a similar fashion by assembling a conductance matrix (G) and solving Eq. (8.5), to obtain voltages. Once the electric field is computed, the electroosmotic flow is calculated using the relation

$$\vec{u}_{eof} = \omega_{eof} \nabla \cdot V \tag{8.11}$$

where ω_{eof} is the electroosmotic mobility, and the arrow represents the vector quantity. However, in our calculation, only the axial component of the velocity plays a significant role. For fluid flow in the linear regime[25] (low Reynolds number flows), it has been previously shown[25] that the cross-sectional velocity profile in a microchannel resulting from the combined pressure-driven and electroosmotic flows could be obtained by superimposing individual profiles for steady state laminar flow. This assumption is used in the derivation of the analytical solution for analyte dispersion based on the "method of moments".

2.4 Governing Equations and Solution Methodology for Analyte Dispersion

The model developed to study the analyte dispersion includes the effects of pressure-driven (parabolic velocity profile) and electroosmotic (plug profile) flows in straight channels and bends (geometry induced dispersion). The transport of the analyte is described by the advection-diffusion equation:

$$\frac{\partial c}{\partial t} + \vec{u} \cdot \nabla c = D \nabla^2 c \tag{8.12}$$

where c is the analyte concentration, D is the diffusion coefficient, and analyte velocity $\vec{u}$ contains contributions from pressure-driven as well as electrokinetic (electroosmosis and electrophoresis) effects. The derivation of the analytical solution consists of recasting (8.12) in a moving coordinate system and calculating moments of the analyte distribution. The pressure-induced dispersion (Taylor dispersion) is calculated using the mass flow rate computed from (8.2). This mass flow rate is used to compute the pressure gradient analytically. The variance and skewness, calculated from the moments of the concentration of the analyte plug, characterizes the dispersion. The solution is derived for a single constant-radius bend channel shown in Fig. 8-2. Here, β (= w/h), r_c, θ, b (= w/r_c), and L (= $r_c\theta$), denote the bend aspect ratio, mean radius, included angle, curvature and mean axial length of the bend, respectively. Extension of the derived solution to a straight channel will be a straightforward application, when we assume bend curvature as zero. The straight and constant-radius bend geometries are elementary constructs that can be used to generate most of the microfluidic layouts studied in the literature[27-29]. The following assumptions are made in deriving the analytical solution:

- The bend curvature (b) is small ($\ll 1$).
- There are no changes in the cross-sectional area or shape of the bend.
- The analytical solution is derived based on the underlying steady state flow and electric field, i.e., migration of the analyte does not induce variations in the fluid properties (dilute approximation).

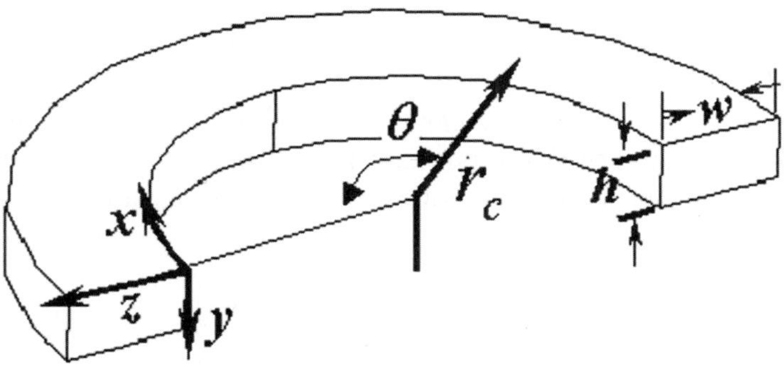

Figure 8-2. Schematic representation of coordinate system and geometry parameters used in the constant radius bend.

2.5 Calculation of Background Flow Field

The analyte flow consists of contributions from the electrokinetic and pressure-driven flows that are additive[25]. In the linear regime and for $b \ll 1$, the apparent axial velocity (u') is expressed as[13,30]:

$$u' = \left(u_{ek} + u_{\mathrm{Pr}}\right) r_c / r \approx \left(u_{ek} + u_{\mathrm{Pr}}\right)\left(1 + b\left(1/2 - z/w\right)\right) \qquad (8.13)$$

where u_{Pr} is the velocity due to the external pressure, u_{ek} is the electrokinetic velocity component, r is the turn radius and r_c/r accounts for the difference in axial travel distance at different locations (z) along the bend width (analyte close to the inside wall transits through a short distance). For the bends considered in this paper (b<<1), $r_c/r \approx \left(1+b\left(1/2-z/w\right)\right)$ is valid in Eq. (8.13). u_{ek} in Eq. (8.13) is given by

$$u_{ek}(z) = \omega_{ek} E = \omega_{ek}\left(-\nabla V\right) = \omega_{ek}\frac{\Delta V}{\theta r_c}\frac{r_c}{r} \approx \left(\omega_{eof}+\omega_{ep}\right)E_t\left(1+b\left(1/2-z/w\right)\right) \tag{8.14}$$

where ω_{ek} is the electrokinetic mobility that has contributions from both electroosmotic (ω_{eof}) and electrophoretic (ω_{ep}) forces, and E_t ($=\Delta V/\theta r_c$) is the axial (tangential) electric field along the mean radius r_c. It is sufficiently accurate to assume E_t to be equal to the average electric field across the cross section, E_{av}[28]. Knowing the electric current, E_{av} can be computed as:

$$E_{av} = \frac{I_{ij}}{whK_e} \tag{8.15}$$

where K_e is the electric conductivity of the buffer solution. The electroosmotic mobility ω_{eof}, if not known *apriori*, may be obtained from the zeta potential (ζ) using the Helmholtz-Smoluchowski equation[31]:

$$\omega_{eof} = \varepsilon\zeta/\mu \tag{8.16}$$

where ε is the permittivity of the buffer and μ is the dynamic viscosity of the fluid.

The flow solver computes only the average values of the mass flow rate and the pressure gradient, across each component. However, the bend curvature induces local variations in the pressure gradients, which may significantly contribute to dispersion. Hence, an analytical solution for u_{Pr} based on the average mass flow rates is derived from the Navier-Stokes equation governing the axial velocity:

$$\mu\left(\frac{\partial^2 u_{\mathrm{Pr}}}{\partial y^2}+\frac{\partial^2 u_{\mathrm{Pr}}}{\partial z^2}\right) = \Pi\left(1+b\left(\frac{1}{2}-\frac{z}{w}\right)\right) \tag{8.17}$$

where $\Pi = dP/d\left(r_c\theta\right)$ refers to the pressure gradient at the mean radius and the term $1+b\left(1/2-z/w\right)$ accounts for the cross-sectional non-uniformity in pressure gradient caused by the turn curvature b. Equation (8.17) can be analytically solved and the solution is

$$u_{\mathrm{Pr}} = \frac{\Pi h^2}{\mu}\left(\left(1+\frac{b}{2}\right)\psi_1 + \frac{b}{\beta}\psi_2\right) \tag{8.18}$$

where

$$\begin{cases} \psi_1 = \sum_{i=1}^{\infty} \frac{2\left(-1+(-1)^i\right)}{i\pi}\left(-\frac{1}{\lambda_i} + \frac{e^{\left(1-\frac{y}{h}\right)\sqrt{\lambda_i}} + e^{\frac{y}{h}\sqrt{\lambda_i}}}{\left(1+e^{\sqrt{\lambda_i}}\right)\lambda_i}\right)\cdot \sin\left(\sqrt{\lambda_i}\frac{z}{h}\right) \\ \psi_2 = \sum_{i=1}^{\infty} \frac{2\beta(-1)^{i+1}}{i\pi\cdot\lambda_i}\left(-1 + \frac{e^{\left(1-\frac{y}{h}\right)\sqrt{\lambda_i}} + e^{\frac{y}{h}\sqrt{\lambda_i}}}{1+e^{\sqrt{\lambda_i}}}\right)\cdot \sin\left(\sqrt{\lambda_i}\frac{z}{h}\right) \end{cases} \tag{8.19}$$

and $\lambda_i = (i\pi/\beta)^2$. Therefore, through cross-sectional integration of Eq. (8.18), pressure gradient Π can be related to the pressure-driven mass flow rate $\dot{m}_{ij}$ by $\Pi = (\dot{m}_{ij}\mu)/\rho\bar{\psi}_1 h^4 \beta$, where the overbar denotes the cross-sectional average. Substituting Eqs. (8.14) and (8.18) into Eq. (8.13), we have

$$u' = \left\{\omega_{ek}E_{av}\left(1+b\left(\frac{1}{2}-\frac{z}{w}\right)\right) + \frac{\Pi h^2}{\mu}\left(\left(1+\frac{b}{2}\right)\psi_1 + \frac{b}{\beta}\psi_2\right)\right\}\left(1+b\left(\frac{1}{2}-\frac{z}{w}\right)\right) \tag{8.20}$$

With $b \ll 1$, all the terms of second order (b^2) in Eq. (8.20) are neglected during its expansion, then Eq. (8.20) can be reduced to

$$u' = \omega_{ek}E_{av}\left(1+b\psi_4\right) + \frac{\Pi h^2}{\mu}\left((1+b)\psi_1 + \frac{b}{\beta}\psi_2 - \frac{b}{\beta}\psi_3\right) \tag{8.21}$$

with

$$\begin{cases} \psi_3 = \frac{z}{h}\psi_1 \\ \psi_4 = \left(1-\frac{2z}{w}\right) \end{cases} \tag{8.22}$$

The residence time Δt of the species band's centroid within a bend is then given by:

$$\Delta t = \frac{\theta r_c}{\left(\overline{u_{ek} + u_{\Pr}}\right)} \tag{8.23}$$

2.6 Calculation of Analyte Dispersion

In our model formulation, we assume that the analyte consists of any number of analytes, each described by its own transport equation. The three-dimensional transient conservation equation for analyte transport [Eq. (8.12)] is recast on a moving coordinate system to derive the analytical solution. The normalized moving spatial and temporal reference coordinate system is defined as follows.

$$\xi = (x - U't)/h,\ \eta = y/h,\ \zeta = z/h \text{ and } \tau = Dt/h^2 \tag{8.24}$$

where, U' refers to the cross-sectional average of u', which can be calculated from Eq. (8.21) as

$$U' = \int_0^\beta \int_0^1 u' d\eta d\zeta = \omega_{ek} E_{av} + \frac{\Pi h^2}{\mu}\left\{(1+b)\overline{\psi}_1 + \frac{b}{\beta}\overline{\psi}_2 - \frac{b}{\beta}\overline{\psi}_3\right\} \tag{8.25}$$

where $\overline{\psi}_4 = 0$ is applied. A normalized apparent velocity χ with respect to U' is defined by:

$$\chi(\eta,\zeta) = (u' - U')/U' \tag{8.26}$$

which can be obtained from Eqs. (8.21) and (8.25):

$$\chi = \frac{\omega_{ek} E_{av} b\psi_4 + \frac{\Pi h^2}{\mu}\left\{(1+b)(\psi_1 - \overline{\psi}_1) + \frac{b}{\beta}(\psi_2 - \overline{\psi}_2) - \frac{b}{\beta}(\psi_3 - \overline{\psi}_3)\right\}}{\omega_{ek} E_{av} + \frac{\Pi h^2}{\mu}\left\{(1+b)\overline{\psi}_1 + \frac{b}{\beta}\overline{\psi}_2 - \frac{b}{\beta}\overline{\psi}_3\right\}} \tag{8.27}$$

In Eq. (8.27), the terms in curly brackets represent the velocity contribution from pressure-driven flow, while the rest (those containing E_{av}) represent the contributions from the electrokinetic flow.

Equation (8.12) can then be rewritten as:

$$\frac{\partial c}{\partial \tau}=\frac{\partial^2 c}{\partial \xi^2}+\frac{\partial^2 c}{\partial \eta^2}+\frac{\partial^2 c}{\partial \zeta^2}-\mathrm{Pe}\,\chi\frac{\partial c}{\partial \xi} \tag{8.28}$$

with the boundary conditions:

$$\partial c/\partial\eta\big|_{\eta=0,1}=0,\quad \partial c/\partial\zeta\big|_{\zeta=0,\beta}=0,\quad c\big|_{\tau=0}=c(\xi,\eta,\zeta,0)$$

where $\mathrm{Pe}=U'h/D$ is the Peclet number representing the ratio of the convective transport rate to the diffusive transport rate along the *depth* of the bend. Using the method of moments, Eq. (8.28) can be formulated in terms of spatial moments of the analyte concentration[13,32]. If the analyte band is entirely contained in the bend, Eq. (8.28) satisfies the condition of $c\to 0$ as $\xi\to\pm\infty$. We define

$$c_p(\eta,\zeta,\tau)=\int_{-\infty}^{\infty}\xi^p c(\xi,\eta,\tau)d\xi \tag{8.29}$$

and

$$m_p(\tau)=\overline{c}_p=\frac{1}{\beta}\int_0^{\beta}\int_0^1 c_p\,d\eta\, d\zeta\quad (p=0,1,2,...) \tag{8.30}$$

where c_p is the pth moment of the concentration in an axial filament of the analyte band that intersects the cross sections at η and ζ, and m_p is the pth moment of the average concentration across the cross-section. Then Eq. (8.28) is reformulated in terms of the moments along with its initial condition and boundary condition.

$$\begin{cases}\dfrac{\partial c_p}{\partial \tau}=\dfrac{\partial^2 c_p}{\partial \eta^2}+\dfrac{\partial^2 c_p}{\partial \zeta^2}+p(p-1)c_{p-2}+p\,\mathrm{Pe}\,\chi c_{p-1}\\ \partial c_p/\partial\eta\big|_{\eta=0,1}=0,\quad \partial c_p/\partial\zeta\big|_{\zeta=0,\beta}=0\\ c_p\big|_{\tau=0}=c_{p0}(\eta,\zeta)=\int_{-\infty}^{\infty}c(\xi,\eta,\zeta,0)\xi^p d\xi\end{cases} \tag{8.31}$$

and

$$\frac{dm_p}{d\tau} = p(p-1)\overline{c_{p-2}} + p\,\mathrm{Pe}\,\overline{\chi c_{p-1}}$$
$$m_p(0) = m_{p0} = \frac{1}{\beta}\int_0^\beta \int_0^1 c_{p0}(\eta,\zeta)\,d\eta\,d\zeta \qquad (8.32)$$

In both Eqs. (8.31) (8.32), any term that contains c_i with $i < 0$ is set to zero. To determine the broadening of the analyte band, these equations are solved to obtain moments up to the second order. Specifically, $c_0(\eta,\zeta,\tau)$ is the analyte mass in the axial filament intersecting η and ζ, and $m_0(\tau)$ is the total analyte mass in the bend. Next, $c_1(\eta,\zeta,\tau)$ is the ξ-coordinate of the centroid of the axial analyte filament, and hence indicates the skew of the analyte band. The cross-sectional average of c_1, $m_1(\tau)$, is the ξ-coordinate of the centroid of the entire analyte band. Finally, $m_2(\tau)$ is used to determine the variance of the analyte band[13,14]. The dispersion parameters describing the shape of the analyte band, such as skewness and variance are obtained from the solution of Eqs. (8.31) and (8.32). Details are outlined elsewhere[13,14] with the difference being that χ in the present work includes both pressure-driven and electrokinetic contributions and varies in both cross-sectional dimensions. Specifically, on substituting $p = 1$ into Eq. (8.31), the skewness of the analyte band is expressed as

$$c_1(\eta,\zeta,\tau) = \sum_{m=0}^{\infty}\sum_{n=0}^{\infty} S_{nm}(\tau)\cos(n\pi\eta)\cos\left(\frac{m\pi\zeta}{\beta}\right) \qquad (8.33)$$

where $S_{nm}(\tau)$ is defined as the skew coefficient, the Fourier series coefficient of c_1, and is given by

$$S_{nm}(\tau) = \begin{cases} S_{nm}(0)e^{-\lambda_{nm}\tau} + \dfrac{\mathrm{Pe}\chi_{nm}\left(1-e^{-\lambda_{nm}\tau}\right)}{\lambda_{nm}} & \text{if } n+m \geq 1 \\ S_{00}(0) & \text{if } n+m = 0 \end{cases} \qquad (8.34)$$

Here $S_{nm}(0)$ is the skew coefficient at the inlet of the bend, χ_{nm} is the Fourier coefficient for the normalized velocity χ and $\lambda_{nm}=(n\pi)^2+(m\pi/\beta)^2$ [$n\geq0$ and $m\geq0$]. The two terms on the right hand side of Eq. (8.34) represent contributions from the skewness at the bend inlet, and the non-uniformity in velocity profiles and axial geometry respectively. Similarly, the variance of the analyte band is expressed as follows.

$$\sigma^2(\tau) - \sigma^2(0) = 2h^2\tau + \frac{2\mathrm{Pe}h^2}{\beta}\sum_{m=0}^{\infty}\sum_{n=0}^{\infty}\frac{\chi_{nm}}{\nu_{nm}\lambda_{nm}}S_{nm}(0)\left(1 - e^{-\lambda_{nm}\tau}\right) + \frac{2\mathrm{Pe}^2h^2}{\beta}\sum_{m=0}^{\infty}\sum_{n=0}^{\infty}\frac{\chi_{nm}^2}{\nu_{nm}\lambda_{nm}^2}\left(e^{-\lambda_{nm}\tau} + \lambda_{nm}\tau - 1\right) \quad (8.35)$$

where ν_{nm} is defined as $\nu_{00}=1/\beta$, $\nu_{0m}=2/\beta$, $\nu_{n0}=2/\beta$ and $\nu_{nm}=4/\beta$ for $n>1$ and $m>1$, $\sigma^2(0)$ represents the variance of the analyte band at the bend inlet. Equation (8.35) shows that the spreading of the analyte band occurs due to three effects: molecular diffusion [the first term on the right hand side of Eq. (8.35)], the initial skewness (the second term) and the convective dispersion associated with the non-uniformities in velocity profile and axial travel distance (the third term). Neglecting terms involving pressure-driven flow will recover dispersion effects due to electrokinetic flow alone[13]. In addition, as bend curvature (b) tends toward zero, Eqs. (8.34) and (8.35) reduce to the special case of a straight channel. Assuming a Gaussian distribution for analyte concentration at the terminals of a component, we can obtain the amplitude of the analyte band as:

$$A(\tau)/A(0) = \sqrt{\sigma^2(0)/\sigma^2(\tau)} \quad (8.36)$$

A simplified injector model has been used to initiate the dispersion calculation. It injects a uniform analyte bands [$S_{nm}(0) = 0$] with an initial variance and amplitude.

2.7 Development of System-level Models

The above solution methodology describes the analyte band characteristics within a component (bend) as a function of time. We assume that the analyte band characteristics are propagated in the network as a signal flow. This means that the parameters associated with the band shape (e.g., variance, time of arrival, skew and amplitude) at the component's outlet are calculated based on the corresponding values at the inlet and contributions from the component itself. This information is then assigned as the input to the next component downstream, locations of which are determined based on the net direction of the analyte migration. The limitation of this approach is that error might be introduced when the analyte band spans two components involving different background velocity profiles.

Knowing the dimensionless residence time of the analyte plug ($\tau_R = \Delta t \cdot D/h^2$), the band characteristics at the outlet of the component are computed by substituting τ_R into Eqs. (8.34)-(8.36), which yields

$$t_{i,out} = t_{i,in} + \Delta t \tag{8.37}$$

$$S_{nm}^{i,out} = \begin{cases} S_{nm}^{i,in} e^{-\lambda_{nm}\tau_R} + \dfrac{\mathrm{Pe}\chi_{nm}\left(1-e^{-\lambda_{nm}\tau_R}\right)}{\lambda_{nm}} & \text{if } n+m \geq 1 \\ S_{nm}^{i,in}(0) & \text{if } n+m=0 \end{cases} \tag{8.38}$$

$$\begin{aligned} &\sigma_{i,out}^2 - \sigma_{i,in}^2 = 2h^2\tau_R + \\ &\frac{2\mathrm{Pe}h^2}{\beta}\sum_{m=0}^{\infty}\sum_{n=0}^{\infty}\frac{\chi_{nm}}{\nu_{nm}\lambda_{nm}}S_{nm}^{i,in}\left(1-e^{-\lambda_{nm}\tau_R}\right) + \\ &\frac{2\mathrm{Pe}^2h^2}{\beta}\sum_{m=0}^{\infty}\sum_{n=0}^{\infty}\frac{\chi_{nm}^2}{\nu_{nm}\lambda_{nm}^2}\left(e^{-\lambda_{nm}\tau_R} + \lambda_{nm}\tau_R - 1\right) \end{aligned} \tag{8.39}$$

$$A_{i,out}\big/A_{i,in} = \sqrt{\sigma_{i,in}^2\big/\sigma_{i,out}^2} \tag{8.40}$$

where index i denotes the ith component in the network, *in* and *out* represent the quantities at the inlet and outlet of the component represented by that component. The values of the band characteristics at the outlet given in the above equations are then assigned as the input to the downstream component, that is, $t_{i,out} = t_{j,in}$, $s_{nm}^{i,out} = s_{nm}^{j,in}$, $\sigma_{i,out}^2 = \sigma_{j,in}^2$ and $A_{i,out} = A_{j,in}$, where j is the downstream component of i. This protocol enables the transmission of the band characteristics within the entire network from the one component to the next.

3. APPLICATIONS

We validate the system solver for pressure-driven as well as electroosmotic flows. Estimates for analyte migration and subsequent separation in pressure-driven as well as electroosmotic flow are validated by comparison with detailed 3-D simulations using a commercial software CFD-ACE+™ (ESI CFD Inc., Huntsville, AL)[33]. The validity of the system solver will be demonstrated using the separation chip shown in Fig. 8-1. The geometry has a cross configuration, with an injection channel (L3, L11, L10 and L1) and a serpentine-shaped separation channel. The injection channel is connected to the analyte reservoir (W1) and a waste reservoir (W3). The separation channel is connected to the buffer reservoir (W2) and a waste reservoir (W4). The analyte and waster reservoirs (W1–W4) act as boundary conditions for the model (with fixed pressure and/or voltage). The effect of momentum loss due to changes in the orientation of the geometry (for example between components L3 and L11 of the injection channel) is neglected.

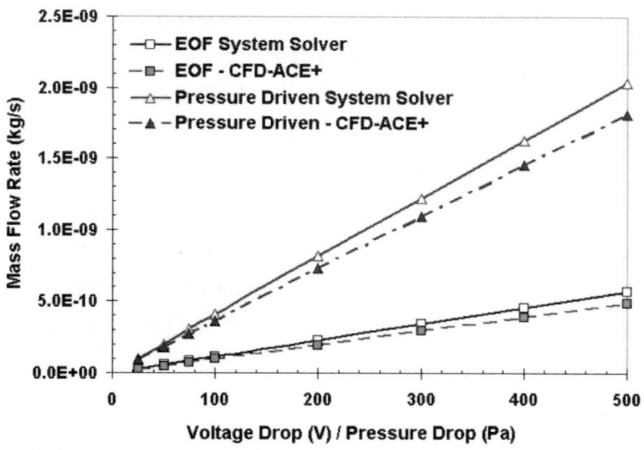

Figure 8-3. Validation of system solver for pressure-driven and electroosmotic flows.

3.1 Validation of Pressure Driven and Electroosmotic Flow

A fixed value of voltage (or pressure) is applied at the components corresponding to wells W1, W2, W3 and W4. The waste well W4 is maintained at the datum (voltage = 0V, or static pressure = 0Pa), while the other three wells are maintained at a higher voltage (or pressure). The voltage (and pressure) at all three wells W1, W2 and W3 is equal. All channels are assumed to be 50 microns wide and 25 microns deep. The channel lengths are chosen as follows: L1 = L3 = 2200 μm, L10 = L11 = 200μm, L4 = 1800μm, L5 = L6 = L7 = 1500μm, L8 = 2000 μm, L9 = 1000 μm. All four bends, B1, B2, B3 and B4 have a mean turn radius of 250 μm. The cross-junction C and detector D are assumed to be point entities that do not have any electrical or hydrodynamic resistance). The buffer is assumed to have the following properties: density = 1000 kg/m^3, kinematic viscosity = 1.0×10^{-6} m^2/s, and electroosmotic mobility = 1.0×10^{-8} m^2/(V s). The net mass flow rate into the waste reservoir is validated against data obtained from the 3-D simulation. The comparison is shown in Fig. 8-3. As expected, the variation of mass flow rate at the outlet boundary (waste well W4) due to pressure-driven or electroosmotic flow is linear with respect to the applied inlet pressure or voltage respectively. Over the range of flow rates studied (Re = 0.001-0.5 where Re is the Reynolds number), the maximum error obtained is less than 10%. The magnitude of the error between compact models used by the system solver and the 3-D model increases with increasing Reynolds number. This is primarily due to the pressure and voltage losses caused by changes in the orientation and intersection of channels (junction component C has been assumed to merely function as a flow distributor, without actually contributing to a pressure or a voltage drop). Typical Reynolds numbers encountered in microfluidic chips are usually less than one, and hence, the agreement is considered to be acceptable. On the other hand, the system solver allows for a speed up of simulation time by more than two orders of magnitude for the present network. An average simulation of electroosmotic and pressure-driven flow using the system solver takes less than 1 second, while the 3-D model takes about 270 seconds on an MS-Windows workstation with an AMD Athlon CPU (2GHz, 1GB RAM).

3.2 Validation of Analyte Dispersion Due to Electrophoresis

In order to validate the dispersion model, we have investigated separation of a sample into its constituent analytes. The sample, consisting of four different positively charged analytes of unit valence with different properties (mobility and diffusivity) is injected using the cross junction and detected at the end of the serpentine section using the detector (D). The properties of the analytes used in the simulation are summarized in Table 8-1.

Table 8-1. Analyte properties used in simulations

Analyte	**Diffusivity (m^2/s)**	**Electrokinetic Mobility [m^2/(V s)]**
Analyte_1	1.0×10^{-10}	3.0×10^{-8}
Analyte_2	3.0×10^{-10}	3.27×10^{-8}
Analyte_3	6.0×10^{-11}	2.73×10^{-8}
Analyte_4	3.0×10^{-11}	2.5×10^{-8}

The sample is transported by electrophoresis alone at E_{av}=300 V/cm (no external pressure applied). The buffer properties are the same as those described in the previous example. During the CFD-ACE+ simulations, the species bands with Gaussian concentration distributions were initially injected at a location 1800 μm upstream of the first turn. The detection of normalized concentration vs. separation time (electropherogram) is taken 2000 μm downstream of the last turn [Fig. 8-4(a)]. Fig. 8-4(b) shows the comparison of the detected electropherograms between the system simulator and CFD-ACE+. Due to the difference in analyte mobility and diffusivity, all bands exhibit different shape and dispersion behavior. Fig. 8-4(a) shows a snapshot of the analyte bands, clearly indicating the differential migration velocities of the analyte bands.

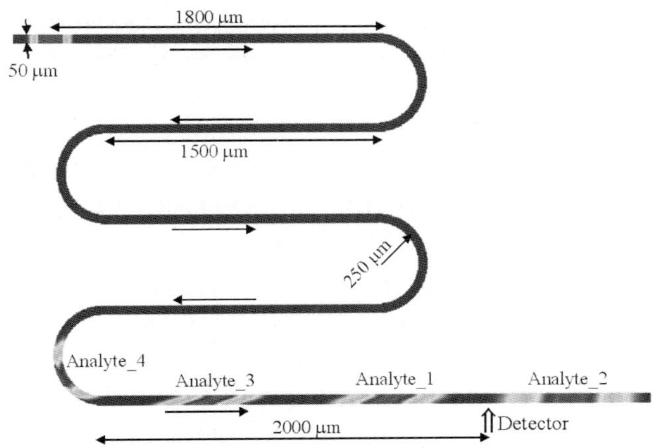

(a)

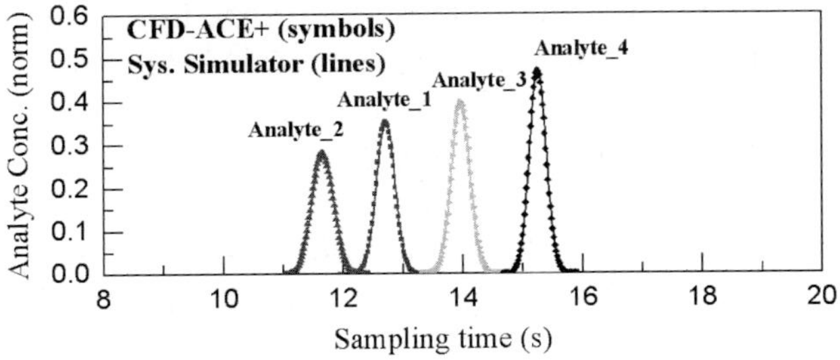

(b)

Figure 8-4. Dispersion due to electrophoresis in serpentine channel: (a) System description; (b) Electropherogram (comparison of system solver with CFD-ACE+).

Figure 8-4(b) shows the analyte concentrations recorded at the detector. Excellent agreement (error < 6.5%) has been obtained for all analytes, which were clearly separated and resolved during their migration through the channels. This includes the extent to which each analyte band gets distorted as it moves through the bends. This simulation demonstrates how the system solver accurately captures the effects of electric field, analyte properties, chip topology and channel dimensions on the dispersion. This methodology also saves computational time (from hours to seconds, Table 8-2), and hence, is suitable for iterative simulation-based design of lab-on-a-chip systems. The high-fidelity model referred in Table 8-2 has sufficient grid density needed to capture the dispersion effects (shown in Fig. 8-4).

Table 8-2. Performance Comparison

Simulation Method	Computational Time
High Fidelity Multiphysics	48 hours
System Simulator	< 5 seconds

3.3 Validation of Analyte Dispersion in Combined Pressure-driven and Electrokinetic Flow

Due to the non-uniformity in the velocity profile, band broadening in pressure-driven flow requires a 3D analysis. Accurate 3-D and transient CFD simulation of analyte transport in a full microfluidic network such as the one described in Fig. 8-1 is prohibitively expensive. In order to overcome this problem, we have setup a simplified geometry that will validate the analytical model. The microfluidic channels are shown in Fig. 8-5(a), along with the channel dimensions. The sample consists of the same four analytes with properties as described previously (Table 8-1). Also shown in the Fig. 8-5 (a) is the state of the analyte (Analyte_1) plug at various locations. The analyte band is initially injected at the middle of the first channel, where both pressure-driven and electrokinetic flow act in the same direction. Obvious analyte band distortion due to pressure-driven flow (Taylor dispersion) can be observed (parabolic interface matching with the parabolic flow profile typical of a pressure-driven laminar flow) even within the straight channel. In the turn, the analyte molecules at the inside wall move faster and travel a shorter distance compared to those at the outside wall (Racetrack effect[28]), leading to a very sharp skew. Fig. 8-5(b) shows the effect of transit time on the variance of the band as it moves through the microchannel network. The grey bars refer to the residence of the analyte bands within the bends. As anticipated, the variance of the analyte band increases as the sample transits the network. The results from the system simulator are compared with CFD-ACE+ for two different values of pressure gradient (Π = 18000 and 30000 Pa/m), while keeping the electric field (E_{av} = 300 V/cm) constant. Good agreement (error < 5%) is observed. Larger pressure gradients lead to higher flow rates and large velocity non-uniformity , which results in greater dispersion of the analyte band. Lower velocities are favorable, but the resulting increase in residence time can lead to increased dispersion from molecular diffusion. Therefore, care must be taken when designing microfluidic networks to keep dispersion to a minimum.

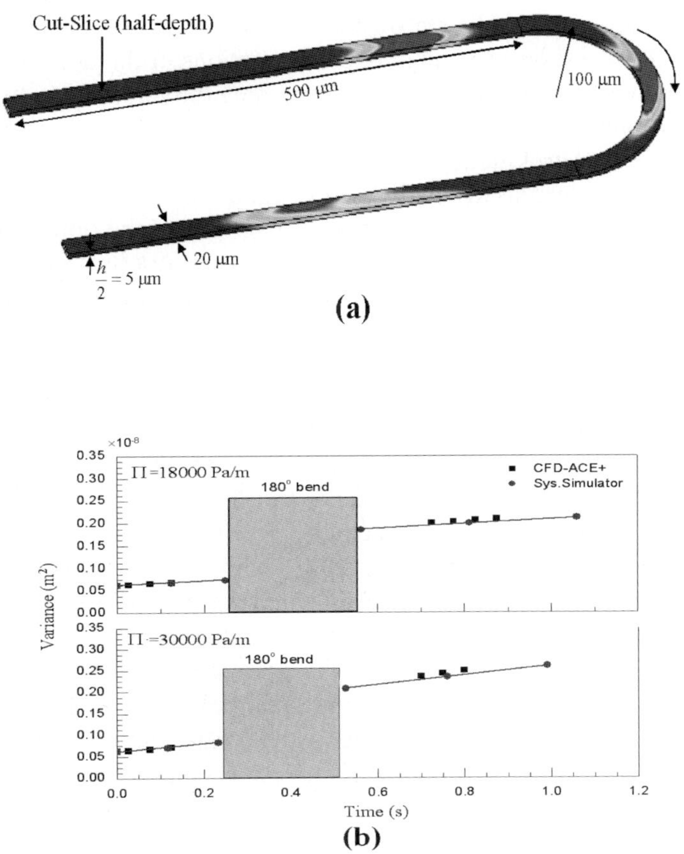

Figure 8-5. Dispersion under combined pressure-driven and electrokinetic flow: (a) System description and contour plot of analyte concentration obtained from 3D CFD-ACE+ simulations (Analyte_1); (b) Increase in band variance with time (comparison of system simulator with CFD-ACE+).

3.4 Application: Electrokinetic Separation Chip

An important application of modeling flow and analyte dispersion in combined pressure-driven and electrokinetic flow is in understanding the effect of unintentional pressure gradients arising from differences in the hydrostatic head of end-channel reservoirs. These gradients can arise as a result of sample loading, evaporation (due to Joule heating and exposure to air) and depletion. In the separation chip shown in Fig. 8-1, an electric field of 300 V/cm is applied across the separation channel, while the injection channel is floated. A channel depth of 25 μm, with a cross section aspect ratio of 2 (i.e., $\beta = 2$) is assumed. In the following discussion, a difference of 2 mm is assumed in the levels of aqueous buffer in the buffer and waste wells. Fig. 8-6(a) shows the analyte concentrations at the detector when the induced pressure-driven flow supplements (same direction, i.e. liquid level in buffer reservoir is higher than that in the waste well) or opposes (reverse direction) the electrokinetic flow. Both situations lead to larger band broadening and decreased analyte concentration peak as compared to electrokinetic flow alone [Fig. 8-4 (b)]. This is due to the non-uniformity in the velocity profile, caused by the external pressure gradient and can lead to potential issues regarding separation resolution and detectability of the band. Convective analytes (i.e. with higher Pe, such as Analyte_3 and Analyte_4 having Pe of 341.25 and 625 respectively) are more sensitive to the induced non-uniform velocity profile and disperse more than diffusive analytes (i.e. with lower Pe, e.g. Analyte_3 and Analyte_4 have Pe of 225 and 82 respectively). This is because for more diffusive analytes, the transverse diffusion can rapidly smear out the skew, and, hence diminish the dispersion effect caused by the non-uniform velocity profile [see the third term in Eq. (35)]. Additionally, the opposing case exhibits a slightly worse band spreading than the supplementing case, which can be attributed to the longer residence time of the analyte band within the channels.

To suppress the dispersion induced by pressure driven flow, the chip can be designed with a higher hydrodynamic resistance (for reducing velocity) using an elongated channel or a higher aspect ratio (β) in the cross section. However, longer channels lead to a larger residence time of the analyte band and can potentially affect the analyte detectability. We select a channel with $\beta = 4$ (depth, $h = 12.5$ μm) and consider the case of the pressure driven flow supplementing the electrokinetic flow. Fig. 8-6(b) shows that this channel significantly reduces the dispersion due to pressure-driven flow. Both the concentration and width of the analyte band remain nearly the same as seen in Fig. 8-4(b), where only electrokinetic flow is involved. In addition, several other parameters (such as channel length, operating voltages, bend curvature etc.) can also be modified to achieve a similar result. The model presented in the present study may be used as a quick method for analyzing the role of these parameters and obtaining an optimal chip design.

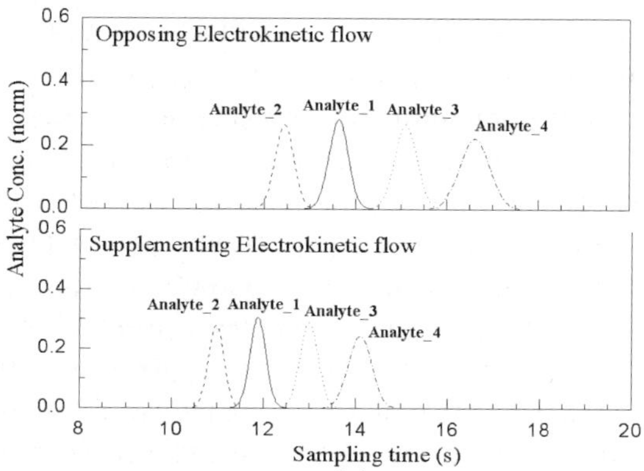

(a)

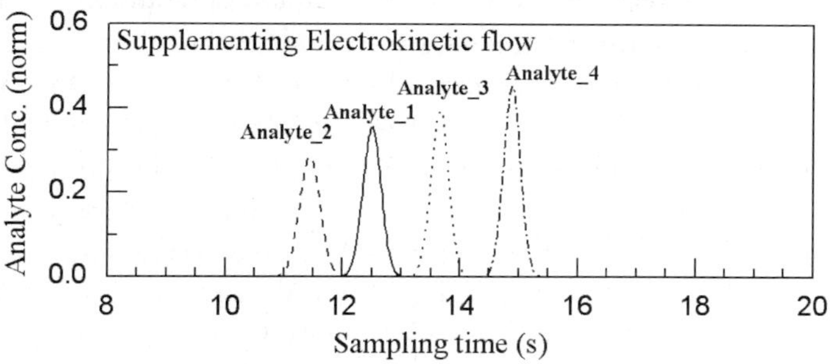

(b)

Figure 8-6. Dispersion due to combined pressure and electrokinetic flow (a) Pressure-driven flow supplementing and opposing electrokinetic flow for $\beta = 2$; (b) Pressure driven flow supplementing electrokinetic flow for $\beta = 4$.

4. CONCLUSION

We have presented a framework for system-level simulation of lab-on-a-chip systems. The compact models are used to compute the flow (pressure- and electrokinetic-driven), and are based on the integral formulation of the mass, momentum and current conservation equations. An analytical model based on the method of moments approach has been developed to characterize the dispersion induced by combined pressure and electrokinetic driven flow. The methodology has been validated against detailed 3D simulations obtained using the commercial software CFD-ACE+. The system solver has been used to analyze hydrostatic pressure effects in electrophoretic separation chips. The pressure gradients due to difference in liquid levels in buffer and waste wells are often created unintentionally but can have an adverse effect on the analyte transport by introducing additional dispersion. This can lead to decreased analyte concentration and potential issues with the limits of detection. Using simulations, we demonstrated that increasing the aspect ratio of the channel cross-section could potentially reduce the dispersion. Overall, a 100-fold improvement in the computational time without significantly compromising the accuracy (error < 10%) has been demonstrated. These benefits will allow for its use for iterative simulation-based design of lab-on-a-chip systems and enable rapid concept evaluation. Previously demonstrated algorithms that optimize chip topology[34] can be incorporated in this framework to allow layout optimization.

The methodology demonstrated has two major limitations: (a) handling of large analyte bands spanning over more than two components or over two components involving different background velocity profiles, and (b) analyte band behavior at junctions and microfluidic components with non-uniform cross-sections. The former is commonly observed, especially in the pressure-driven flow, where Taylor dispersion can cause the band to spread significantly and span multiple components. The latter is important in lab-on-chip systems designed specifically for multiplexed assays. Future work will be aimed at resolving these two issues. In addition, the present framework can be extended to include reaction models for biochemical assays. Compartment models have been well studied[35,36] in literature, and are amenable to coupling with the method of moments approach.

ACKNOWLEDGEMENT

This work was supported by the National Aeronautics and Space Administration under Grant NNC04CA05C.

REFERENCES

1. S. C. Jacobson, T. E. McKnight, and J. M. Ramsey, "Microfluidic devices for electrokinetically driven parallel and serial mixing," *Analytical Chemistry*, vol. 71, pp. 4455-4459, (1999).
2. D. R. Reyes, D. Iossifidis, P.-A. Auroux, and A. Manz, "Micro total analysis systems. 1. Introduction, theory, and technology," *Analytical Chemistry*, vol. 74, pp. 2623, (2002).
3. P. Mitchell, "Microfluidics - downsizing large-scale biology," *Nature Biotechnology*, vol. 19, pp. 717-721, (2001).
4. A. S. Bedekar, J. J. Feng, K. Lim, S. Krishnamoorthy, G. T. R. Palmore, and S. Sundaram, "Computational analysis of microfluidic biofuel cells," Austin, TX, United States, 2004.
5. J. J. Feng, S. Krishnamoorthy, and S. Sundaram, "Numerical and analytical studies of electrothermally induced flow and its application for mixing and cleaning in microfluidic systems," Boston, MA, United States, 2004.
6. M. Sigurdson, C. Meinhart, D. Wang, X. Liu, J. J. Feng, S. Krishnamoorthy, and S. Sundaram, "AC electrokinetics for microfluidic immunosensors," Washington, DC, United States, 2003.
7. S. Krishnamoorthy, J. J. Feng, and V. B. Makhijani, "Analysis of sample transport in capillary electrophoresis microchip using full-scale numerical analysis," Hilton Head Island, SC, United States, 2001.
8. M. G. Giridharan, S. Krishnamoorthy, and A. Krishnan, "Computational simulation of microfluidics, electrokinetics and particle transport in biological MEMS devices," *Proceedings of SPIE - The International Society for Optical Engineering*, vol. 3680, pp. 150-160, (1999).
9. D. Erickson, "Towards numerical prototyping of labs-on-chip: Modeling for integrated microfluidic devices," *Microfluidics and Nanofluidics*, vol. 1, pp. 301, (2005).
10. D. C. Duffy, J. C. McDonald, O. J. A. Schueller, and G. M. Whitesides, "Rapid prototyping of microfluidic systems in poly(dimethylsiloxane)," *Analytical Chemistry*, vol. 70, pp. 4974, (1998).
11. J. Ng, I. Gitlin, A. Stroock, and G. Whitesides, "Components for integrated poly(dimethylsiloxane) microfluidic systems.," *Electrophoresis*, vol. 23, pp. 3461–3473, (2002).
12. R. Qiao and N. R. Aluru, "A compact model for electroosmotic flows in microfluidic devices," *Journal of Micromechanics and Microengineering*, vol. 12, pp. 625-635, (2002).
13. Y. Wang, Q. Lin, and T. Mukherjee, "System-oriented dispersion models of general-shaped electrophoresis microchannels," *Lab on a Chip*, vol. 4, pp. 453-463, (2004).
14. Y. Wang, Q. Lin, and T. Mukherjee, "A model for Joule heating-induced dispersion in microchip electrophoresis," *Lab on a Chip*, vol. 4, pp. 625-631, (2004).
15. Y. Wang, Q. Lin, and T. Mukherjee, "A model for laminar diffusion-based complex electrokinetic passive micromixers," *Lab on a chip*, vol. 5, pp. 877-887, (2005).
16. M. J. Mitchell, R. Qiao, and N. R. Aluru, "Meshless analysis of steady-state electro-osmotic transport," *Journal of Microelectromechanical Systems*, vol. 9, pp. 435-449, (2000).
17. R. Qiao and N. R. Aluru, "Transient analysis of electro-osmotic transport by a reduced-order modelling approach," *International Journal for Numerical Methods in Engineering*, vol. 56, pp. 1023-1050, (2003).
18. R. Qiao and N. R. Aluru, "Mixed-domain and reduced-order modeling of electroosmotic transport in Bio-MEMS," presented at Proceedings of the IEEE/ACM International Workshop on Behavior Modeling and Simulation, 2000.

19. R. Magargle, J. F. Hoburg, and T. Mukherjee, "Microfluidic Injector Models Based on Neural Networks," presented at Technical Proceedings of the 2005 NSTI Nanotechnology Conference and Trade Show (Nanotech 2004), 2005.
20. O. Mikulchenko, A. Rasmussen, and K. Mayaram, "A Neural Network Based Macromodel for Microflow Sensors," presented at Technical Proceedings of 2000 International Conference on Modeling and Simulation of Microsystems (MSM'00), San Diego, CA, 2000.
21. A. N. Chatterjee and N. R. Aluru, "Combined circuit/device modeling and simulation of integrated microfluidic systems," *Journal of Microelectromechanical Systems*, vol. 14, pp. 81-95, (2005).
22. T. H. Zhang, K. Chakrabarty, and R. B. Fair, "Behavioral modeling and performance evaluation of microelectrofluidics-based PCR systems using SystemC," *IEEE Transactions on Computer-Aided Design of Integrated Circuits and Systems*, vol. 23, pp. 843-858, (2004).
23. Y. Wang, R. M. Magargle, Q. Lin, J. F. Hoburg, and T. Mukherjee, "System-oriented modeling and simulation of biofluidic lab-on-a-Chip," presented at Proceedings of the 13th International Conference on Solid-State Sensors and Actuators, Seoul, Korea, 2005.
24. S. V. Patankar, *Numerical heat transfer and fluid flow*. Washington New York: Hemisphere Pub. Corp. McGraw-Hill, 1980.
25. A. Ajdari, "Steady flows in networks of microfluidic channels: building on the analogy with electrical circuits," *Comptes Rendus Physique*, vol. 5, pp. 539-546, (2004).
26. F. M. White, *Viscous fluid flow*, 2nd ed. New York: McGraw-Hill, 1991.
27. B. M. Paegel, L. D. Hutt, P. C. Simpson, and R. A. Mathies, "Turn geometry for minimizing band broadening in microfabricated capillary electrophoresis channels," *Analytical Chemistry*, vol. 72, pp. 3030-3037, (2000).
28. C. T. Culbertson, S. C. Jacobson, and J. M. Ramsey, "Dispersion sources for compact geometries on microchips," *Analytical Chemistry*, vol. 70, pp. 3781-3789, (1998).
29. C. T. Culbertson, S. C. Jacobson, and J. M. Ramsey, "Microchip devices for high-efficiency separations," *Analytical Chemistry*, vol. 72, pp. 5814-5819, (2000).
30. S. K. Griffiths and R. H. Nilson, "Band spreading in two-dimensional microchannel turns for electrokinetic species transport," *Analytical Chemistry*, vol. 72, pp. 5473-5482, (2000).
31. R. F. Probstein, *Physicochemical Hydrodynamics: An Introduction*, 2nd ed. New York: John Wiley & Sons, 2003.
32. R. Aris, "On the Dispersion of a Solute in a Fluid Flowing through a Tube," *Proceedings of the Royal Society of London Series a-Mathematical and Physical Sciences*, vol. 235, pp. 67-77, (1956).
33. E. C. Inc., "CFD-ACE+ Theory Manual," (2004).
34. A. J. Pfeiffer, T. Mukherjee, and S. Hauan, "Design and optimization of compact microscale electrophoretic separation systems," *Industrial & Engineering Chemistry Research*, vol. 43, pp. 3539-3553, (2004).
35. K. Sigmundsson, G. Masson, R. H. Rice, N. Beauchemin, and B. Obrink, "Determination of active concentrations and association and dissociation rate constants of interacting biomolecules: An analytical solution to the theory for kinetic and mass transport limitations in biosensor technology and its experimental verification," *Biochemistry*, vol. 41, pp. 8263-8276, (2002).
36. D. G. Myszka, X. He, M. Dembo, T. A. Morton, and B. Goldstein, "Extending the range of rate constants available from BIACORE: Interpreting mass transport-influenced binding data," *Biophysical Journal*, vol. 75, pp. 583-594, (1998).

Chapter 9

MICROFLUIDIC INJECTOR MODELS BASED ON ARTIFICIAL NEURAL NETWORKS

Ryan Magargle, James F. Hoburg, and Tamal Mukherjee
Carnegie Mellon University, 5000 Forbes Ave., Pittsburgh, 15213, USA

Abstract: Lab-on-a-chip systems can be functionally decomposed into their basic operating devices. Common devices are mixers, reactors, injectors, and separators. In this work, the injector device is modeled using artificial neural networks trained with finite element simulations of the underlying mass transport PDE's. This technique is used to map the injector behavior into a set of analytical performance functions parameterized by the system's physical variables. The injector examples shown are the cross, double-tee, and gated-cross. The results are four orders of magnitude faster than numerical simulation and accurate with mean square errors on the order of 10^{-4}. The resulting neural network training data compares favorably with experimental data from a gated-cross injector found in the literature.

Keywords: microfluidic, electrokinetic, lab-on-a-chip, injector, neural network, simulation

1. INTRODUCTION

Microfluidic lab-on-a-chip (LoC) systems have been studied for more than a decade and have many applications in biology, medicine, and chemistry[1,2]. They generally perform chemical analysis involving sample preparation, mixing, reaction, injection, separation analysis, and detection. Compared to traditional analytical labs, LoC has the significant advantage of increased analysis speed, parallel processing, and high integration and automation.

The simulation of complex LoC systems requires computationally expensive numerical solutions to partial differential equations. As *design* of LoC systems requires many repeated simulations, iterative design using numerical simulation is computationally infeasible. A much more efficient alternative involves functional decomposition into a series of interconnected blocks, as previously proposed for the mixer, injector, and separator[3-6], which

K. Chakrabarty and J. Zeng (eds.),
Design Automation Methods and Tools for Microfluidics-Based Biochips, 215–233.

can be used to compose an entire LoC[7]. Behavioral artificial neural networks (hereafter referred to as simply neural network, or NN) work to create models for use in the decomposition when first principle models are not possible. For the mixer and separator, band shape assumptions are used with analytical techniques to simplify the partial differential equations into several ordinary differential equations. For the injector, the resulting modeled outputs are described by a finite number of performance functions; hence we use numerical NN techniques to describe these functions in a finite domain.

A diagram representing a typical LoC consisting of a mixer, reactor, injector, and separation system is shown in Figure 9-1. There are two stages to its operation. In the first stage, a voltage is applied to nodes 1, 2, 3, 4, and 5. The resulting electric fields induce electrokinetic flow[8] which causes the fluid to be pumped through the channels. This dilutes the sample from node 2 with buffer from node 1 in the mixer before it is reacted with a catalyst and ultimately focused by nodes 4 and 5. In the second stage, called electrophoretic separation, new voltages are applied to nodes 1-5, which causes a band of the reacted analyte to be injected into the separation system. Because the analyte can be composed of multiple species of different charge, the individual species migrate at different speeds in the electric field causing them to separate from each other.

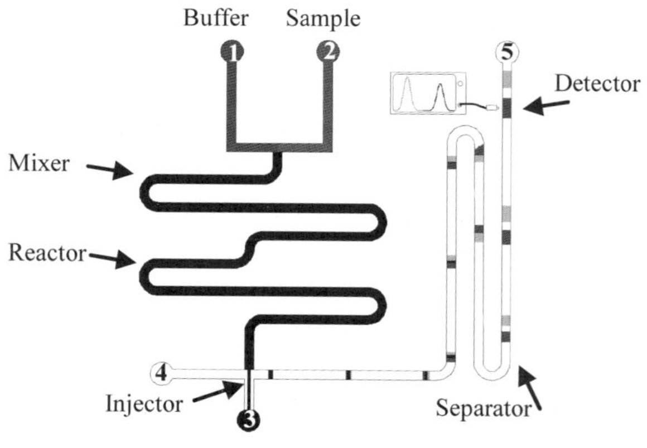

Figure 9-1. Diagram of a canonical lab-on-a-chip (LoC).

The importance of the injector as a component in a microfluidic separation system derives from the fact that it defines the shape and quantity of analyte that will be used for separation and analysis. Various forms of microfluidic electrokinetic injectors such as the tee, double-tee, cross, double-cross, and gated-cross have been introduced[9-13]. A first generation injector model produced by the authors was specific to the cross injector and was defined by a two-dimensional parameter space[4]. This work improves on previous generation injector modeling by using neural network behavioral modeling concepts to create a library of models for injectors, including the cross, double-tee, and gated-cross, each defined by a four-dimensional parameter space. The methodology is not specific to any one injector topology and has been used in macromodels that have many more dimensions[14].

The methodology described here involves an exploration of a relevant portion of the injector physical parameter space using numerical solutions of the convection diffusion equation. These solutions are then used to train a neural network that describes the performance of the injector. The behavioral output parameters are sufficient to construct a Gaussian distribution, approximating the width-averaged plug concentrations as a function of longitudinal position, for the input to the separation channel. Once the neural network has defined the behavioral model, an explicit analytic equation defining the neural network is created and can easily be ported into any modeling environment. This modeling method is related to the work done by Mikulchenko[15], however in our method we have not utilized quasi-random training sequences because our training data sets are not large enough to achieve the asymptotic low-discrepancy distribution[16]. Also, our system has a larger number of physical variables making the computational cost of our network training greater, thus we use the Buckingham-π theorem to reduce the size of the variable space needed for training simulations[17].

The scope of this work is primarily focused on the application of composable system simulation for continuous flow microfluidic LoC systems. The broad range of macromodeling application of neural networks from VLSI CAD[14], microfluidics[15], and RF MEMS[18] suggests that further application in microfluidics, such as digital microfluidics, should be possible, but the exploration of such possibilities are beyond the scope of this work.

The injector topologies will be introduced in the next section. Section 3 will introduce the method used to model the injector device. The physical variables of the injectors and their non-dimensional reduction will be shown in section 3.1 and section 3.2, and the parameters governing the creation of the neural network model will be shown in section 3.3 and section 3.4. Finally, the impact of the resulting model on a composable system simulation will be shown in section 3.5, with conclusions in section 4.

2. INJECTOR TOPOLOGIES

The three injectors considered in this paper are: The cross, double-tee, and gated-cross. Each injector is defined by its geometry and operating procedure. The geometries and operating steps are shown in Figure 9-2.

The cross injector, Figure 9-2a, is a simple intersection of two channels of the same width. The injector is operated by first loading the sample into the injection chamber, followed by a change in voltage that dispenses a portion of the injected flow into the separation channel. In both the loading and dispensing stages, accessory fields can be used to manipulate the shape of the injected plug. In the loading stage, accessory pinching (from E_{2L} and E_{4L}) can be used to achieve a more compact plug and to reduce electrokinetic biasing[9]. In the dispensing stage, accessory pullback (from E_{1D} and E_{3D}) prevents leakage into the separation channel. The double-tee injector, Figure 9-2b, is a more general form of the cross injector where the loading channels are offset to produce a larger injection plug. The operation scheme is the same as with the cross, with two stages of operation and optional accessory fields for plug shaping.

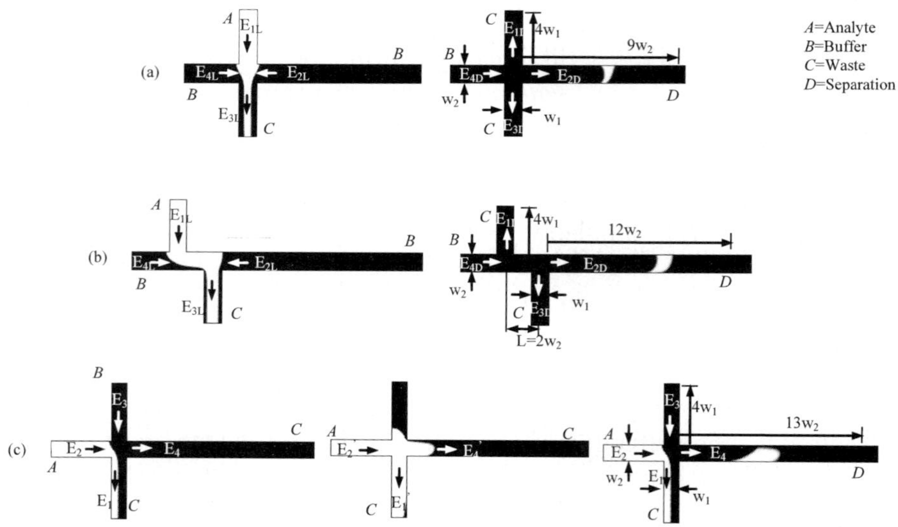

Figure 9-2. Example injections and geometries for cross (a), double-tee (b), and gated-cross (c). The light shades indicate regions of high analyte concentration.

The gated-cross, Figure 9-2c, has the same geometry as the cross, but uses a three stage operating scheme. The gated-cross control scheme allows for continuous flow injection, so that new sample can be loaded at the same time that previously dispensed plugs are run through the separation channel. The first step of operation involves creating the gate in the injection chamber by counterflowing an analyte stream against a buffer stream. The second step involves removing the applied potential from the buffer port, which allows a portion of analyte to overflow the injection chamber. The final step is a return to the applied potentials of the first stage, which reestablishes the gate, while simultaneously injecting the overflown analyte into the separation channel. Since the action taken in the second stage involves floating only one node while all others remain unchanged, no independent parameters are introduced by this stage. The fields and flow patterns are completely determined by the parameters set in the first stage.

3. MODELING METHODOLOGY

The complexity of the geometry and resulting transitional electric field structure[19] make first principles models of the detailed physical performance of the injectors very difficult. The modeling approach used in this paper avoids the complexity of the underlying physical structure, while maintaining an accurate description of the injector performance based upon the key characteristics that define the input/output mapping.

The injector modeling methodology consists of four steps: (1) The Buckingham-π theorem is used to reduce the order of the space of physical parameters describing the behavior of the injector to a reduced set of non-dimensional parameters. In all three of the injectors shown in this paper, the number of non-dimensional parameters describing the injector's operation is reduced to four, which is relatively small compared to the dimensions of typical VLSI analog design spaces[14]. (2) Numerical simulations are carried out at points in the non-dimensional parameter space. The goal is to use as few numerical simulations as possible, since they are computationally expensive. (3) A neural network is constructed to analytically describe the parameter space. Functions available in MATLAB 7[20] were used to train the neural network. A feed-forward back-propagation neural network topology was chosen for its strong non-linear regression capabilities and its ability to be converted to an explicit algebraic function. The neural network as a regressor has the ability to handle high dimensional spaces with relatively sparse data sets[21,22]. The neural network also 'learns' the functional mapping without any knowledge of the underlying physics or basis functions. (4) The trained network is converted to an explicit algebraic function, suitable for use in any software environment for simulation or synthesis. This conversion is

To solve for the electrokinetic flow, first Laplace's equation is solved for the electric potential,

$$\nabla^2\Phi = 0 \tag{9.2}$$

which is used to determine the velocity, which is proportional to the electric field through:

$$\vec{u} = -\mu\vec{\nabla}\Phi \tag{9.3}$$

where μ is the electrokinetic mobility, and the negative gradient of the electric potential, Φ, is the electric field[24].

In all simulations, the channels are long enough to contain the transitions in the electric field structure. The buffer, analyte, and waste channels were all of the same length, $4w$, for each injector topology. The separation channel length, set so it can encompass the entire injected band, is $9w$ for the cross, $12w$ for the double-tee, and $13w$ for the gated-cross. This will allow the numerical simulation data to be used to train the neural network to contain a highly accurate description of the transitional electric field structure and its effect on the dispersion of the analyte.

For each injector topology, using conservation of current and assuming that all channel widths are the same, a relationship between the steady electric fields can be determined. As in prior work on the cross injector[4], the electric field magnitudes in the cross and double-tee are related for the loading stage by:

$$E_{3L} = E_{1L} + 2E_{2,4L} \tag{9.4}$$

and for the dispensing stage by:

$$E_{4D} = E_{2D} + 2E_{1,3D} \tag{9.5}$$

where it is assumed that the accessory fields are applied symmetrically, i.e., $E_{2L}=E_{4L}=E_{2,4L}$ and $E_{1D}=E_{3D}=E_{1,3D}$. The fields for the gated-cross are related by:

$$E_2 + E_3 = E_4 + E_1 \tag{9.6}$$

Using Eqs. (9.4, 9.5) and (9.6) and the conversion matrix introduced by Magargle[4], it is possible to calculate the electric potentials for the boundary conditions, as they change throughout the π-space.

To perform, on average, 200-300 simulations per injector, a FEMLAB script was written to automate the setup and execution of the simulations. The π-space domains for each of the injectors are defined in Table 9-2. A parallel computing cluster was utilized, which reduced several weeks of computation to several days on a shared 100 node Beowulf cluster. This is a one-time computational expense used to create the much more efficient neural network description, which 'learns' from the results of the simulations.

Table 9-2. π-space domains for the training simulations, covering a wide range of physically viable values.

	π_1	π_2	π_3	π_4
Cross	[0:1/2]	[0:2]	[50:5000]	[50:5000]
Double-Tee	[0:1/4]	[0:4]	[50:5000]	[50:5000]
Gated-Cross	[1/10:1]	[1:4]	[50:5000]	[1/20:1/3]

The FEMLAB algorithms have been validated against many experiments found in the literature demonstrating microfluidic mixing, joule heating, injection, and separation[3,5,25]. To validate the simulations for the new gated-cross topology, a test was run to measure the leakage of the injector during the loading stage, as was first done by Ermakov[13]. Figure 9-3 shows the results of the simulations at the same parameters used by Ermakov. The resulting iso-concentration contours agree very well with Ermakov's numerical and experimental results.

After numerically simulating points throughout the parameter space, the form of the output of the behavioral model is chosen. Analysis of the numerical results can be done to extract any desired behavioral outputs. In this work, the outputs were chosen to be the peak height and variance of an effective Gaussian describing the transversely averaged concentration profile output from the injector, such that the area and variance of the original distribution are conserved. The resulting form with non-dimensional outputs is:

$$\bar{f}(\pi_{1(k)}, \pi_{2(k)}, \pi_{3(k)}, \pi_{4(k)}) = \begin{bmatrix} \sigma^2 / w^2 \\ C_{eff} / C_o \end{bmatrix}_{(k)} \tag{9.7}$$

where $k \in [C, D, G]$, and C_o is the value of the uniform concentration of the input species from the reactor.

While the neural network is inherently capable of describing multiple-input, multiple-output models, we use a multiple-input, single-output neural network to train each output separately.

Using effective Gaussian outputs as the interface parameters for a system simulation accomplishes two purposes: (9.1) The system simulation is efficient, and (9.2) while there is sufficient information to accurately describe the band after it has traveled a short distance down the separation channel, there is not enough information to describe the non-Gaussian shape of the band with high accuracy immediately after it leaves the injector.

As long as diffusion transversely homogenizes the band before detection occurs, or before another source of dispersion, such as a channel bend, is introduced, the approximation will provide accurate results. By requiring the time it takes particles to diffuse across the channel to be shorter than the time it takes the particles to reach the first channel bend or detector, a bound is defined:

$$\tau = \frac{Pe_w}{L / w} < 2 \tag{9.8}$$

where Pe_w is the Peclet number based on the channel width w, and L is the length of straight channel before the detector or first channel bend. This expression is conservative in that it requires the particles travel across the entire channel width to homogenize the band. In many cases, as seen in Figure 9-2, the band coming out of the injector is axially compact and fairly uniform across the channel and so homogenization occurs much more rapidly.

An example for a double-tee injection is shown in Figure 9-4. The picture on the left shows the band immediately after the injector, and the picture on the right shows the band after traveling 4.7 mm where from Eq. (9.8), $\tau = 2$. As an example of an actual experimental system, the LoC developed by Chiem and Harrison[26] has $\tau = 0.2$, which is an order of magnitude within the bounds defined by Eq. (9.8).

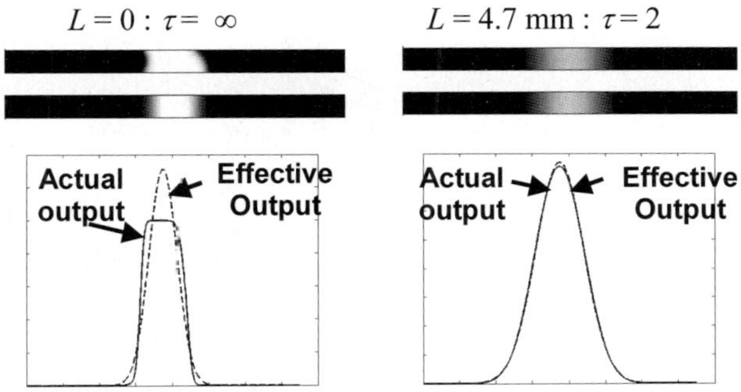

Figure 9-4. Actual output of double-tee injector (top channel) compared to effective parameterized output of analytical model (bottom channel). The band on the right traveled 4.7 mm, for $\tau = 2$, and a channel width of 50um. In these simulations $\pi_{1D} = 186$, $\pi_{2D} = 186$, $\pi_{3D} = 1/8$, $\pi_{4D} = 0.57$.

3.3 Training the Neural Network

A neural network is a mathematical structure that is adept at learning non-linear functional relations and complex item categorizations[22]. The general principle of operation of the feed-forward neural network is shown in Figure 9-5a, and the topology of the two layer neural network used to model the injectors for this work is shown in Figure 9-5b. The feed-forward neural network is a set of nodes that are connected only to the layers above and below. Each node takes, as the argument of an activation function, the weighted sum of the outputs of the layer below. To create a neural network model, the number of layers, nodes per layer, node transfer function, and interconnecting weights must be selected.

It has been shown that to perform non-linear regression to an arbitrary degree of accuracy, where the number of nodes is not an issue, two layers are sufficient (one hidden, one output)[27]. Likewise with classification networks, it has been shown that a decision boundary of arbitrary complexity can be defined using only three layers (two hidden, one output)[22]. Therefore our regression models use two layer topologies, and our classification networks use three layer topologies.

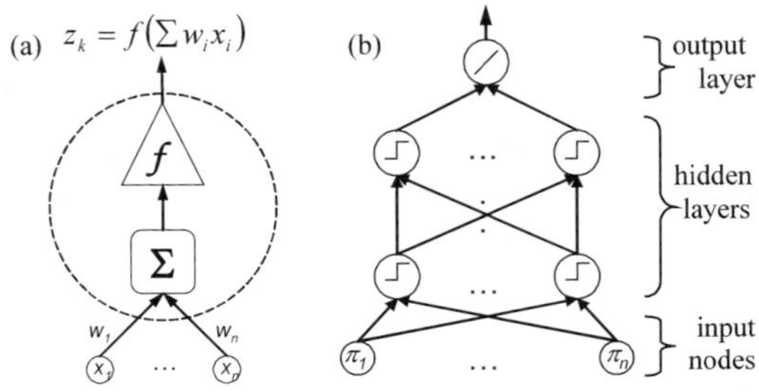

Figure 9-5. (a) The activation, z_k, of a node is a function of the weighted inputs of the previous layer's activations. (b) General feedforward neural network topology utilizing multiple hidden layers with an unspecified number of nodes with non-linear activation functions, and a single output node using a linear activation function.

If the network is to perform non-linear regression, the hidden-layer consists of bound non-linear activation functions, and the output nodes are unbound linear activation functions. If the network makes discrete classifications, then all nodes are bound non-linear functions. In both cases, the actual shape of the non-linear function is not important so long as it is bound[22]. The function used in this work is:

$$f(x) = \frac{2}{1+\exp(-2x)} - 1 \qquad (9.9)$$

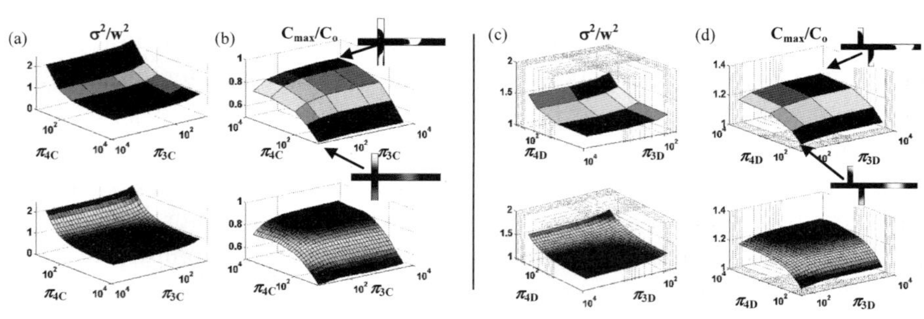

Figure 9-6. (a) Variance and (b) effective peak concentration results for the cross, where π_{1C} and π_{2C} fixed at 0.2. (c) Variance and (d) effective peak concentration results for the double-tee, where π_{1D} and π_{2D} fixed at 0.05 and 0.3 respectively. The top plots show the FEMLAB simulations at a coarse set of points and the bottom plots show the much faster and higher resolution evaluations produced by the NN.

The number of input nodes is determined by the number of independent parameters in the model and the number of nodes in the output layer is determined by the number of dependent parameters in the model. The number of nodes in the hidden layer is chosen to give the network a complexity that minimizes the degree of underfitting (error to the given training data) and overfitting (error to any other points not given in the training set). This error is measured by the generalization error[21], which can be approximated using a k-fold cross validation technique (KFCV)[21]. The generalization error is measured for networks with varying numbers of hidden nodes. The number of nodes that minimizes the generalization error is then chosen for the final network topology. The hidden layers of the injector neural networks typically contain less than 30 nodes.

After the topology and node count have been determined, the network is trained on the FEMLAB simulation data. The process of training a neural network involves modifying the weights between layers until the outputs of the network sufficiently match the simulation training data. This takes the form of an optimization problem. The Levenberg-Marquardt algorithm included in MATLAB was used due to its speed for networks of complexity similar to the ones shown here.

For the models in this paper, no more than 350 iterations were used to achieve a mean squared error (MSE) of less than 1×10^{-4}. To improve the network's convergence rate, the range of the input data is prescaled. In particular, the Peclet number is varied on a logarithmic rather than linear scale, and the variance training data for the gated-cross injector is normalized to unity. To validate the neural network models, first the MSE and KFCV are examined; the results are summarized in Table 9-3, with very small validation parameters throughout. As a visual confirmation, response surfaces are shown for portions of the π-space for each injector. Figure 9-6 shows the results for the cross and double-tee injector, and Figure 9-7 shows the results for the gated-cross injector.

The graphs on top represent the sparse set of points simulated in FEMLAB, while those on the bottom show the densely simulated points using the much faster neural network after being trained on the FEMLAB data. The neural network operates in a fraction of a second per point, more than four orders of magnitude faster than numerical simulation.

Figure 9-6a and 9-6b show the variance and effective peak concentration results, respectively, for the cross, and Figure 9-6c and 9-6d show the same results for the double-tee. When the cross and double-tee are operated with the same accessory fields, the cross has a smaller variance and a smaller peak concentration.

Figure 9-7a shows the feasibility regions for the gated-cross, and Figure 9-7b and 9-6c show the variance and peak concentration results for the gated-cross injector, respectively. The feasibility space is independent of π_{4G} because the leakage is determined in the loading stage independently of how the band is dispensed. The most notable feature of the feasibility space is that as π_{3G}, the Peclet number, becomes smaller, the amount of feasible space decreases significantly. This is due to the fact that regardless of how much the gate is closed, when diffusion is large the diffusive flux leaking into the separation channel is large. Figure 9-7b shows the variance results in the π-space of the gated-cross, and Figure 9-7c shows the effective peak concentration results. The missing surface areas in Figure 9-7b and 9-7c represent infeasible regions as described in section 3.1.

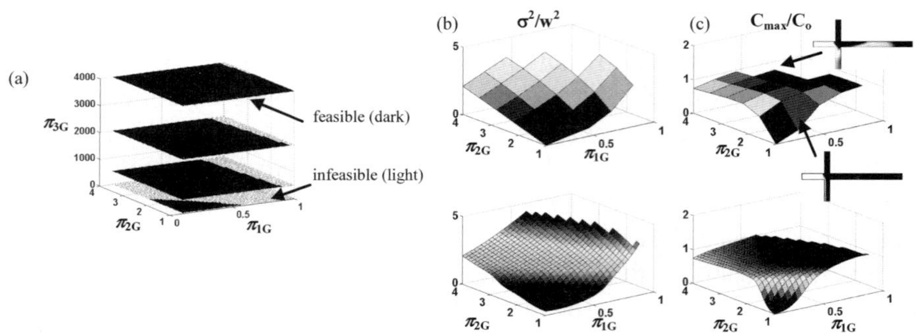

Figure 9-7. (a) Gated-cross feasibility space (dark is feasible). (b) Variance and (c) effective peak concentration results for the gated-cross, where π_{3G} and π_{4G} are fixed at 232 and 0.1208 respectively.

Table 9-3. Neural Network MSE and Cross Validation: For all three injectors, both validation parameters are very small.

	MSE		KFCV		
	σ^2/w^2	C_{max}/C_o	k	σ^2/w^2	C_{max}/C_o
Cross	1.2e-4	2e-5	10	2e-6	9e-5
Double-Tee	1.8e-5	1e-5	8	3e-6	2e-5
Gated-Cross	4e-5	4.2e-5	10	1.1e-4	2.4e-4

3.4 Extracting an Analytic Equation

Since the neural network is connected in a feed-forward configuration, it is straightforward to identify the explicit relation for the output node in terms of all the nodes in the previous layers. The general explicit form for the two-layer topology with a single output node is:

$$y_k = g_{out}\left[\sum_{j=0}^{M} v_{kj} g_{hid}\left[\sum_{i=0}^{N} w_{ji} x_i\right]\right] \tag{9.10}$$

where y_k is the output of the network, $g_{out}[\xi]$ is the linear output node activation function, $g_{hid}[\zeta]$ is the non-linear hidden node activation function from Eq. (9.9), and v_{kj} and w_{ji} are the weights between hidden-output and input-hidden layers, respectively.

Once the network is trained and the weights are fixed, a simple algorithm can be written to extract the network data from the MATLAB data structures. The result is then used in VerilogA to perform analytical system simulations.

3.5 Results

As a simple example to illustrate the benefit of using the accurate neural network model, a standard cross separation is performed in Figure 9-8. Comparison is made between numerical simulation of the separation using FEMLAB, analytical simulation using our VerilogA simulator with the device models of the injector and separation channel[28] (another simulator has been built in Matlab[29]), and a hand calculation using a simplified injector model with diffusion-based dispersion. The simplified injector model assumes that the resulting injection will be a rectangular plug made using error functions with concentration, C_o, and resulting standard deviation of w/sqrt(12)[30].

To compare the simulation results, the resolution and total injected mass for the separation are calculated. Resolution is defined as:

$$R = \frac{(d_2 - d_1)}{4\sigma_{ave}} \tag{9.11}$$

where d_2 and d_1 are the centers of the two bands in time, respectively, and σ_{ave} is the average standard deviation of the two bands in time. These quantities are meaningful when the electropherogram peaks are approximated to be Gaussian, as they are in this case. Four times the standard deviation therefore accounts for more than 99.99% of the total mass of the band. A desirable resolution is usually at least 1.5, so the ease of the separation in the example of Figure 9-8 is illustrated by its very high resolution number.

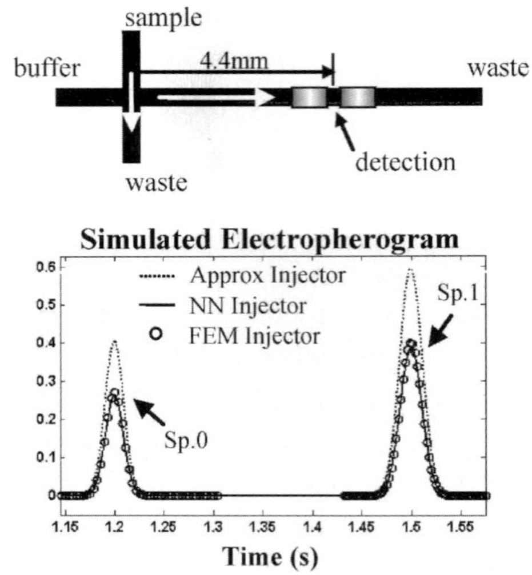

Simulation Properties

	Sp. 0	Sp. 1
π_{1c}	500	469
π_{2c}	500	469
π_{3c}	.5	.5
π_{4c}	2	2
C_o	0.7	1

Resolution Results

	Res
Approx	7.3
NN	7.5
FEM	7.6

Mass Results

	Mass
Approx	0.027
NN	0.017
FEM	0.017

Figure 9-8. Simulations of the cross separation system using component separation system model and (1) an approximate injector as described in the text, (2) the neural network injector, (3) full finite element method (FEM). Results shown for resolution and total mass of species 0 (Sp. 0) and 1 (Sp. 1).

The results show the simulation with the NN injector model in close agreement with the numerical simulation. The calculation using the approximate injector model, shows a respectable 4% difference in resolution, but more than 58% difference in total mass of both bands. In this example, the large differences in the described amount of injected mass could significantly affect the results of modeled quantification assays such as immunoassays[7,26]. This necessitates the use of the more accurate injector models described in this work.

4. CONCLUSION

This work has shown a methodology that was used to model the injector component of a microfluidic LoC, and examples were shown for the cross, double-tee, and gated-cross injector designs. The key contributions of this work were the use of non-dimensionalized reduced variable sets for the numerical simulation, and the use of simplified Gaussian interface parameters to describe the injection output to the downstream separation channel models. The impact of the neural network based injector model was seen to be significant compared to a simple approximate injector model and could become a significant design factor in realistic systems separating 10's to 100's of species. These injector models combined with other component models allow for a functional block decomposition of the system for efficient simulation. The speed and accuracy of these analytic block models present a far more feasible method of CAD than using numerical solutions of partial differential equations for whole complex systems.

This work lays the foundation for future enhancements that include the geometric π-space variables for a more complete synthesis formulation. Due to its ability to model arbitrary functional forms with an arbitrary number of inputs and outputs, with further investigation, the methodology described here can be used to model other microfluidic components for use in composable system simulation.

ACKNOWLEDGEMENTS

This research effort is sponsored by the Defense Advanced Research Projects Agency under the Air Force Research Laboratory, Air Force Material Command, USAF, under grant number F30602-01-2-0587 and the NSF ITR program under grant number CCR-0325344. Computing resources were made possible by NSF equipment grant CTS-0094407 and Intel Corporation.

REFERENCES

1. D.R. Reyes, D. Iossifidis, P.-A. Auroux, and A. Manz, "Micro Total Analysis Systems. Introduction, Theory, and Technology," Vol. 74, *Anal. Chem.* 2002, pp. 2623-2636.
2. P-A. Auroux, D. Iossifidis, D.R. Reyes, and A. Manz, "Micro Total Analysis Systems. Analytical Standard Operations and Applications," Vol. 74, *Anal. Chem.* 2002, pp. 2637-2652.
3. Y. Wang, Q. Lin, and T. Mukherjee, "Applications of Behavioral Modeling and Simulation on Lab-on-a-Chip: Micro-Mixer and Separation System," *BMAS* 2004, pp. 1-6.

4. R. Magargle, J.F. Hoburg, and T. Mukherjee, "An Injector Component Model For Complete Microfluidic Electrokinetic Separation Systems," *Proc. NanoTech* 2004, pp 77-80.
5. Y. Wang, Q. Lin, and T. Mukherjee, "System-Oriented Dispersion Models of General-Shaped Electrophoresis Microchannels," Vol. 4, *Lab-on-a-Chip* 2004, pp. 453-463.
6. Y. Wang, Q. Lin, T. Mukherjee, "Composable Behavioral Models and Schematic-Based Simulation of Electrokinetic Lab-on-a-Chip Systems," *IEEE Trans. on Computer Aided Design* 2005, DOI: 10.1109/TCAD.2005.855942.
7. Y. Wang, R. Magargle, Q. Lin, J.F. Hoburg, and T. Mukherjee, "System-Oriented Modeling and Simulation of Biofluidic Lab-on-a-Chip," Vol. 2, *Proc. Transducers* 2005, pp. 1280-1283.
8. R.F. Probstein, *Physicochemical Hydrodynamics : An Introduction*, 2nd ed. New York: John Wiley & Sons, 1994.
9. S.C. Jacobson, R. Hergenröder, L.B. Koutny, and R.J. Warmack, "Effects of Injection Schemes and Column Geometry on the Performance of Microchip Electrophoresis Devices," Vol. 66, *Anal. Chem.* 1994, pp. 1107-1113.
10. S.V. Ermakov, S.C. Jacobson, and J.M. Ramsey, "Computer Simulations of Electrokinetic Transport in Microfabricated Channel Structures," Vol. 70, *Anal. Chem.* 1998, pp. 4494-4504.
11. L.L. Shultz-Lockyear, C.L. Colyer, Z.H. Fang, K.I. Roy, and D.J. Harrison, "Effects of Injector Geometry an Sample Matrix on Injection and Sample Loading in Integrated Capillary Electrophoresis Devices," Vol. 20, *Electrophoresis* 1999, pp. 529-538.
12. M. Deshpande, K.B. Greiner, J. West, and J.R. Gilbert, "Novel Designs for Electrokinetic Injection in μTAS," *Micro Total Analysis Systems* 2000, pp. 339-342.
13. S.V. Ermakov, S.C. Jacobson, and J.M. Ramsey, "Computer Simulations of Electrokinetic Injection Techniques in Microfluidic Devices," Vol. 72, *Anal. Chem.* 2000, pp. 3512-3517.
14. H. Liu, A. Singhee, R.A. Rutenbar, and L.R. Carley, "Remembrance of Circuits Past: Macromodeling by Data Mining in Large Analog Design Spaces," *DAC* 2002, pp. 437-442.
15. O. Mikulchenko, A. Rasmussen, and K. Mayaram, "A neural network based macromodel for microflow sensors," *Proc. MSM* 2000, pp. 540-543.
16. P. Bratley and B.L. Fox, "Implementing Sobol's quasirandom sequence generator," *ACM Trans. Math. Softw.* 1998, Vol. 14, pp. 88-100.
17. S. Rudolph, "On Topology, Size and Generalization of Non-Linear Feed-Forward Neural Networks," *Neurocomputing* 1997, Vol. 16, pp. 1-22.
18. Y. Lee, Y. Park, F. Niu, B. Bachman, and D. Filipovic, "Computer Aided Design and Optimization of Integrated Circuits with RF MEMS Devices by an ANN Based Macro-Modeling Approach," Vol. 3, *Proc. MSM* 2005, pp. 565-568.
19. N.A. Patankar and H.H. Hu, "Numerical Simulation of Electroosmotic Flow," Vol. 70, *Anal. Chem.* 1998, pp. 1870-1881.
20. The Mathworks, 2005, http://www.mathworks.com
21. T. Hastie, R. Tibshirani, and J.H. Friedman, *The elements of statistical learning : data mining, inference, and prediction*. New York: Springer, 2001.
22. C.M. Bishop, Neural Networks for Pattern Recognition. Oxford University Press, New York, 1995.
23. Comsol, 2005, http://www.comsol.com

24. E.B. Cummings, S.K. Giffiths, R.H. Nilson, and P.H. Paul, "Conditions for Similitude between the Fluid Velocity and Electric Field in Electroosmotic Flow", Vol. 72, *Anal. Chem.* 2000, pp. 2526 2532.
25. Y. Wang, Q. Lin, and T. Mukherjee, "A Model for Joule Heating-Induced Dispersion in Microchip Electrophoresis," Vol.4, *Lab-on-a-chip* 2004, pp. 625-631.
26. N.H. Chiem and D.J. Harrison, "Microchip systems for immunoassay: an integrated immunoreactor with electrophoretic separation for serum theophylline determination," Vol. 44, *Clin. Chem.* 1998, pp. 591-598.
27. A.R. Barron, "Universal Approximation Bounds For Superpositions of a Sigmoidal Function," Vol. 39, *IEEE Trans. on Information Theory* 1993, pp. 930-945.
28. Y. Wang, Q. Lin, and T. Mukherjee, "Composable Behavioral Models and Schematic-Based Simulation of Electrokinetic Lab-on-a-Chip Systems," in: *Design Automation Methods and Tools for Microfluidics-Based Biochips*, eds. K. Chakrabarty and J. Zeng, Springer, Norwell, MA, 2006.
29. A.J. Pfeiffer, X. He, T. Mukherjee, S. Hauan., "A Modular Simulation Framework for Microfluidic Chips", *American Institute of Chem. Eng.* annual meeting, (AIChE-2005) Nov. 2005.
30. J.C. Sternberg, "Extracolumn Contributions to Band Broadening," Vol. 2, *Adv. Chrom.* 1966, pp. 205-270.

Chapter 10

COMPUTER-AIDED OPTIMIZATION OF DNA ARRAY DESIGN AND MANUFACTURING

Andrew B. Kahng

CSE and ECE Department, University of California at San Diego
La Jolla, CA 92093-0114, USA
abk@ucsd.edu

Ion I. Măndoiu

CSE Department, University of Connecticut
371 Fairfield Rd., Unit 2155, Storrs, CT 06269-2155, USA
ion@engr.uconn.edu

Sherief Reda

CSE Department, University of California at San Diego
La Jolla, CA 92093-0114, USA
sreda@cs.ucsd.edu

Xu Xu

CSE Department, University of California at San Diego
La Jolla, CA 92093-0114, USA
xuxu@cs.ucsd.edu

Alex Z. Zelikovsky

CS Department, Georgia State University
University Plaza, Atlanta, Georgia 30303, USA
alexz@cs.gsu.edu

K. Chakrabarty and J. Zeng (eds.),
Design Automation Methods and Tools for Microfluidics-Based Biochips, 235–269.

Abstract: *DNA probe arrays*, or *DNA chips*, have emerged as a core genomic technology that enables cost-effective gene expression monitoring, mutation detection, single nucleotide polymorphism analysis and other genomic analyses. DNA chips are manufactured through a highly scalable process, called *Very Large-Scale Immobilized Polymer Synthesis (VLSIPS)*, that combines photolithographic technologies adapted from the semiconductor industry with combinatorial chemistry. As the number and size of DNA array designs continues to grow, there is an imperative need for highly-scalable software tools with predictable solution quality to assist in the design and manufacturing process. In this chapter we review recent algorithmic and methodological advances forming the foundation for a new generation of DNA array design tools. A recurring motif behind these advances is exploiting the analogy between silicon chip design, pointing to the value of technology transfer between the 40-year old VLSI CAD field and the newer DNA array design field.

Keywords: DNA arrays, computer-aided design, design flow, border length minimization, probe placement, probe embedding, algorithms

1. INTRODUCTION

DNA probe arrays – DNA arrays or DNA chips for short – have recently emerged as one of the core genome technologies. They provide a cost-effective method for obtaining fast and accurate results in a wide range of genomic analyses, including gene expression monitoring, mutation detection, and single nucleotide polymorphism analysis (see [45] for a survey). The number of applications is growing at an exponential rate [28], [55], already covering a diversity of fields ranging from health care to environmental sciences and law enforcement. The reasons for this rapid acceptance of DNA arrays are a unique combination of robust manufacturing, massive parallel measurement capabilities, and highly accurate and reproducible results.

Today, most DNA arrays are manufactured through a highly scalable process, referred to as *Very Large-Scale Immobilized Polymer Synthesis (VLSIPS)*, that combines photolithographic technologies adapted from the semiconductor industry with combinatorial chemistry [1], [2], [25]. Similar to Very Large Scale Integration (VLSI) circuit manufacturing, multiple copies of a DNA array are simultaneously synthesized on a *wafer*, typically made out of quartz. To initiate synthesis, linker molecules including a photo-labile protective group are attached to the wafer, forming a regular 2-dimensional pattern of synthesis sites. Probe synthesis then proceeds in successive steps, with one nucleotide (A, C, T, or G) being synthesized at a selected set of sites in each step. To select which sites will receive nucleotides, photolithographic *masks* are placed over the wafer. Exposure to light de-protects linker molecules at the non-masked sites. Once the desired sites have been activated in this way, a solution containing a single type of nucleotide (which bears its own photo-labile protection group to prevent the probe from growing by more than one nucleotide) is flushed over the wafer's surface. Protected nucleotides attach to the unprotected linkers, initiating the probe synthesis process. In each subsequent step, a new mask is used to enable selective de-protection and single-nucleotide synthesis. This cycle is repeated until all probes have been fully synthesized.

As the number of DNA array designs is expected to ramp up in coming years with the ever-growing number of applications [28], [55], there is an urgent need for high-quality software tools to assist in the design and manufacturing process. The biggest challenges to rapid growth of DNA array technology are the drastic increase in design sizes with simultaneous decrease of array cell sizes – next-generation designs are envisioned to have hundreds of millions of cells of sub-micron size [2], [45] – and the increased complexity of the design process, which leads to unpredictability of design quality and design turnaround time.

In this chapter we review recent algorithmic and methodological advances addressing these challenges and already proved to yield significant solution quality and scalability improvements over existing methods. A recurring motif behind these advances is exploiting the analogy between silicon chip design and DNA chip design, pointing to the value of technology transfer between the 40-year old VLSI CAD field and the newer DNA array design field.

The organization of the chapter is as follows. In Section 2 we introduce the main steps of the DNA array design flow. In Section 3 we formalize the synchronous and asynchronous array design problems and establish lower bounds on the achievable border length. Algorithms for the two versions of the array design problem are presented in Sections 4 and 5, respectively. In Section 6, we give empirical results comparing the presented algorithms and review novel methodologies for characterizing heuristic suboptimality scaling. Finally, we discuss enhancements of DNA array design flow in Section 7 and conclude in Section 8.

2. DNA ARRAY DESIGN STEPS

In this section we introduce the main steps of a typical design flow for DNA arrays, noting the similarities to the VLSI design flow and briefly reviewing previous work. The application of this flow to the design of a DNA chip for studying gene expression in the Herpes B virus is described in [9]. In Section 7 we will revise this flow and show how it can be enhanced by adding flow-awareness to each optimization step and introducing feedback loops between steps - techniques that have proved very effective in the VLSI design context [22], [51].

2.1 Probe Selection

Analogous to logic synthesis in VLSI design, the probe selection step is responsible for implementing the desired functionality of the DNA array. Although probe selection is application-dependent, several underlying selection criteria are common to all designs, regardless of the intended application [1], [2], [44], [8], [36], [47].

First, in order to meet array functionality, the selected probes must have low hybridization energy for their intended targets and high hybridization energy for all other target sequences. Hence, a standard way of selecting probes is to select a probe of minimum hybridization energy from the set of probes which maximizes the minimum number of mismatches with all other sequences [44]. Second, since selected probes must hybridize under similar operating conditions, they must have similar melting temperatures.[1] Finally, to simplify array design, probes are often constrained to be substrings of a predetermined nucleotide deposition sequence. Typically, there are multiple probe candidates satisfying these constraints.

2.2 Deposition Sequence Design

The number of synthesis steps directly affects manufacturing time and the number of masks in the mask set, as well as the likelihood of manufacturing errors. Therefore, a basic optimization in DNA array design is to minimize the number of synthesis steps. In the simplest model, this optimization has been reformulated as the classical *shortest common supersequence* (SCS) problem [42], [53]: Given a finite alphabet Σ (for DNA arrays $\Sigma = \{A, C, T, G\}$) and a set $P = \{p_1, ..., p_t\} \subseteq \Sigma^n$ of probes, find a minimum-length string $s_{opt} \in \Sigma^*$ such that every string of P is a subsequence of s_{opt}. (A string p_i is a subsequence of s_{opt} if s_{opt} can be obtained from p_i by inserting zero or more symbols from Σ.) The SCS problem has been studied for over two decades from the point of view of computational complexity, probabilistic and worst-case analysis, approximation algorithms and heuristics, experimental studies, etc. (see, e.g., [10], [12], [13], [20], [26], [27], [35], [48]).

The general SCS problem is NP-hard, and cannot be approximated within a constant factor in polynomial time unless $P = NP$ [35]. On the other hand, a $|\Sigma|$-approximation is produced by using the *trivial periodic supersequence* $s = (x_1 x_2 \ldots x_{|\Sigma|})^n$, where $\Sigma = \{x_1, x_2, \ldots, x_{|\Sigma|}\}$ Better results are produced in practice by a simple greedy algorithm usually referred to as the "majority merge" algorithm [26], or variations of it that add randomization, lookahead, bidirectionality, etc. (see, e.g., [42]). Some DNA array design methodologies

[1] At the melting temperature, two complementary strands of DNA are as likely to be bound to each other as they are to be separated. A practical method for estimating the melting temperature is suggested in [36].

bypass the deposition design step and use a predefined periodic deposition sequence such as $ACTGACTG\ldots$ (see, e.g., [42], [53]).

2.3 Design of Control and Test Structures

DNA array manufacturing defects can be classified as *non-catastrophic*, i.e., defects that affect the reliability of hybridization results, but do not compromise chip functionality when maintained within reasonable limits, and *catastrophic*, i.e., defects that render the chip unusable. Non-catastrophic defects are caused by systematic error sources in the VLSIPS manufacturing process, such as unintended illumination due to diffraction, internal reflection, and scattering. Their likelihood can be reduced during the physical design stage, as detailed in next section.

Catastrophic manufacturing defects affect a large fraction of the probes on the chip, and can be caused caused by missing, out-of-order, or incomplete synthesis steps, wrong or misaligned masks, etc. These defects can be detected by incorporating on the chip test structures similar to built-in self-test (BIST) structures in VLSI design. A common approach is to synthesize a small set of test probes (sometimes referred to as *fidelity probes* [33]) on the chip and add their fluorescently labeled complements to the genomic sample that is hybridized to the chip. Multiple copies of each fidelity probe are deliberately manufactured at different locations on the chip using different sequences of synthesis steps. Lack of hybridization at some of the locations where fidelity probes are synthesized can be used not only to detect catastrophic manufacturing defects, but also to identify the erroneous manufacturing steps. Further results on test structure design for DNA chips include those in [7], [17], [49].

2.4 Physical Design

Physical design for DNA arrays is equivalent to the physical design phase in VLSI design. It consists of two steps: *probe placement*, which is responsible for mapping selected probes onto locations on the chip, and *probe embedding*, which embeds each probe into the deposition sequence (i.e., determines synthesis steps for all nucleotides in the probe). The result of probe placement and embedding is the complete description of the reticles used to manufacture the array.

Under ideal manufacturing conditions, the functionality of a DNA array should not be affected by the placement of the probes on the chip or by the probe synthesis schedule. In practice, the manufacturing process is prone to synthesis errors that are highly sensitive to the actual probe placement and synthesis schedule. There are several types of synthesis errors that take place during array manufacturing. First, a probe may not loose its protective group when exposed to light, or the protective group may be lost but the nucleotide to be synthesized may not attach to the probe. Second, due to diffraction, internal reflection, and scattering, unintended illumination may occur at sites that are geometrically close to intentionally exposed regions. The first type of manufacturing errors can be effectively controlled by carefully choosing manufacturing process parameters, e.g., by properly controlling exposure times and by inserting correction steps that irrevocably end synthesis of all probes that are unprotected at the end of a synthesis step [1]. Errors of the second type result in synthesis of unforeseen sequences in masked sites and can compromise interpretation of hybridization intensities. To reduce such uncertainty, one can exploit the freedom available in assigning probes to array sites during placement and in choosing among multiple probe embeddings, when available. The objective of probe placement and embedding algorithms is therefore to minimize the sum of border lengths in all masks, which directly corresponds to the magnitude of the unintended illumination effects. Reducing these effects improves the signal to noise ratio in image analysis after hybridization, and thus permits smaller array sites or more probes per array [34].[2]

Let $S = e_1 e_2 \dots e_K$ denote the nucleotide deposition sequence, i.e., $e_i \in \{A, C, T, G\}$ denotes the nucleotide synthesized in the ith synthesis step. Clearly, every probe in the array must be a subsequence of S. When a probe corresponds to multiple subsequences of S, one such subsequence (embedding of the probe into S) must be chosen as the schedule for synthesizing the probe. Clearly, the geometry of the masks is uniquely determined by the placement of the probes on the array and the synthesis schedule used for each probe.

More formally, the border minimization problem is equivalent to finding a *three-dimensional placement* of the probes [37, 41]: two dimensions represent the site array, and the third dimension represents the nucleotide deposition sequence S (see Figure 10-1). Each layer in the third dimension corresponds to a mask that induces deposition of a particular nucleotide (A, C, G, or T); while columns correspond to embedded probes. The border length of a given mask is computed as the number of *conflicts*, i.e., pairs of adjacent transparent and masked sites in the mask. Given two adjacent embedded probes p and p', the *conflict distance* $d(p, p')$ is the number of conflicts between the corresponding columns. The total border length of a three-dimensional placement is the sum of conflict distances between adjacent probes, and the *border minimization problem (BMP)* seeks to minimize this quantity.

We distinguish two types of DNA array synthesis. In *synchronous* synthesis, the ith period $(ACGT)$ of the periodic nucleotide deposition sequence S synthesizes a single nucleotide (the ith) in each probe. This corresponds to a unique (and trivially computed) embedding of each probe p in the sequence S;

[2]Unfortunately, the lack of publicly available information about DNA array manufacturing yield makes it impossible to assign a concrete economic value to decreases in total border length.

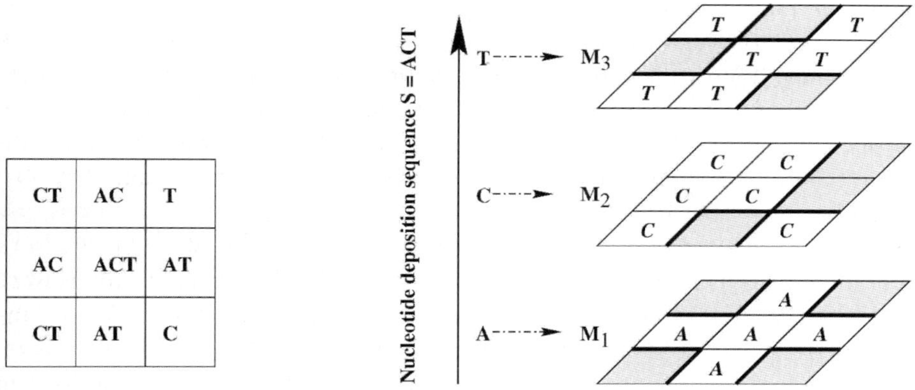

Figure 10-1. (a) Two-dimensional probe placement. (b) Three masks corresponding to nucleotide deposition sequence $S = (ACT)$. Masked sites are shaded, and borders between transparent and masked sites are thickened.

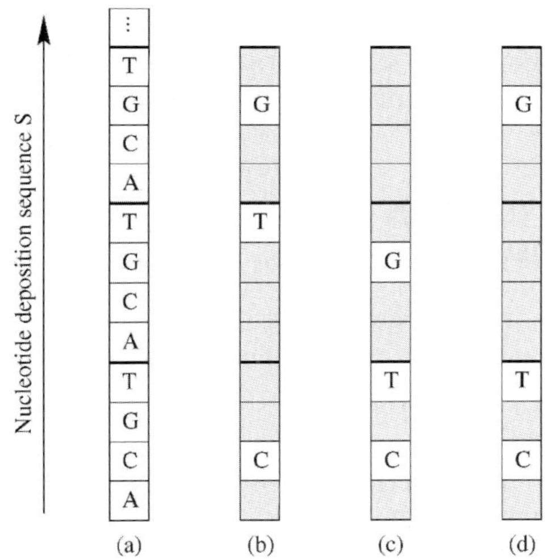

Figure 10-2. (a) Periodic nucleotide deposition sequence S. (b) Synchronous embedding of probe CTG into S; the shaded sites denote the masked sites in the corresponding masks. (c-d) Two different asynchronous embeddings of the same probe.

see Figure 10-2(a-b). On the other hand, *asynchronous* array synthesis permits arbitrary embeddings, as shown in Figure 10-2(c-d).

The border minimization problem was first considered for *uniform arrays* (i.e., arrays containing all possible probes of a given length) by Feldman and Pevzner [23], who proposed an optimal solution based on 2-dimensional Gray codes. Hannenhalli et al. [29] gave heuristics for the special case of synchronous synthesis. In this case the border-length contribution from two probes

p and p' placed next to each other (in the synchronous synthesis regime) is twice the Hamming distance between them, i.e., twice the number positions in which they differ. Hence, BMP reduces to finding a two-dimensional placement of the probes that minimizes the sum of Hamming distances between adjacent probes. The method of [29] is to order the probes in a traveling salesman problem (TSP) tour that heuristically minimizes the total Hamming distance between neighboring probes. The tour is then *threaded* into the two-dimensional array of sites, using a technique similar to one previously used in VLSI design [43]. For the same synchronous context, improved probe placement algorithms were proposed in [37, 38, 40, 41]. These algorithms, drawing on techniques borrowed from the VLSI circuit placement literature, such as epitaxial growth, recursive partitioning, or window-based local re-optimization, are discussed in detail in Section 4.

The general border minimization problem, which allows *asynchronous* probe embeddings, was introduced by Kahng et al. [37], who proposed a dynamic programming algorithm that embeds a given probe optimally with respect to fixed embeddings of the probe's neighbors, and used it to decrease border length by iteratively re-embedding array probes after placing them using a synchronous placement algorithm. Asynchronous probe placement and embedding algorithms in [37] and subsequent improvements in [38, 41] are discussed in Section 4.

3. ARRAY DESIGN PROBLEM FORMULATIONS AND LOWER BOUNDS

Following [37, 41], in this section we give graph-theoretical formulations and theoretical lower bounds for the synchronous and asynchronous variants of BMP.

Let $G_1(V_1, E_1, w_1)$ and $G_2(V_2, E_2, w_2)$ be two edge-weighted graphs with weight functions w_1 and w_2. (In the following, any edge not explicitly defined is assumed to be present in the graph with weight zero.) A bijective function $\phi : V_2 \rightarrow V_1$ is called a *placement* of G_2 on G_1. The cost of the placement is defined as

$$cost(\phi) = \sum_{x,y \in V_2} w_2(x,y) w_1(\phi(x), \phi(y)).$$

The *optimal placement problem* is to find a minimum cost placement of G_2 on G_1.

The border minimization problem for synchronous array design can be cast as an optimal placement problem. In this case we let G_2 be a *two-dimensional grid graph* corresponding to the arrangement of sites in the DNA array, i.e., $V(G_2)$ has $N \times N$ vertices corresponding to array sites, and $E(G_2)$ has edge weights of 1 for every vertex pair corresponding to adjacent sites, and edge weights of 0 otherwise. Also, let H be the *Hamming graph* defined by the set of probes, i.e., the complete graph with probes as vertices and each edge weight equal to twice the Hamming distance between corresponding probes. The border minimization problem for synchronous array design can then be formulated as follows:

Synchronous Array Design Problem (SADP). Find a minimum-cost placement of the Hamming graph H on the two-dimensional grid graph G_2.

For asynchronous array design, formalizing BMP is more involved. Conceptually, asynchronous design consists of two steps: (i) embedding each probe p into the nucleotide deposition sequence S, and (ii) placing the embedded probes into the $N \times N$ array of sites. Let H' be the complete graph with vertices corresponding to the embedded probes and with edge weights equal to the Hamming distance between them.[3] The border minimization problem for asynchronous array design can then be formulated as follows:

Asynchronous Array Design Problem (AADP). Find embeddings into the nucleotide deposition sequence S for all given probes and a placement of the corresponding graph H' on the two-dimensional grid graph G_2 such that the cost of the placement is minimized.

Let L be the directed graph over the set of probes obtained by including arcs from each probe to the 4 closest probes with respect to Hamming distance, and then deleting the heaviest $4N$ arcs. Since the total weight of L cannot exceed the conflict cost of any valid placement of H on the grid graph $G2$, it follows that:

THEOREM 10.1 *[37, 41] The total arc weight of L is a lower bound on the cost of the optimum SADP solution.*

In order to obtain non-trivial lower-bounds on the cost of the optimum AADP solution, it is necessary to establish a lower-bound on the conflict distance between two probes independent of their embedding into S. We get such a lower-bound by observing that the number of nucleotides (mask steps) common to two embedded probes cannot exceed the length of the *longest common subsequence* (LCS) of the two probes. Define the *LCS distance* between probes p and p' by $lcsd(p, p') = k - |LCS(p, p')|$, where $k = |p| = |p'|$, and let L' be the directed graph over the set of probes obtained by including arcs from each probe to the 4 closest probes with respect to LCS distance, and then deleting the heaviest $4N$ arcs. Similar to Theorem 10.1, it follows that:

[3]Recall that embedded probes are viewed as sequences of length $K = |S|$ over the alphabet $\{A, C, G, T, b\}$ such that the jth letter is either b or s_j. Thus, conflicts between two adjacent embedded probes occur only on positions where a nucleotide in one probe corresponds to a blank in the other.

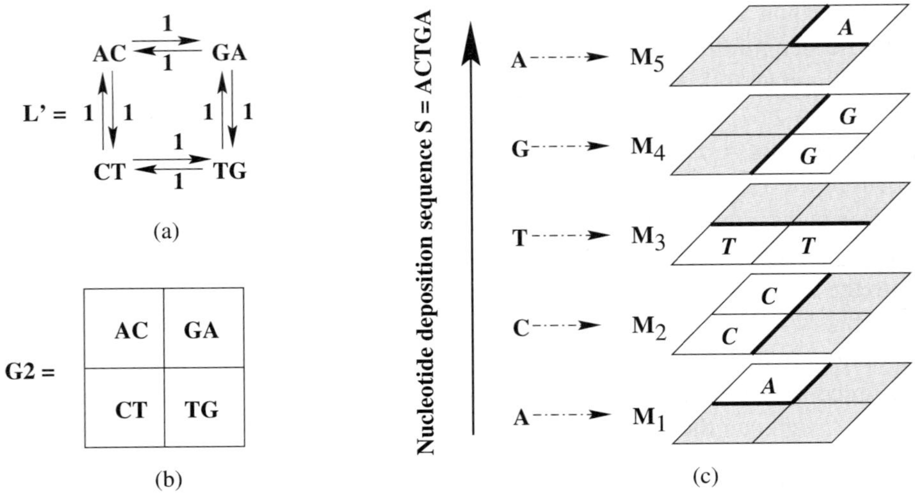

Figure 10-3. (a) Lower-bound digraph L' for the probes AC, GA, CT, and TG. The arc weight of L' is 8. (b) Optimum two-dimensional placement of the probes. (c) Optimum embedding of the probes into the nucleotide deposition supersequence $S = ACTGA$. The optimum embedding has 10 conflicts, exceeding the lower bound by 2.

THEOREM 10.2 *[37, 41] The total arc weight of L' is a lower bound on the cost of the optimum AADP solution.*

The weight of L' may be smaller than the optimum cost, since the embeddings needed to achieve LCS distance between pairs of adjacent probes may not be compatible with each other. Figure 10-3 gives one such example consisting of four dinucleotide probes, AC, GA, CT, and TG, which must be placed on a 2×2 grid. In this case, the lower bound on the number of conflicts is 8 while the optimum number of conflicts is 10.

4. SCALABLE ALGORITHMS FOR SADP

In this section, we review recent highly-scalable heuristics for synchronous probe placement. We first describe the epitaxial growth algorithm in [37] and its highly scalable row-epitaxial version in [41]. Finally, we describe the sliding window matching heuristic for synchronous placement improvement [38] (based on optimally re-placing an independent set of probes via a reduction to minimum cost assignment), and the partition based synchronous probe placement in [40]. A recurring motif behind these algorithms is the technology transfer between the 40-year VLSI design literature and the newer field of DNA chip design.

Input: Set P of N^2 probes, scaling coefficients k_i, $i = 1, \ldots, 3$
Output: Assignment of the probes to the sites of an $N \times N$ grid

---

1. Mark all grid sites as empty
2. Assign a randomly chosen probe to the center site and mark this site as full
3. While there are empty sites, do

If there exists an empty site c with all 4 neighbors full, then

Find probe $p(c) \in P$ with minimum sum of Hamming distances to the neighboring probes

Assign probe $p(c)$ to site c and mark c as full

Else

For each empty site c with $i > 0$ adjacent full sites, find probe $p(c) \in P$ with minimum sum S of Hamming distances to the probes in full neighbors, and let $norm\_cost(c) = k_i S/i$.

Let c^* be the site with minimum $norm\_cost$

Assign probe $p(c^*)$ to site c^* and mark c^* as full

Figure 10-4. The Epitaxial Algorithm.

4.1 Epitaxial Growth SADP Algorithms

In this section, we describe the so-called *epitaxial growth* approach to SADP and discuss some efficient implementation details [37, 41]. Epitaxial, or seeded crystal growth, placement is a technique that has been well-explored in the VLSI circuit placement literature [46, 50]. The technique essentially grows a two-dimensional placement around a single starting seed.

The algorithm in [29], which finds a TSP tour and then threads it into the array, optimizes directly only half of the pairs of adjacent probes in the array (those corresponding to tour edges). Intuitively, the epitaxial algorithm (see Figure 10-4) attempts to make full use of the available information during placement. The algorithm places a random probe at the center of the array, and then iteratively places probes in sites adjacent to already-placed probes so as to greedily minimize the average number of conflicts induced between *all* newly created pairs of neighbors. Sites with more filled neighbors have higher priority to be filled; in particular, sites with 4 known neighbors have the highest priority. To avoid repeated distance computations, the algorithm maintains for each border site a list of probes sorted by normalized cost. For each array site, this list is computed at most four (and on the average two) times, i.e., when one of the neighboring sites is being filled while the site is still empty.

While the epitaxial algorithm achieves good results, it runtime becomes impractical for DNA chips with dimensions of 300×300 or more. Any synchronous placement method can be trivially scaled by partitioning the set of probes and the probe array into K subsets ("chunks"), then solving K independent placement problems. While this ensures linear scaling of runtime, two types of losses are incurred: (i) from lack of freedom of a probe to move anywhere other than its subset's assigned chunk of array sites, and (ii) lack of optimization on borders between chunks. In [41] it is noted that better solution quality is achieved for a different scalable variant of the epitaxial algorithm called the *row-epitaxial* algorithm. There are three main distinguishing features of the row-epitaxial variant:

(1) It re-shuffles an existing pre-optimized placement rather than starting with an empty placement;

(2) The sites are filled with crystallized probes in a predefined order, namely, row by row and within a row from left to right;

(3) The probe filling each site is chosen as the best candidate not among *all* remaining ones, but among a bounded number of them (the not yet "crystallized" probes within the next k_0 rows, where k_0 is a parameter of the algorithm).

Feature (1) is critical for compensating the loss in solution quality due to the reduced search space imposed by (2) and (3). Since the initial placement must be very fast to compute, one cannot afford using any two-dimensional placement based on computing all pairwise distances between probes (such as TSP-based placement in [29]). Possible initial placement algorithms can be based on space-filling curve (e.g., Gray code) ordering [11]; indeed such orderings have had success in the VLSI context [5]. As noted in [41], an excellent initial placements is obtained by simply ordering the probes lexicographically (this can be done in linear time by radix sort) and then threading them as in [29]. Features (2) and (3) speed-up the algorithm significantly, with the number k_0 of look-ahead rows allowing a fine tradeoff between solution quality and runtime.

4.2 Highly Scalable Algorithms for Synchronous Placement Improvement

In the early VLSI placement literature, iterative placement improvement methods relied on weak neighborhood operators such as pair-swap, leveraged by meta-heuristics such as simulated annealing. More recently, strong neighborhood operators have been proposed which improve larger portions of the placement. For example, the DOMINO approach [21] iteratively determines an optimal reassignment of all objects within a given window of the placement. The end-case placer of [14] uses branch and bound to optimally reorder small sub-rows of a row-based placement. Extending such improvement operators to full-chip scale, such that placeable objects can eventually migrate to good locations within practical runtimes, is typically achieved by shifting a fixed-size sliding window [21] around the placement; cf. cycling and overlapping [32], row-ironing [14], etc.

For DNA arrays, an initial placement (and embedding) of probes in array sites may be improved by changing the placement and/or the embedding of individual probes. However, randomly chosen pairs of probes are extremely unlikely to be swappable with reduction in border cost. On the other hand, optimal probe re-placement of an entire window of probes is not practical even for very small window sizes. However, as noted in [38], optimal probe re-placement of large sets of *independent* (i.e., non-adjacent) probes reduces to computing a minimum cost assignment, where the cost of assigning a probe p to a cell c is given by the sum of Hamming distances between p and the probes placed in the four cells adjacent to c. For a set of t independent cells, computing the minimum cost assignment requires $O(t^3)$ time. Full-chip application with practical runtime is achieved by iteratively choosing the independent set from a sliding window that is moved around the array; this approach is a reminiscence of early work on electronic circuit placement by [3, 52].

Following extensive algorithm engineering, the following implementation of the sliding window method was found to work best [38]. (1) First, radix-sort all probes lexicographically and then perform 1-threading as in [29]. (2) For each sliding $W_0 \times W_0$ window, choose one random maximal independent set of sites and determine the cost of (asynchronous) reassignment of each associated probe to each site, then reassign probes according to the minimum weight perfect matching in the resulting weighted bipartite graph. (3) The window slides in rows, beginning in the top-left corner of the array; at each step, it slides horizontally to the right as far as possible while maintaining a prescribed amount of *window overlap*. After the right side of the array is reached, the window returns to the left end of the next row while maintaining the prescribed overlap with the preceding row. When the bottom side of the array is reached, the window returns to the top-left corner. The experiments in [38] have shown that an overlap equal to half the window size gives best results. (4) The window-sliding continues until an entire pass through the array results in less than 0.1% reduction of border cost. Figure 10-5 illustrates the heuristic tuning with respect to varying window sizes.

4.3 Partition Based Probe Placement

Recursive partitioning has been the basis of numerous successful VLSI placement algorithms [6], [15], [54] since it produces placements with acceptable wirelength within practical runtimes. The main goal of partitioning in VLSI is to divide a set of cells into two or four sets with minimum edge or hyper-edge cut between these sets. The min-cut goal is typically achieved through the use of the Fiduccia-Mattheyses procedure [24], often in a multilevel framework [15].

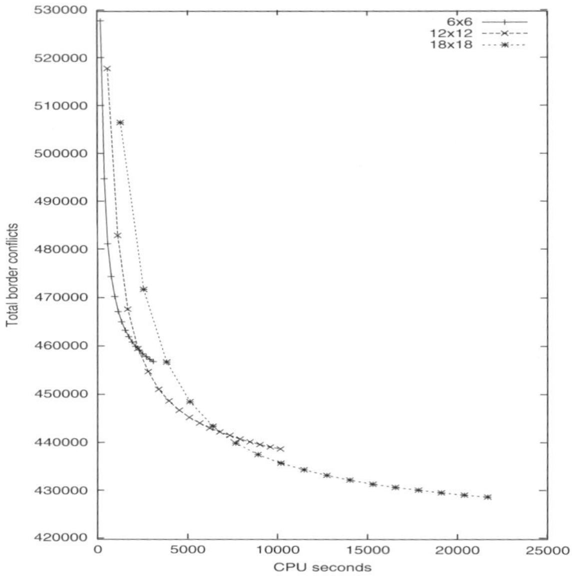

Figure 10-5. Solution quality vs. runtime for Synchronous Sliding-Window Matching with varying window size; array size = 100×100.

Unfortunately, direct transfer of the recursive min-cut placement paradigm from VLSI to VLSIPS is blocked by the fact that the possible interactions between probes must be modeled by a complete graph and, furthermore, the border cost between two neighboring placed partitions can only be determined after the detailed placement step which finalizes probe placements at the border between the two partitions. In this section we describe the *centroid-based quadrisection* method proposed in [40], which applies the recursive partitioning paradigm to DNA probe placement.

Assume that at a certain depth of the recursive partitioning procedure, a probe set R is to be *quadrisectioned* into four equally sized subsets R_1, R_2, R_3 and R_4. The probes in R are partitioned among R_i's by picking a *representative*, or *centroid*, probe C_i for each R_i, and then iteratively assigning remaining probes to the subset R_i whose representative is closest in Hamming distance. The procedure for selecting the four centroids, reminiscent of the k-center approach to clustering studied by Alpert et al. [4] and of methods used in large-scale document classification [19], is described in Figure 10-6.

The complete partitioning-based placement algorithm for DNA arrays is given in Figure 10-7. The algorithm recursively quadrisects every partition at a given level, assigning the probes so as to minimize distance to the centroids

Input: Partition (set of probes) R
Output: Probes C_0, C_1, C_2, C_3 to be used as centroids for the 4 sub-partitions

---

Randomly select probe C_0 in R
Choose $C_1 \in R$ maximizing $d(C_1, C_0)$
Choose $C_2 \in R$ maximizing $d(C_2, C_0) + d(C_2, C_1)$
Choose $C_3 \in R$ maximizing $d(C_3, C_0) + d(C_3, C_1) + d(C_3, C_2)$
Return (C_0, C_1, C_2, C_3)

Figure 10-6. The **SelectCentroid()** procedure for selecting the centroid probes of sub-partitions.

Input: Chip size $N \times N$; set R of DNA probes
Output: Probe placement which heuristically minimizes total conflicts

---

Let $l = 0$ and let $L =$ maximum recursion depth
Let $R^l_{1,1} = R$
For $l = 0$ to $L - 1$
    For $i = 1$ to 2^l
        For $j = 1$ to 2^l
            $(C_0, C_1, C_2, C_3) \leftarrow$ *SelectCentroid*($R^l_{i,j}$)
            $R^{l+1}_{2i-1,2j-1} \leftarrow \{C_0\}$; $R^{l+1}_{2i-1,2j} \leftarrow \{C_1\}$; $R^{l+1}_{2i,2j-1} \leftarrow \{C_2\}$;
                $R^{l+1}_{2i,2j} \leftarrow \{C_3\}$
            For each probe $p \in R^l_{i,j} \setminus \{C_0, C_1, C_2, C_3\}$
                Insert p into the yet-unfilled partition of $R^l_{i,j}$ whose centroid has minimum distance to p
For $i = 1$ to 2^L
    For $j = 1$ to 2^L
        Reptx($R^L_{i,j}$, $R^L_{i,j+1}$)

Figure 10-7. Partitioning-based DNA probe placement heuristic.

of sub-partitions.[4] A multi-start heuristic in the innermost of the three nested *for* loops of Figure 10-7, whereby r different random probes are tried as seed C_0, and the result that minimizes the total distance to the centroids is selected. Within the innermost of the three nested *for* loops, our implementation actually performs, and benefits from, a *dynamic update* of the partition centroid whenever a probe is added into a given partition.[5]

Once the maximum level L of the recursive partitioning is attained, detailed placement is executed via a modified version of the row-epitaxial algorithm. Since the basic implementation of the row-epitaxial algorithm treats the last locations within a region "unfairly" (e.g., only one candidate probe will remain available for placing in a region's last location), the modified algorithm permits "borrowing" probes from the next region. The modified row-epitaxial algorithm is also "border-aware", that is, it takes into account Hamming distances to the already placed probes in adjacent regions.

5. IN-PLACE OPTIMIZATION OF PROBE EMBEDDINGS

Experiments in [37, 38] indicate that separate optimization of probe placement and embedding yields better results for AADP than simultaneous optimization of the two degrees of freedom. For example, the asynchronous version of the epitaxial algorithm [37] and the asynchronous version of sliding-window matching [38] are both dominated by algorithms implementing the following two-step flow:

Step (i). Find a two-dimensional placement based on synchronous embedding for the probes (using, e.g., the row-epitaxial and sliding-window matching algorithms discussed in the previous section, or the TSP+1-Threading of [29]).

Step (ii). Iteratively optimize probe embeddings, *without changing their location on the array.*

In this section we consider the second step of the above flow. We first present a dynamic programming algorithm for optimally embedding a single probe with respect to its neighbors, as well as a lower-bound on the optimum

[4]The variables i and j index the row and column of a given partition within the current level's array of partitions.

[5]Details of the dynamic centroid update, reflecting an efficient implementation, are as follows. The "pseudo-nucleotide" at each position t (e.g., $t = 1, \ldots, 25$ for probes of length 25) of the centroid C_i can be represented as $C_i[t] = \bigcup\limits_s \frac{N_{s,t}}{N_i} \cdot s$, where N_i is the current number of probes in the partition R_i and $N_{s,t}$ is the number of probes in the partition having the nucleotide $s \in \{A, T, C, G\}$ in t-th position. The Hamming distance between a probe p and C_i is $d(p, C_i) = \frac{1}{N_i} \sum\limits_t \sum\limits_{s \neq p[t]} N_{s,t}$.

Input: Nucleotide deposition sequence $S = s_1 s_2 \ldots s_K$, $s_i \in \{A, C, G, T\}$; set X of probes already embedded into S; and unembedded probe $p = p_1 p_2 \ldots p_k$, $p_i \in \{A, C, G, T\}$
Output: The minimum number of conflicts between an embedding of p and probes in X, along with a minimum-conflict embedding

---

1. For each $j = 1, \ldots, K$, let x_j be the number of probes in X which have a non-blank letter in j^{th} position.
2. $cost(0,0) = 0$; For $i = 1, \ldots, k$, $cost(i, 0) = \infty$
3. For $j = 1, \ldots, K$ do

$cost(0, j) = cost(0, j-1) + x_j$

For $i = 1, \ldots, k$ do

If $p_i = s_j$ then
$cost(i, j) = \min\{cost(i, j-1) + x_j,\ cost(i-1, j-1) + |X| - x_j\}$
Else $cost(i, j) = cost(i, j-1) + x_j$

4. Return $cost(k, K)$ and the corresponding embedding of s

Figure 10-8. The Single Probe Alignment Algorithm.

border cost for in-place probe embedding [37]. We then present three methods for iterative in-place probe embedding optimization [37, 41], and conclude with a useful theoretical bound on the amount of improvement available during this optimization step.

5.1 Optimum Embedding of a Single Probe

The basic operation used by in-place embedding optimization algorithms is to find the optimum embedding of a probe when the adjacent sites contain already embedded probes. In other words, the goal is to simultaneously align the given probe s to its embedded neighboring probes, while making sure this alignment gives a feasible embedding of s in the nucleotide deposition sequence S. In this section we present an efficient dynamic programming algorithm given in [37] for computing this optimum alignment.

The Single Probe Alignment algorithm (see Figure 10-8) essentially computes a shortest path in a specific directed acyclic graph $G = (V, E)$. Let p be the probe to be aligned, and let X be the set of already embedded probes adjacent to p. Each embedded probe $q \in X$ is a sequence of length $K = |S|$ over the alphabet $\{A, C, G, T, b\}$, with the jth letter of q being either a blank or s_j, the jth letter of the nucleotide deposition sequence S. The graph G (see Figure 10-9) has vertex set $V = \{0, \ldots, k\} \times \{0, \ldots, K\}$ (where k is the length of the probe p and K is the length of the deposition sequence S), and edge set $E = E_{horiz} \cup E_{diag}$ where

$$E_{horiz} = \{(i, j-1) \rightarrow (i, j) \mid 0 \leq i \leq k, 0 < j \leq K\}$$

and

$$E_{diag} = \{(i-1, j-1) \to (i, j) \mid 0 < i \leq k, 0 < j \leq K, p_i = s_j\}.$$

The cost of a horizontal edge $(i, j-1) \to (i, j)$ is defined as the number of embedded probes in X which have a non-blank letter on j^{th} position, while the cost of a diagonal edge $(i-1, j-1) \to (i, j)$ is equal to the number of embedded probes of X with a blank on the jth position. The Single Probe Alignment algorithm computes the shortest path from the source node $(0, 0)$ to the sink node (k, K) using a topological traversal of G, which corresponds to the optimum embedding of s:

THEOREM 10.3 *The algorithm in Figure 10-8 returns, in $O(kK)$ time, the minimum number of conflicts between an embedding of s and the adjacent embedded probes X (along with a minimum-conflict embedding of s).*

Proof : Each directed path from $(0, 0)$ to (k, K) in G consists of K edges, k of which must be diagonal. Each such path P corresponds to an embedding of p into S as follows. If the jth arc of P is horizontal, the embedding has a blank in j^{th} position. Otherwise, the jth arc must be of the form $(i-1, j-1) \to (i, j)$ for some $1 \leq i \leq k$, and the embedding of p corresponding to P has $p_i = s_j$ in the jth position. It is easy to verify that the edge costs defined above ensure that the total cost of P gives the number of conflicts between the embedding of p corresponding to P and the set X of embedded neighbors.

Remarks. The above dynamic programming algorithm can be easily extended to find the optimal simultaneous embedding of $n > 1$ probes. The corresponding directed acyclic graph G consist of $k^n K$ nodes $(i_1, \ldots, i_n, j)$, where $0 \leq i_l \leq k$, $1 \leq j \leq K$. All arcs into $(i_1, \ldots, i_n, j)$ come from nodes $(i'_1, \ldots, i'_n, j-1)$, where $i'_l \in \{i_l, i_l - 1\}$. Therefore the in-degree of each node is at most 2^n. The weight of each edge is defined as above such that each finite weight path defines embeddings for all probes and the weight equals the number of conflicts. Finally, computing the shortest path between $(0, \ldots, 0)$ and $(k, \ldots, k, K)$ can be done in $O(2^n k^n K)$ time. The probe alignment algorithm can also be extended to handle practical concerns such as pre-placed control probes, presence of polymorphic probes, unintended illumination between non-adjacent array sites, and position-dependent border conflict weights, we refer the reader to [38, 41] for details.

5.2 Algorithms for Iterative In-Place Embedding Optimization

Batched Greedy [37]. A natural greedy algorithm is to find a probe that offers largest cost gain from optimum re-embedding with respect to the (fixed) embeddings of its neighbors, perform this re-embedding, and repeat these steps until no further improvement is possible. The *batched* version of the greedy algorithm (see Figure 10-10) trades some gain in re-embedding steps for faster runtime. During each *batched* phase the algorithm attempts to re-embed all probes in the order given by cost gains at the beginning of the phase. The algorithm gains in efficiency by avoiding to update probe gains after each probe re-embedding.

Chessboard Optimization [37]. The main idea behind our so-called "Chessboard" algorithm is to maximize the number of *independent* re-embeddings, where two probes are independent if changing the embedding of one does not affect the optimum embedding of the other. It is easy to see that if we bicolor our grid as we would a chessboard, then all white (resp. black) sites will be independent and can therefore be simultaneously, and optimally, re-embedded. The Chessboard Algorithm (see Figure 10-11) alternates re-embeddings of black and white sites until no improvement is obtained.

A 2×1 version of the Chessboard algorithm partitions the array into iso-oriented 2×1 tiles and bicolors them. Then using the multi-probe alignment algorithm (see the remark in Section 5.1) with $n = 2$ it alternatively optimizes the black and white 2×1 tiles.

Sequential Probe Re-Embedding [41]. In this method, probes are sequentially re-embedded optimally with respect to their neighbors in a row-by-row order. A shortcoming of the Batched Greedy and Chessboard algorithms is that, by always re-embedding *independent* sets of probes, it takes longer to propagate the effects of a new embedding. Performing the re-embedding sequentially permits faster propagation and convergence to a better local optimum.

5.3 A Lower Bound for In-Place Probe Re-Embedding

Let LG be a grid graph with weights on edges equal to the LCS distance between endpoint probes. The following theorem gives a lower bound that is very useful in assesing the quality of in-place probe re-embedding algorithms:

THEOREM 10.4 *[37] The total edge weight of the graph LG is a lower bound on the optimum AADP solution cost with a given placement.*

6. EMPIRICAL RESULTS

In this section we give experimental results comparing algorithms introduced in previous sections. Unless otherwise specified, experiments reported in this chapter were performed on test cases obtained by generating each probe candidate uniformly at random and reported numbers are averages over 10 random instances. The probe length was set to 25, which is the typical value for commercial arrays [1]. We used the canonical periodic deposition sequence, $(ACTG)^{25}$. All reported runtimes are for a 2.4 GHz Intel Xeon server with 2GB of RAM running under Linux.

Table 10-1. Total border cost, gap from the synchronous placement lower-bound (in percents), and CPU time (in seconds) for the TSP threading (TSP+1Thr), the row-epitaxial (Row-Epitaxial), and sliding-window matching (SWM) heuristics, and the simulated annealing algorithm (SA).

Chip	LB	TSP+1Thr			Row-Epitaxial			SWM			SA		
Size	Cost	Cost	Gap	CPU	Cost	Gap	CPU	Cost	Gap	CPU	Cost	Gap	CPU
100	0.41M	0.55M	35.3	113	0.50M	22.5	108	0.60M	47.7	2	0.58M	42.4	20769
200	1.51M	2.14M	41.6	1901	1.91M	26.6	1151	2.36M	56.1	8	2.42M	59.9	55658
300	3.23M	4.67M	44.3	12028	4.18M	29.4	3671	5.19M	60.6	19	5.50M	70.2	103668
500	8.46M	12.70M	50.1	109648	11.18M	32.2	10630	13.75M	62.5	50	15.43M	82.5	212390

In a first set of experiments we compare synchronous probe placement heuristics: the TSP 1-threading heuristic of [29] (TSP+1Thr), the Row-Epitaxial and sliding-window matching (SWM) heuristics of [38], a simulated annealing algorithm (SA), and the partitioning based algorithm. These experiments use an upper bound of 20,000 on the number of candidate probes in Row-Epitaxial, and 6×6 windows with overlap 3 for SWM. The SA algorithm starts by sorting the probes and threading them onto the chip. It then slides a 6×6 window over the chip in the same way as the SWM algorithm (with overlap 3). For every window position, SA picks 2 random probes in the window and swaps them with probability 1 if the swap improves total border cost. If the swap increases border cost by δ, the swap is performed only with probability $e^{-\delta/T}$, where T is the current temperature. A number of 6^3 SA iterations was performed for every window position.

Table 10-1 shows that among TSP+1Thr, Row-Epitaxial, SWM, and SA heuristics, Row-Epitaxial is the algorithm with highest solution quality (i.e., lowest border cost), while SWM is the fastest, offering competitive solution quality with much less runtime. SA takes the largest amount of time, and also gives the worse solution quality.

Table 10-2 gives results for the recursive partitioning algorithm (RPART) with recursion depth L varying between 1 and 3. Compared to Row-Epitaxial, recursive partitioning based placement achieves improved runtime and similar or better solution quality.

In a second set of experiments we compared the three probe embedding algorithms in Section 5 on random instances with chip sizes between 100 and 500 and an initial two-dimensional placement obtained using TSP+1-threading. All algorithms were stopped when the improvement cost achieved in one iteration over the whole chip drops below 0.1% of the total cost. The results in Table 10-3 show that sequential re-embedding of the probes in a row-by-row order yields the smallest border cost with a runtime similar to that of the other methods.

In another series of experiments, we ran complete placement and embedding flows obtained by combining each of the five two-dimensional placement

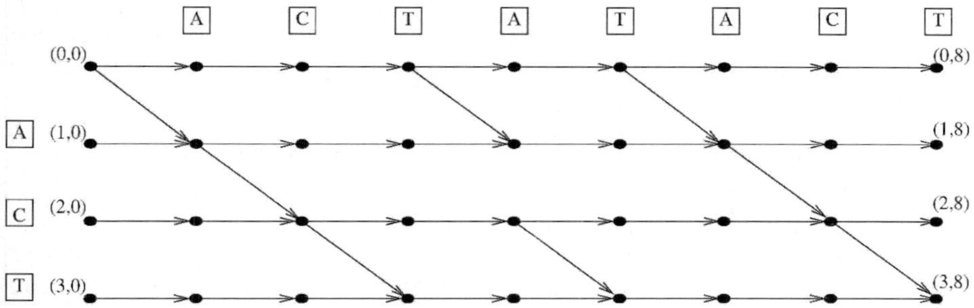

Figure 10-9. Directed acyclic graph G_1 representing possible embeddings of probe $p = ACT$ into the nucleotide deposition sequence $S = ACTATACT$.

Table 10-2. Total border cost, gap from the synchronous placement lower-bound (in percents), and CPU time (in seconds) for the recursive partitioning algorithm with $r = 10$ and recursion depth L varying between 1 and 3.

Chip	LB	RPART $L = 1$			RPART $L = 2$			RPART $L = 3$		
Size	Cost	Cost	Gap	CPU	Cost	Gap	CPU	Cost	Gap	CPU
100	0.41M	0.48M	16.1	69	0.49M	20.0	24	0.50M	23.1	10
200	1.51M	1.81M	19.9	992	1.87M	23.4	283	1.92M	27.2	81
300	3.23M	4.14M	27.9	3529	4.07M	26.0	1527	4.18M	29.1	240
500	8.46M	11.28M	33.4	10591	11.05M	30.6	9678	11.13M	31.6	3321

Table 10-3. Total border cost, gap from the synchronous placement lower-bound (in percents), and CPU time (in seconds) for the batched greedy, chessboard, and sequential in-place re-embedding algorithms.

Chip	LB	Batched Greedy			Chessboard			Sequential		
Size	Cost	Cost	Gap	CPU	Cost	Gap	CPU	Cost	Gap	CPU
100	0.36M	0.45M	25.7	40	0.44M	20.5	54	0.44M	19.9	64
200	1.43M	1.80M	26.3	154	1.72M	20.9	221	1.72M	20.3	266
300	3.13M	3.97M	26.7	357	3.80M	21.5	522	3.77M	20.6	577
500	8.59M	10.92M	27.1	943	10.43M	21.4	1423	10.38M	20.9	1535

Input: Feasible AADP solution, i.e., placement in $G2$ of probes embedded in S
Output: A heuristic low-cost feasible AADP solution

While there exist probes which can be re-embedded with gain in cost do

Compute gain of the optimum re-embedding of each probe.

Unmark all probes

For each unmarked probe p, in descending order of gain, do

Re-embed p optimally with respect to its four neighbors

Mark p and all probes in adjacent sites

Figure 10-10. The Batched Greedy Algorithm.

Input: Feasible AADP solution, i.e., placement in $G2$ of probes embedded in S
Output: A heuristic low-cost feasible AADP solution

Repeat until there is no gain in cost

For each site (i, j), $1 \leq i, j \leq N$ with $i + j$ even, re-embed probe optimally with respect to its four neighbors

For each site (i, j), $1 \leq i, j \leq N$ with $i + j$ odd, re-embed probe optimally with respect to its four neighbors

Figure 10-11. The Chessboard Algorithm.

Table 10-4. Total border cost, gap from the synchronous placement lower-bound (in percents), and CPU time (in seconds) for the TSP threading (TSP+1Thr), the row-epitaxial (Row-Epitaxial), and sliding-window matching (SWM) heuristics, and the simulated annealing algorithm (SA) followed by sequential in-place probe re-embedding.

Chip	LB	TSP+1Thr			Row-Epitaxial			SWM			SA		
Size	Cost	Cost	Gap	CPU	Cost	Gap	CPU	Cost	Gap	CPU	Cost	Gap	CPU
100	0.22M	0.44M	99.5	113	0.42M	88.3	161	0.44M	99.8	93	0.46M	107.6	11713
200	0.80M	1.72M	115.8	1901	1.61M	101.4	1368	1.72M	115.6	380	1.84M	130.9	42679
300	—	3.80M	—	12028	3.53M	—	3861	3.80M	—	861	4.16M	—	101253
500	—	10.43M	—	109648	9.46M	—	12044	10.16M	—	2239	11.57M	—	222376

algorithms compared above with the sequential in-place re-embedding algorithm. Results are given in Tables 10-4–10-5. Again, SA and TSP+1Thr are dominated by both REPTX and SWM in both conflict cost and running time. REPTX produces less conflicts than SWM but SWM is considerably faster. Recursive partitioning consistently outperforms the other flows with similar or lower runtime.

Table 10-5. Total border cost, gap from the synchronous placement lower-bound (in percents), and CPU time (in seconds) for the recursive partitioning algorithm with $r = 10$ and recursion depth L varying between 1 and 3, followed by sequential in-place probe re-embedding.

Chip	LB	RPART $L = 1$			RPART $L = 2$			RPART $L = 3$		
Size	Cost	Cost	Gap	CPU	Cost	Gap	CPU	Cost	Gap	CPU
100	0.22M	0.39M	78.3	123	0.40M	81.1	44	0.41M	86.2	10
200	0.80M	1.52M	90.9	1204	1.55M	93.5	365	1.57M	97.0	101
300	—	3.49M	—	3742	3.41M	—	1951	3.43M	—	527
500	—	9.55M	—	11236	9.36M	—	10417	9.30M	—	3689

6.1 Quantified Suboptimality of Placement Algorithms

As noted in the introduction, next-generation of DNA probe arrays will contain up to one hundred million probes, orders of magnitude more than current designs. Thus, it is of interest to study not only runtime scaling for available heuristics, but also the scaling of their suboptimality. Following [40], we present next an experimental framework for quantifying suboptimality of probe placement heuristics This framework, inspired by similar studies in the area of VLSI placement [30],[16],[18], comprises two basic types of instance scaling.

- **Instances with known optimum solution.** These instances consist of all 4^k probes of length k padded with the same prefix up to the prescribed probe length. These instances are are placeable such that every probe has border cost of 2 to each of its neighboring probes using 2-dimensional Gray codes [23] (see Figure 10-12).
- **Instances with known suboptimal solutions.** Because constructed instances with known optimum solutions are not representative of "real" instances, we also apply a technique of [30] that allows real instances to be scaled, such that they offer insights into scaling of heuristic suboptimality. The technique is applied as follows. Beginning with a problem instance I, we construct three isomorphic versions of I by three distinct mappings of the nucleotide set $\{A, C, G, T\}$ onto itself. Each mapping yields a new probe set that can be placed with optimum border cost exactly equal to the optimum border cost of I. Our scaled instance I' consists of the union of the original probe set and its three isomorphic copies. Observe that one placement solution for I' is to optimally place I and its isomorphic copies as individual chips, and then to adjoin these placements as the four quadrants of a larger chip. Thus, an *upper bound* on the optimum border cost for I' is 4 times the optimum border cost for I, plus the border cost between the copies of I; see Figure 10-13. If a heuristic H places I' with cost $c_H(I') \geq 4 \cdot c_H(I)$, then we may infer

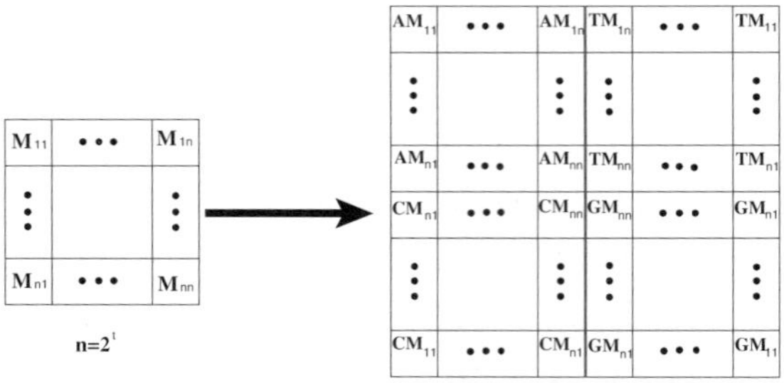

Figure 10-12. 2-dimensional Gray code placement.

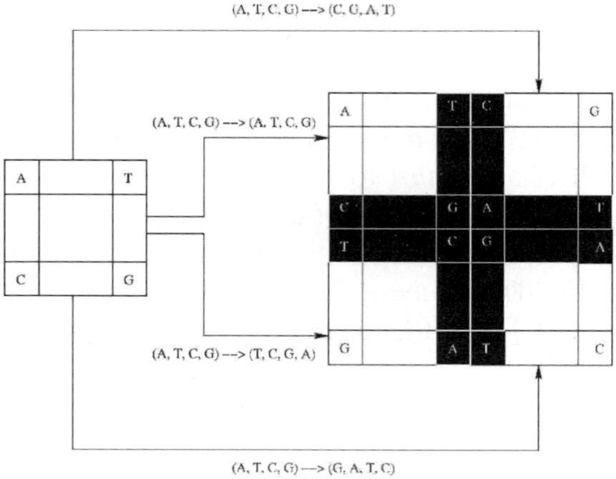

Figure 10-13. Scaling construction used in the suboptimality experiment.

that the heuristic's suboptimality is growing by at least a factor $\frac{c_H(I')}{4 \cdot c_H(I)}$. On the other hand, if $c_H(I') < 4 \cdot c_H(I)$, then the heuristic's solution quality would be said to scale well on this class of instances.

Table 10-6 shows results from executing the various placement heuristics on instances with known optimum solution. We see from these results that sliding-window matching is closest to the optimum, with a suboptimality gap of 4-30%. Overall, DNA array placement algorithms appear to be performing better than their VLSI counterparts [16] when it comes to results on special-case instances with known optimal cost. Of course, results from placement algorithms (whether for VLSI or DNA chips) on special benchmark instances should not be generalized to arbitrary benchmarks. In particular, our results show that algorithms that perform best for arbitrary benchmarks are not necessarily the best performers for specially constructed benchmarks.

Table 10-7 shows the results obtained by running the synchronous placement heuristics on scaled versions of random DNA probe sets, with original instances

Table 10-6. Comparison of placement algorithms performance on instances with known optimal solution. SW matching is using a window size of 20 x 20 and an overlap of 10. Row-epitaxial uses $10,000/chipsize$ lookahead rows.

Chip Size	Optimal Cost	TSP+1Thr Cost	TSP+1Thr Gap	Row-Epitaxial Cost	Row-Epitaxial Gap	SWM Cost	SWM Gap	RPART Cost	RPART Gap
16	960	1380	44	960	0	992	4	1190	24
32	3968	6524	65	5142	30	4970	25	5210	31
64	16128	27072	68	16128	0	19694	22	21072	31
128	65024	111420	71	92224	42	86692	33	88746	36
256	261120	457100	75	378612	45	325566	25	359060	37
512	1046528	1844244	76	1573946	50	1414154	35	1476070	41

Table 10-7. Suboptimality of placement algorithm performance on scaled benchmarks. SW matching is using a window size of 20 x 20 and a step of 10. Row epitaxial uses $10,000/chipsize$ lookahead rows.

Original Size	Row Epitaxial U-Bound	Row Epitaxial Actual	Row Epitaxial Ratio	SWM U-Bound	SWM Actual	SWM Ratio	RPART U-Bound	RPART Actual	RPART Ratio
100	2024464	1479460	0.73	2203132	1999788	0.91	1919328	1425806	0.73
200	7701848	6379752	0.83	8478520	6878096	0.81	7497520	6107394	0.82
300	16817110	12790186	0.76	18645122	13957686	0.75	16699806	12567786	0.75
400	29239934	24621324	0.84	32547390	26838164	0.82	30450780	24240850	0.80
500	44888710	38140882	0.85	49804320	41847206	0.84	47332142	37811712	0.80

ranging in size from 100 x 100 to 500 x 500. The results show that in general, placement algorithms for DNA arrays offer excellent scaling suboptimality. We believe that this is primarily due to the already noted fact that algorithm quality (as reflected by normalized border costs) improves with instance size. The larger number of probes in the scaled instances gives more freedom to the placement algorithms, leading to heuristic placements that have scaling suboptimality factor well below 1.

7. FLOW ENHANCEMENTS

As noted in [39], the basic DNA array design flow described in Section 2 can be significantly improved by introducing flow-aware problem formulations, adding feedback loops between optimization steps, and/or integrating multiple optimizations. These enhancements, which are represented schematically in Figure 10-14 by the dashed arcs, are similar to flow enhancements that have proved very effective in the VLSI design context [22], [51]. In this section

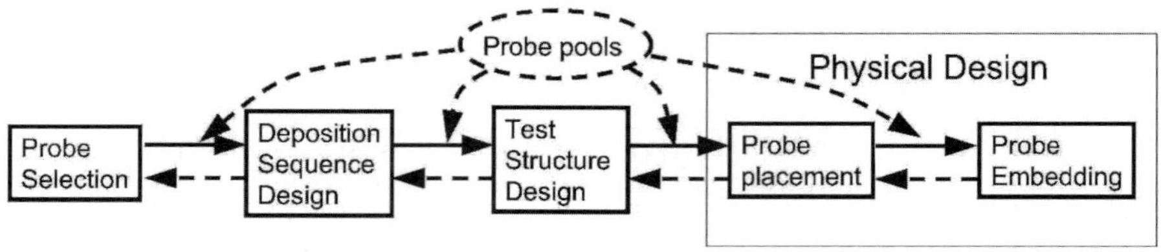

Figure 10-14. A typical DNA array design flow with solid arcs and proposed enhancements represented by dashed arcs.

we describe two such enhancements, both aiming at further reductions in total border length. The first enhancement is a tighter integration between probe placement and embedding. The second enhancement is the integration between physical design and probe selection, which is achieved by passing the entire pools of candidates available for each probe to the physical design step. These enhancements yield significant improvements (up to 15%) in border length compared to best flows in [38, 41]. [6]

7.1 Problem Formulation for Integrated Probe Selection and Physical Design

To integrate probe selection and physical design, one can pass the entire pools of candidates for each probe to the physical design step (Figure 10-14). As discussed in Section 2, candidate probes are selected so that they have similar hybridization properties (e.g., melting temperatures), and can thus be used interchangeably. The availability of multiple probe candidates gives additional freedom during placement and embedding, and may potentially reduce final border cost. DNA array physical design with probe pools is captured by the following problem formulation [39]:

Integrated DNA Array Design Problem

Given:

- Pools of candidates $P_i = \{p_{ij} \mid j = 1, \ldots, l_i\}$ for each probe $i = 1, \ldots, N^2$, where $N \times N$ is the size of array
- The number of masks K

Find:

1 Probes $p_{ij} \in P_i$ for every $i = 1, \ldots, N^2$,

[6] This formulation also integrates deposition sequence design. For simplicity, we leave out design of control and test sequences.

2 A deposition sequence $S = s_1, \ldots, s_K$ which is a supersequence of all selected probes p_{ij},

3 A placement of the selected probes p_{ij} into an $N \times N$ array,

4 An embedding of the selected probes p_{ij} into the deposition sequence S

Such that:

- The total number of conflicts between adjacent embedded probes is minimized

Although the flow in Figure 10-14 suggests a particular order for making the choices 1-4, the integrated formulation above allows interleaving these decisions. The following two algorithms capture key optimizations and will be used as building blocks for constructing integrated optimization flows. They are "probe pool" versions of the Row-epitaxial and re-embedding algorithms described in previous sections, and degenerate to the latter ones in the case when each probe pool contains a single candidate.

- The *Pool Row-Epitaxial algorithm* (Pool-REPTX) is the extension to probe pools of the REPTX probe placement algorithm. Pool-REPTX performs choices 1 and 3 for given choices 2 and 4, i.e., it simultaneously chooses already embedded candidates from the respective pools and places them on the array. The input of Pool-REPTX consists of probe candidates p_{ij} embedded in the deposition sequence S. Each such embedding is written as a sequence of length $K = |S|$ over the alphabet $\{A, C, T, G, Blank\}$, where A, C, T, G denote embedded nucleotides and $Blank$'s denote positions of S left unused by the embedded candidate probe. Pool-REPTX consists of the following steps: (1) Lexicographic sorting of the pools (based on the first candidate, when more than one candidate is available in the pool); (2) Threading the sorted pools in row-by-row order; (3) Traversing array cells, again in row-by-row order, and placing at each location the best probe candidate – i.e., the candidate having the minimum number of conflicts with already placed neighbors – within a prescribed lookahead region.

- The *sequential in-place pool re-embedding* algorithm is the extension to probe pools of the sequential probe re-embedding algorithm given in Section 5. It complements Pool-REPTX by iteratively modifying candidate selections within each pool and their embedding (choices 2 and 4) as follows. In row-by-row order, for each array cell and for each candidate p_{ij} from the associated pool of probe candidates, an embedding having minimum number of conflicts with the existing embeddings of the neighbors is computed, and then the best embedded candidate probe is used to replace the current one.

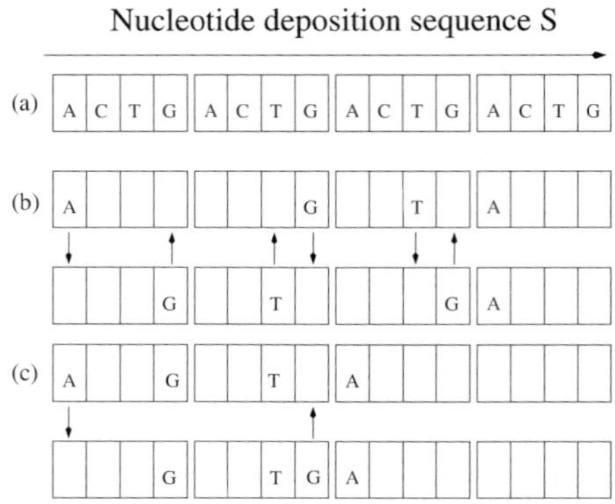

Figure 10-15. (a) Periodic deposition sequence. (b) Synchronous embedding of the probes $AGTA$ and $GTGA$ gives 6 border conflicts (indicated by arrows). (c) "As soon as possible" asynchronous embedding of the probes $AGTA$ and $GTGA$ gives only 2 border conflicts.

7.2 Improved Integration of Probe Placement and Embedding

As noted in [37], allowing asynchronous embeddings leads to further reductions in border length compared to synchronous embedding (e.g., contrast (b) and (c) in Figure 10-15). An interesting question is finding the best order in which the placement and embedding degrees of freedom should be exploited. Existing methods [37, 38, 41] can be divided into two main classes: (1) methods that perform placement and embedding decisions simultaneously, and (2) methods that exploit the two degrees of freedom one at a time. Currently, best methods in the second class (e.g., synchronous row-epitaxial followed by chessboard/sequential in-place probe re-embedding [38, 41]) outperform the methods in the first class (e.g., the asynchronous epitaxial algorithm in [37]) in terms of both runtime and solution quality.

Methods in the second class perform synchronous probe placement followed by iterated in-place re-embedding of the probes (with locked probe locations). More specifically, these methods perform the following 3 steps:

- Synchronous embedding of the probes.
- Probe placement with costs given by the Hamming distance between the synchronous probe embeddings.
- Iterated sequential probe re-embedding.

In [39] we noted that significant reductions in border cost are possible by performing the placement based on asynchronous, rather than synchronous, embeddings of the probes, and therefore proposed a modified scheme as follows:

- Asynchronous embedding of the probes.

- Placement with costs given by the Hamming distance between the fixed asynchronous probe embeddings.
- Iterated sequential probe re-embedding.

Since solution spaces for placement and embedding are still searched independently of one another, and the computation of an initial asynchronous embedding does not add significant overhead, the proposed change is unlikely to adversely affect the runtime. However, because placement optimization is now applied to embeddings more similar to those sought in the final optimization stage, there is significant potential for improvement.

In the implementation proposed in [39] the first step consists of embedding each probe using an "as soon as possible," or *ASAP*, synthesis schedule (see Figure 10-15(c)). Under ASAP embedding the nucleotides in a probe are embedded sequentially by always using the earliest available synthesis step. The intuition behind using ASAP embeddings is that, since ASAP embeddings are more densely packed, the likelihood that two neighboring probes will both use a synthesis step increases compared to synchronous embeddings. This translates directly into reductions in the number of border conflicts.

Indeed, consider two random probes p, p' picked from the uniform distribution. When performing synchronous embedding, the length of the deposition sequence is $4 \times 25 = 100$. The probability that any one of the 100 synthesis step is used by one of the random probes and not the other is $2 \times (1/4) \times (3/4)$, and therefore the expected number of conflicts is $100 \times 2 \times (1/4) \times (3/4) = 37.5$. Assume now that the two probes are embedded using the ASAP algorithm. Notice that for every $0 \leq i \leq 3$ the ASAP algorithm will leave a gap of length i with probability 1/4 between any two consecutive letters of a random probe. This results in an average gap length of 1.5, and an expected number of synthesis steps of 25 + 24*1.5 = 61. Assuming that p and p' are both embedded within 61 steps, the number of conflicts between their ASAP embeddings is then approximately $61 \times 2 \times (25/61) \times ((61 - 25)/61) \approx 29.5$. Although in practice many probes require more than 61 synthesis steps when embedded using the ASAP algorithm, they still require much less than 100 steps and result in significantly fewer conflicts compared to synchronous embedding.

We compared ASAP and synchronous initial embeddings on test cases ranging in size from 100×100 to 500×500. For both embedding strategies, the second and third steps are implemented using REPTX and sequential in-place probe re-embedding, respectively. Tables 10-8 and 10-9 give the border-length and CPU time (in seconds) after each flow step. Remarkably, the simple switch from synchronous to ASAP initial embedding results in 5-7% reduction in total border-length. Furthermore, the runtimes for the two methods are comparable. In fact, sequential re-embedding becomes faster in the ASAP-based method compared to the synchronous-based one since fewer iterations are needed to

Table 10-8. Total border cost (averages over 10 random instances) for synchronous and ASAP initial probe embedding followed by row-epitaxial and sequential in-place probe re-embedding.

Chip	Synchronous Initial Embedding			ASAP Initial Embedding			%
Size	Sync.	REPTX	Sequential	ASAP	REPTX	Sequential	Impr.
100	619153	502314	415227	514053	393765	389637	5.2
200	2382044	1918785	1603745	1980913	1496937	1484252	6.7
300	5822857	4193439	3514087	4357395	3273357	3245906	6.9
500	18786229	11203933	9417723	11724292	8760836	8687596	7.0

Table 10-9. CPU seconds (averages over 10 random instances) for synchronous and ASAP initial probe embedding followed by row-epitaxial and sequential in-place probe re-embedding.

Chip	Synchronous Initial Embedding			ASAP Initial Embedding		
Size	Sync+REPTX	Sequential	Total	ASAP+REPTX	Sequential	Total
100	166	81	247	188	29	217
200	1227	340	1567	1302	114	1416
300	3187	748	3935	2736	235	2971
500	8495	2034	10529	6391	451	6842

converge to a locally optimal solution (the number of iterations drops from 9 to 3 on the average).

7.3 Integrated Probe Selection and Physical Design

Two different methods for exploiting the availability of multiple probe candidates during placement and embedding were proposed in [39]. A first method uses the pool versions of the row-epitaxial and sequential in-place probe re-embedding algorithms described above. This method is an instance of integration between multiple flow steps, since probe selection decisions are made during probe placement and can be further changed during probe re-embedding. The detailed steps are as follows:

- Perform ASAP embedding of all probe candidates.
- Run the Pool-REPTX (or a pool version of the recursive-partitioning placement algorithm) using border costs given by the Hamming distance between the ASAP embeddings.
- Run the pool version of the sequential in-place re-embedding algorithm.

The second method preserves the separation between candidate selection and placement+embedding. However, probe selection is modified to make its results more suitable for the subsequent placement and embedding optimizations. Building on the observation that shorter probe embeddings lead to improved border length, the modified probe selection algorithm picks from the available candidates the one that embeds in the *least* number of steps of the standard periodic deposition sequence using ASAP:

- Perform ASAP embedding of all probe candidates.
- Select from each pool of candidates the one that embeds in the least number of steps using ASAP.
- Run the REPTX or recursive-partitioning placement algorithm using only the selected candidates and border costs given by the Hamming distance between the ASAP embeddings.
- Run the iterated sequential in-place probe re-embedding algorithm, again using only selected candidates.

Table 10-10 gives the border-length and the runtime (in CPU seconds) for the two methods (each number represents the average over 10 test cases of the given size). The pool version of the recursive-partitioning uses $L = 3$. In these experiments, the number of candidates available for each probe is varied between 1 and 16; probe candidates were generated uniformly at random.

As expected, for each method and chip size, the improvement in solution quality grows monotonically with the number of available candidates. The improvement is significant (up to 15% when running the first method on a 100×100 chip with 16 candidates per probe), but varies non-uniformly with the method and chip size. For small chips the first method gives better solution quality than the second. For chips of size 200×200 the two methods give comparable solution quality, while for chips with size 300×300 or larger the second method is better (by over 5% for 500×500 chips with 8 probe candidates). The second method is faster than first for all chip sizes. The speedup factor varies between $5\times$ and $40\times$ when the number of candidates varies between 2 and 16. Interestingly, the runtime of the second method is slightly improving with the number of candidates, the reason being that the number of iterations of sequential re-embedding decreases when the length of the ASAP embedding of the selected candidates decreases.

8. CONCLUSIONS

In this chapter we have reviewed several recent algorithmic and methodological advance in DNA array design, focusing on minimizing the total mask border length during probe placement and embedding. Unlike VLSI placement, where placer suboptimality generally increases with instance size, empirical experimental results suggest that the opposite trend holds for DNA array placement: current algorithms are able to find DNA array placements with smaller normalized border cost when the number of probes in the design grows. Second, the lower bounds for DNA probe placement and embedding appear to be tighter

Table 10-10. Total border cost and runtime (averages over 10 random instances) for the two methods of combining probe placement and embedding with probe selection. The improvement (in percents) is relative to the single-candidate version of the respective method.

Chip	Pool	Multi-Candidate						ASAP-Based Selection					
		Row-Epitaxial			Partitioning			Row-Epitaxial			Partitioning		
Size	Size	Border	CPU	%	Border	CPU	%	Border	CPU	%	Border	CPU	%
100	1	0.39M	217	–	0.38M	115	–	0.39M	217	–	0.38M	115	–
	2	0.37M	1040	4.3	0.37M	676	0.9	0.38M	212	3.2	0.36M	114	3.6
	4	0.36M	1796	8.2	0.36M	1274	5.0	0.36M	193	6.6	0.35M	127	7.0
	8	0.34M	3645	11.8	0.34M	2605	8.7	0.35M	191	9.8	0.34M	109	9.4
	16	0.33M	7315	15.2	0.33M	5003	12.2	0.34M	185	12.8	0.33M	121	11.6
200	1	1.48M	1416	–	1.45M	1012	–	1.48M	1416	–	1.45M	1012	
	2	1.44M	6278	3.1	1.44M	7281	0.6	1.44M	1176	3.3	1.41M	946	2.5
	4	1.39M	12750	6.6	1.39M	13231	4.1	1.39M	1189	6.6	1.36M	932	5.9
	8	1.33M	27382	10.1	1.33M	26413	7.7	1.34M	1121	9.9	1.31M	957	9.2
	16	1.28M	44460	13.5	1.28M	52400	11.2	1.29M	1117	13.1	1.28M	971	11.7
300	1	3.25M	2971	–	3.22M	2975	–	3.25M	2971	–	3.22M	2975	–
	2	3.19M	14956	1.9	3.18M	13161	1.1	3.14M	2724	3.2	3.05M	2134	5.4
	4	3.09M	26514	4.7	3.09M	24671	3.9	3.02M	2771	7.0	2.94M	2118	8.8
	8	2.99M	51226	8.0	2.99M	45607	7.3	2.92M	2603	10.0	2.83M	2079	12.0
	16	2.88M	98189	11.3	2.88M	85311	10.6	2.84M	2760	12.6	2.71M	2247	15.9
500	1	8.69M	6842	–	8.65M	5608	–	8.69M	6842	–	8.65M	5608	–
	2	8.61M	51847	0.9	8.61M	41409	0.4	8.41M	6090	3.2	8.27M	5468	4.3
	4	8.48M	86395	2.4	8.48M	94566	1.9	8.11M	6709	6.7	7.96M	5591	8.0
	8	8.25M	161651	5.1	8.25M	213264	4.6	7.81M	6085	10.1	7.64M	5782	11.7
	16	–	–	–	–	–	–	7.52M	5986	13.5	7.45M	5601	13.9

than those available in the VLSI placement literature. Developing even tighter lower bounds is, of course, an important open problem. Other direction of future research is to find formulations and algorithms for integrated optimization of test structure design and physical design. Since test structures are typically pre-placed at sites uniformly distributed across the array, integrated optimization can have a significant impact on the total border length.

REFERENCES

[1] http://www.affymetrix.com

[2] http://www.perlegen.com

[3] S. Akers, "On the Use of the Linear Assignment Algorithm in Module Placement," *Proc. 1981 ACM/IEEE Design Automation Conference (DAC'81)*, pp. 137–144.

[4] C.J. Alpert and A.B. Kahng, "Geometric Embeddings for Faster (and Better) Multi-Way Netlist Partitioning" *Proc. ACM/IEEE Design Automation Conf.*, 1993, pp. 743-748.

[5] C.J. Alpert and A.B. Kahng, "Multi-Way Partitioning Via Spacefilling Curves and Dynamic Programming," *Proc. 1994 ACM/IEEE Design Automation Conference (DAC'94)*, pp. 652-657.

[6] C.J. Alpert and A.B. Kahng, "Recent directions in netlist partitioning: A survey", *Integration: The VLSI Jour.* 19 (1995), pp. 1-81.

[7] N. Alon, C.J. Colbourn, A.C.H. Lingi and M. Tompa, "Equireplicate Balanced Binary Codes for Oligo Arrays", *SIAM Journal on Discrete Mathematics* 14(4) (2001), pp. 481-497.

[8] A.A. Antipova, P. Tamayo and T.R. Golub, " A strategy for oligonucleotide microarray probe reduction", *Genome Biology* 2002 3(12):research0073.1-0073.4

[9] M. Atlas, N. Hundewale, L. Perelygina and A. Zelikovsky, "Consolidating Software Tools for DNA Microarray Design and Manufacturing", *Proc. International Conf. of the IEEE Engineering in Medicine and Biology (EMBC'04)*, 2004, pp. 172-175.

[10] K. Nandan Babu and S. Saxena, "Parallel algorithms for the longest common subsequence problem", *Proc. 4th Intl. Conf. on High-Performance Computing*, Dec. 1997, pp. 120-125.

[11] J.J. Bartholdi and L.K. Platzman, "An O(N log N) Planar Travelling Salesman Heuristic Based On Spacefilling Curves," *Operations Research Letters* 1 (1982), pp. 121-125.

[12] J. Branke and M. Middendorf, "Searching for shortest common supersequences", *Proc. Second Nordic Workshop on Genetic Algorithms and Their Applications*, 1996, pp. 105-113.

[13] J. Branke, M. Middendorf and F. Schneider, "Improved heuristics and a genetic algorithm for finding short supersequences", *OR Spektrum* 20(1) (1998), pp. 39-46.

[14] A.E. Caldwell, A.B. Kahng and I.L. Markov, "Optimal Partitioners and End-Case Placers for Standard-Cell Layout," *Proc. ACM 1999 International Symposium on Physical Design (ISPD'99)*, pp. 90-96.

[15] A. Caldwell and A. Kahng and I. Markov, "Can Recursive bisection Produce Routable Designs?", *DAC*, 2000, pp.477-482.

[16] C.C. Chang, J. Cong and M. Xie, "Optimality and Scalability Study of Existing Placement Algorithms", *Proc. Asia South-Pacific Design Automation Conference*, Jan. 2003.

[17] C.J. Colbourn, A.C.H. Lingi and M. Tompa, "Construction of optimal quality control for oligo arrays", *Bioinformatics* 18(4) (2002), pp. 529–535.

[18] J. Cong, M. Romesis and M. Xie, "Optimality, Scalability and Stability Study of Partitioning and Placement Algorithms", *Proc. ISPD*, 2003, pp. 88-94.

[19] D.R. Cutting, D.R. Karger, J.O. Pederson and J.W. Tukey, "Scatter/Gather: A Cluster-Based Approach to Browsing Large Document Collections", (15th Intl. ACM/SIGIR Conference on Research and Development in Information Retrieval) *SIGIR Forum* (1992), pp. 318–329.

[20] V. Dancik, "Common subsequences and supersequences and their expected length", *Combinatorics, Probability and Computing* 7(4) (1998), pp. 365-373.

[21] K. Doll, F.M. Johannes, K.J. Antreich, "Iterative Placement Improvement by Network Flow Methods", *IEEE Transactions on Computer-Aided Design* 13(10) (1994), pp. 1189-1200.

[22] J.J. Engel et al. "Design methodology for IBM ASIC products", *IBM Journal for Research and Development* 40(4) (1996), pp. 387.

[23] W. Feldman and P.A. Pevzner, "Gray code masks for sequencing by hybridization", *Genomics*, 23 (1994), pp. 233–235.

[24] C.M. Fiduccia and R.M. Mattheyses, "A Linear-Time Heuristic for Improving Network Partitions", *Proc. Design Automation Conference (DAC 1982)*, pp. 175–181.

[25] S. Fodor, J. L. Read, M. C. Pirrung, L. Stryer, L. A. Tsai and D. Solas, "Light-Directed, Spatially Addressable Parallel Chemical Synthesis", *Science* 251 (1991), pp. 767–773.

[26] D.E. Foulser, M. Li and Q. Yang, "Theory and algorithms for plan merging", *Artificial Intelligence* 57(2-3) (1992), pp. 143-181.

[27] C.B. Fraser and R.W. Irving, "Approximation algorithms for the shortest common supersequence", *Nordic J. Computing* 2 (1995), pp. 303-325.

[28] D.H. Geschwind and J.P. Gregg (Eds.), *Microarrays for the neurosciences: an essential guide*, MIT Press, Cambridge, MA, 2002.

[29] S. Hannenhalli, E. Hubbell, R. Lipshutz and P.A. Pevzner, "Combinatorial Algorithms for Design of DNA Arrays," in *Chip Technology* (ed. J. Hoheisel), Springer-Verlag, 2002.

[30] L.W. Hagen, D.J. Huang and A.B. Kahng, "Quantified Suboptimality of VLSI Layout Heuristics", *Proc. ACM/IEEE Design Automation Conf.*, 1995, pp. 216–221.

[31] S. A. Heath and F. P. Preparata, "Enhanced Sequence Reconstruction With DNA Microarray Application", *Proc. 2001 Annual International Conf. on Computing and Combinatorics (COCOON'01)*, pp. 64-74.

[32] D. J. Huang and A. B. Kahng, "Partitioning-Based Standard-Cell Global Placement with an Exact Objective", in *Proc. ACM/IEEE Intl. Symp. on Physical Design*, Napa, April 1997, pp. 18-25.

[33] E. Hubbell and P.A. Pevzner, "Fidelity Probes for DNA Arrays", *Proc. Seventh International Conference on Intelligent Systems for Molecular Biology*, 1999, pp. 113-117.

[34] E. Hubbell and M. Mittman , *personal communication* (Affymetrix, Santa Clara, CA), July 2002.

[35] T. Jiang and M. Li, "On the approximation of shortest common supersequences and longest common subsequences", *SIAM J. on Discrete Mathematics* 24(5) (1995), pp. 1122-1139.

[36] L. Kaderali and A. Schliep, "Selecting signature oligonucleotides to identify organisms using DNA arrays", Bioiformatics 18:1340-1349, 2002.

[37] A.B. Kahng, I.I. Măndoiu, P.A. Pevzner, S. Reda, and A. Zelikovsky, "Border Length Minimization in DNA Array Design", *Proc. 2nd International Workshop on Algorithms in Bioinformatics (WABI 2002),* R. Guigó and D. Gusfield (Eds.), Springer-Verlag Lecture Notes in Computer Science Series 2452, pp. 435-448.

[38] A.B. Kahng, I.I. Măndoiu, P.A. Pevzner, S. Reda, and A. Zelikovsky, "Engineering a Scalable Placement Heuristic for DNA Probe Arrays", *Proc. 7th Annual International Conference on Research in Computational Molecular Biology (RECOMB 2003)*, W. miller, M. Vingron, S. Istrail, P. Pevzner and M. Waterman (Eds.), 2003, pp. 148–156.

[39] A.B. Kahng, I.I. Măndoiu, S. Reda, X. Xu, and A. Zelikovsky. Design flow enhancements for DNA arrays. In *Proc. IEEE International Conference on Computer Design (ICCD)*, pages 116–123, 2003.

[40] A.B. Kahng, I.I. Măndoiu, S. Reda, X. Xu, and A. Zelikovsky. Evaluation of placement techniques for DNA probe array layout. In *Proc. IEEE-ACM International Conference on Computer-Aided Design (ICCAD)*, pages 262–269, 2003.

[41] A.B. Kahng, I.I. Măndoiu, P. Pevzner, S. Reda, and A. Zelikovsky. Scalable heuristics for design of DNA probe arrays. *Journal of Computational Biology*, 11(2–3):429–447, 2004.

[42] S. Kasif, Z. Weng, A. Derti, R. Beigel, and C. DeLisi, "A computational framework for optimal masking in the synthesis of oligonucleotide microarrays", *Nucleic Acids Research* vol. 30 (2002), e106.

[43] T. Kozawa et al., "Automatic Placement Algorithms for High Packging Density VLSI", *Proc. 20th Design Automation Conference (DAC 1983)*, pp. 175–181.

[44] F. Li and G.D. Stormo, "Selection of optimal DNA oligos for gene expression arrays," Bioinformatics 17(11):1067-1076, 2001.

[45] R.J. Lipshutz, S.P. Fodor, T.R. Gingeras, D.J. Lockhart, "High density synthetic oligonucleotide arrays," *Nature Genetics* 21 (1999), pp. 20–24.

[46] B. T. Preas and M. J. Lorenzetti (Eds.), *Physical Design Automation of VLSI Systems*, Benjamin-Cummings, 1988.

[47] S. Rahmann. "Rapid large-scale oligonucleotide selection for microarrays", *Proc. IEEE Computer Society Bioinformatics Conference (CSB)*, 2002.

[48] S. Rahmann, "The Shortest Common Supersequence Problem in a Microarray Production Setting," *Bioinformatics* 19 Suppl. 2 (2003), pp. 156-161.

[49] R. Sengupta and M. Tompa, "Quality Control in Manufacturing Oligo Arrays: a Combinatorial Design Approach", *Journal of Computational Biology* 9 (2002), pp. 1–22.

[50] K. Shahookar and P. Mazumder, "VLSI Cell Placement Techniques", *Computing Surveys* 23(2) (1991), pp. 143-220.
[51] N.A. Sherwani, *Algorithms for VLSI Physical Design Automation*, Kluwer Academic Publishers, Norwell, MA, 199
[52] L. Steinberg, "The backboard wiring problem: a placement algorithm", *SIAM Review* 3 (1961), pp. 37–50.
[53] A.C. Tolonen, D.F. Albeanu, J.F. Corbett, H. Handley, C. Henson, and P. Malik, "Optimized in situ construction of oligomers on an array surface", *Nucleic Acids Research* 30 (2002), e107.
[54] M. Wang, X. Yang and M. Sarrafzadeh, "DRAGON2000: Standard-cell Placement Tool For Large Industry Circuits", *Proc. International Conference on Computer-Aided Design (ICCAD 2001)*, pp. 260–263.
[55] J.A. Warrington, R. Todd, and D. Wong (Eds.). *Microarrays and cancer research* BioTechniques Press/Eaton Pub., Westboro, MA, 2002.

Chapter 11

SYNTHESIS OF MULTIPLEXED BIOFLUIDIC MICROCHIPS

Anton J. Pfeiffer
Carnegie Mellon University
Department of Chemical Engineering
pfeiffer@cmu.edu

Tamal Mukherjee
Carnegie Mellon University
Department of Electrical and Computer Engineering
tamal@ece.cmu.edu

Steinar Hauan
Carnegie Mellon University
Department of Chemical Engineering
hauan@cmu.edu

Abstract: Lab-on-a-Chip (LoC) devices are a class of microfluidic chip-based systems that show a great deal of promise for complex chemical and biological sensing and analysis applications. We are developing an approach for full-custom LoC design that leverages optimal design techniques and System on a Chip (SoC) physical design methods. We simultaneously consider both the physical design of the chip and the microfluidic performance to obtain complete LoC layouts. We demonstrate our approach by designing multiplexed capillary electrophoresis (CE) separation microchips. We believe that this approach provides a foundation for future extension to LoC devices in which many different complex chemical operations are performed entirely on-chip.

Keywords: Lab on a Chip design, biofluidic chip layout, microfluidic synthesis

K. Chakrabarty and J. Zeng (eds.),
Design Automation Methods and Tools for Microfluidics-Based Biochips, 271–300.

1. INTRODUCTION

Microchips represent an advantageous platform for microscale chemical sensors and analytical devices. Microfluidic devices that are fast, accurate, portable and readily automated can be fabricated inexpensively using methods adapted from the semiconductor industry. This paper explores model-based design of microfluidic channel systems that fall within a broad category known as *Lab on a Chip* (LoC) [1]. A LoC is essentially a miniaturized, microchip implementation of a macroscale analytical chemistry laboratory. LoC's have already seen a great deal of use within the life-science and biomedical industries for applications in genomics, proteomics and combinatorial chemistry. An exciting direction for LoC development concerns Point of Care (PoC) medical devices. PoC devices must be both highly compact and capable of efficiently performing complex chemical operations. However, the design of LoC devices is complicated by performance limits arising from complex physiochemical phenomena and difficult layout and integration issues.

Design methods for various LoC components have recently been developed. A partitioning based approach has been successfully used to layout regular-arrays of DNA probes on microchips [2]. Shape optimization methods have been employed to generate high performance channel geometries [3]. Both heuristic and optimal design methods for capillary electrophoresis channel systems have been developed [4]. Each of these design methods pertain to sub-components or subsystems within a complete LoC. In addition, a design methodology for reconfigurable droplet-based microfluidic systems has been investigated [5]. In this architecture, droplets rather than continuous streams of fluid are moved throughout the chip.

Here we present our work toward automating the design of complete LoC systems. For instance, for DNA analysis, our methodology would allow for the simultaneous design and layout of the channel network that would be needed to sample body fluid, biochemically pre-process the fluid, and bring the analyte to the DNA sensor array. The DNA array itself could be designed simultaneously with the rest of the chip using [2].

Our approach combines System on a Chip (SoC) circuit floorplanning and routing techniques with rigorous subsystem design optimization to develop methods for full-custom LoC design. We focus on the design of multiplexed systems, in which multiple non-interacting subsystems of similar function are compacted onto a single chip. Multiplexing exploits device level parallelism to increase analysis speed and decrease the required sample size, which provides redundancy, robustness and the potential for combinatorial experiments. This makes multiplexed LoC's applicable to PoC applications.

We demonstrate our approach by designing microchips containing multiplexed capillary electrophoresis (CE) subsystems. CE is a common on-chip

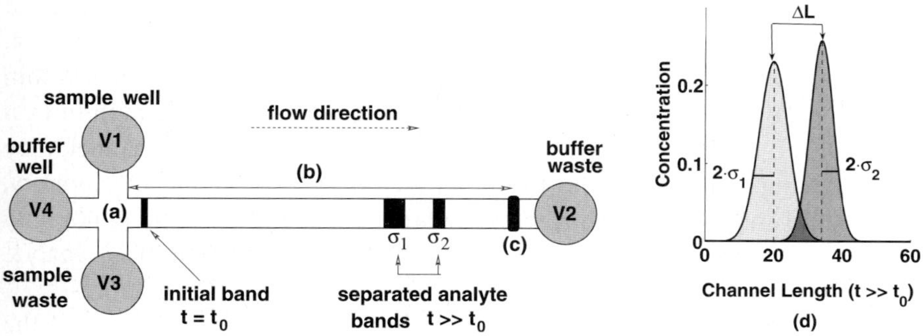

Figure 11-1. Simple capillary electrophoresis schematic: (a) injector (b) separation channel length L (c) detector and (d) the resulting detector output (electropherogram) showing the two species bands as Gaussian distributions.

separation technology that is capable of separating complex biological mixtures [1] and is thus relevant for use in PoC devices.

The organization of the paper is as follows: Section 1.1 presents the background and design considerations of microchip based CE. Section 2 reviews subsystem simulation and optimization. An algorithm for designing and placing subsystems on a chip is introduced in section 3. Section 4 presents an approach for connecting subsystems and fluid I/O structures on the chip. Results and completed designs are discussed in section 5. Finally, extensions and future directions are summarized in section 6.

1.1 Capillary Electrophoresis on Microchips

Figure 11-1 shows the major components of a simple chip-based capillary electrophoretic separation system. The chip is composed of: (a) an injector where the analyte, or mixture of chemical species to be separated, enters the system, (b) the separation channel where the analyte separates into unique species bands and (c) the detector where species bands are detected; typically optically or electrically. The detector output or electropherogram is shown in (d) where the species band concentration profiles are represented as Gaussian distributions. The microchannels are filled with a carrier electrolyte known as a buffer solution. Electrodes positioned in the four wells generate an electric field through the buffer that imparts specific velocities to the species within the channel.

The CE separation process has two phases: A loading phase, where a continuous analyte stream is drawn from the sample well to the sample waste well by applying a DC potential between V1 and V3 and a dispensing stage, where a band of analyte material is injected into the separation channel by applying a potential between V2 and V4 [6].

Electrophoretic separation occurs because of the differential transport of charged species in the presence of an electric field generated between V2 and V4. As an analyte mixture travels down the separation channel, the species within the mixture separate into bands according to their electrophoretic mobilities μ [7]. However, all the while species bands are traveling down the channel, they are broadening or dispersing due to factors such as diffusion, channel geometry, Joule heating, adsorption, and electromigration [7]. Band broadening not only impedes separation but also results in the dilution or reduction of an analyte band's concentration.

The total dispersion for a given analyte band is quantified in terms of the variance σ^2_{tot} of that band's concentration distribution. The total dispersion can be estimated by summing each of the major contributions to dispersion Eq. (11.1).

$$\sigma^2_{tot} = \sigma^2_{inj} + \sigma^2_{diff} + \sigma^2_{geom} + \sigma^2_{JH} + \ \ldots \tag{11.1}$$

In Eq. (11.1) the contributions to dispersion from the injector σ^2_{inj} [8], diffusion σ^2_{diff} [7], channel geometry σ^2_{geom} [9] and Joule heating σ^2_{JH} [10] are all functions of the separation channel length L and the applied voltage V.

The measure that describes the degree of separation between a pair of migrating analytes, i and j, is the resolution $\hat{R}_{i,j}$ Eq. (11.2).

$$\hat{R}_{i,j} = \frac{\Delta L}{2(\sigma_i + \sigma_j)} = \frac{(\mu_i - \mu_j)L}{2\mu_i(\sigma_i + \sigma_j)} \tag{11.2}$$

Here, ΔL is the distance between the peaks of the concentration distributions of bands i and j, (Fig. 11-1) where μ_i and μ_j are the species mobilities with $\mu_i > \mu_j$. The denominator of Eq. (11.2) quantifies the average dispersion of bands i and j in terms of the standard deviations of their concentration distributions, σ_i and σ_j. A value of $\hat{R} \geq 1.5$ implies baseline resolution [7] which we use as a lower bound on separation performance in our studies.

1.2 Design Considerations

The separation channel length L and the applied voltage V are the key design variables we manipulate to achieve separation. We consider channel width ω to be a fixed variable since its feasible range is generally small. The design variables interact in complex ways. For example, for constant L, increasing V will typically reduce diffusional dispersion σ^2_{diff}, but may increase the dispersion resulting from channel geometry σ^2_{geom}, and bulk fluid heating σ^2_{JH}. Likewise, for constant V, increasing L provides more time for analyte bands to move apart, but causes σ^2_{diff} to increase. In this case, σ^2_{geom} and σ^2_{JH} can become negligible when compared to σ^2_{diff}. Additionally, an analyte band's dispersion is inversely related to its concentration. A band's peak concentration

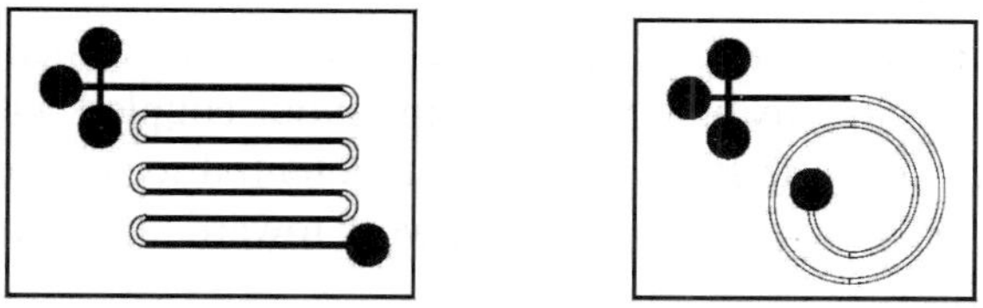

(a) (b)

Figure 11-2. Two common compact channel topologies: (a) Serpentine topology composed of an injection cross with wells (black circles) straight channels (black) and U-bends (white) (b) Spiral topology composed of an injection cross with wells, U-bends and an elbow.

C may drop below the limit of detection for a particular detection technology when dispersion is too great [7].

In practice, for difficult separations ($\mu_i \approx \mu_j$), the available voltage source becomes a limiting factor and long separation channels are required to achieve baseline resolution. However, long channels can only be made to fit on a microchip by adding turns or bends to the design, which results in a phenomenon called turn induced dispersion [11]. The topology of a design, or interconnectivity of channel sections, can also be shown to have a significant impact on the performance of a design [12]. Thus, minimizing design area and maximizing separation performance are conflicting design goals.

Two common topologies found in the literature are the *serpentine* and *spiral* topologies (Figs. 11-2(a) and 11-2(b)). In our view, channel topologies can be built from a library of channel sections consisting of straights, U-bends, elbows, crosses, tee's and wells as described in [13]. For example: Fig. 11-2(a) is composed of a cross injector followed by an alternating series of straight (black) and U-bend (white) channel sections. We use the serpentine topology for every subsystem in our designs because serpentines are compact and allow for easy access to I/O ports in a multiplexed layout.

Our design approach uses accurate parameterized reduced order models of CE that are derived from the partial differential equations that describe the electric field, species transport, and thermal effects within channel sections [13]. These models were verified against experimental designs presented in the literature and against a suite of finite element simulations. They can be connected together to create a flexible CE simulation framework that is amenable to both iterative studies and design optimization.

In LoC devices, the subsystem design and ultimate chip layout are intimately linked due to the influence that channel geometry has on device performance. Therefore, a solution approach that is capable of simultaneously considering both subsystem design and overall system layout is required. Currently, creating multiplexed designs [14] requires a substantial investment of time, manpower

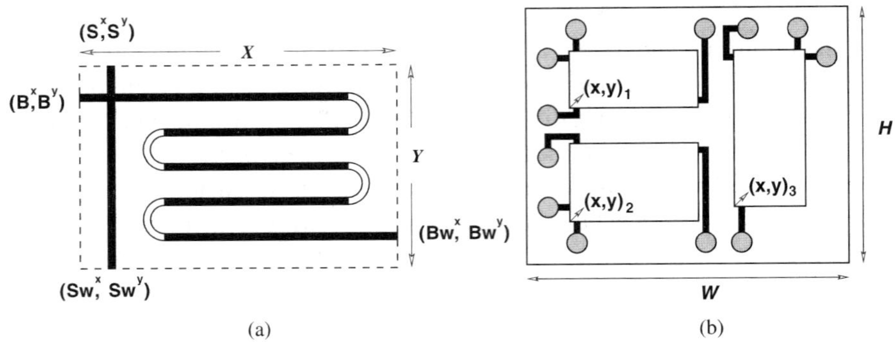

Figure 11-3. (a) Schematic of a serpentine subsystem showing horizontal X and vertical Y dimensions, and port locations (S^x, S^y), (B^x, B^y), (Sw^x, Sw^y), and (Bw^x, Bw^y). (b) Three subsystem chip layout showing the horizontal W and vertical H dimensions and subsystem locations $(x, y)_i$. Fluid I/O wells are shown as circles, and auxiliary channels (black lines) connect wells and subsystems.

that all subsystems will be similar to the topology shown in Fig. 11-3(a), with ports deterministically located one per side. This configuration allows for convenient subsystem I/O and heuristically reduces the potential for routing congestion around each subsystem.

Figure 11-3(b) is an example instance of a design containing three subsystems. Auxiliary channels (black lines) transport fluid between subsystem ports and particular wells (circles). The width and height of the chip are indicated by W and H respectively. The wells form the *world-to-chip* interface and allow fluids to be introduced and removed from the chip. The position of each subsystem and well is defined by an (x, y) coordinate point in its lower left corner. Notice that the wells are relegated to the chip edges. This requirement makes it easier to transport fluids on and off the chip. Figs. 11-3(a) and 11-3(b) give a schematic description of the simultaneous subsystem design and layout problem.

3.1 Floorplan Problem Definition

We define our floorplan problem as follows: given a set of subsystems $\mathcal{N}$ and their associated input and output wells $\mathcal{W}$ obtain a planar arrangement $\mathcal{A}$ of subsystems and wells that does not exceed the total chip height H or width W, and where all subsystems with respect to the constraint set in MSA are feasible. The objective is to choose the arrangement of subsystems and wells that is most compact and where fluid I/O wells are positioned on the edge closest to their associated subsystem ports.

Our problem is similar to a standard VLSI floorplan problem in which chip area and wire length are minimized. However, unlike many typical VLSI floorplan formulations, neither the subsystem aspect ratio, nor total area is known *a priori* since subsystem dimensions are a function of the constraint set shown in MSA. We assume a building-block layout style for our problem as opposed to a grid-point assignment layout style [15] because the dimensions of each subsystem are highly variable. We allow the dimensions of each subsystem to vary so that for a given arrangement $\mathcal{A}$, the optimal values of W and H can be obtained.

3.2 Floorplan Problem Formulation

In previous work, we developed a rigorous floorplanning formulation using a concise modeling technique known as General Disjunctive Programming (GDP) [16]. Typically, GDP's are translated into mixed integer nonlinear programs (MINLP) and solved using deterministic optimization approaches that combine gradient-based search methods with Branch and Bound [17]. However, the floorplanning problem that we have formulated is a rectangle packing problem that has been shown to be NP-hard [18]. In addition, floorplan optimization using Branch and Bound [19] or conventional mixed integer linear programming (MILP) approaches [20] often have difficulty handling problems of realistic size. Since we are coupling rectangle packing with a non-convex (nonlinear) physiochemical problem, our problem should be at least as hard as rectangle packing alone. Our computational experience supports this conjecture. Solving our GDP formulation directly using an MINLP solver requires hours of CPU time for problems of trivial size (i.e. $|\mathcal{N}| \leq 4$). While we consider our GDP formulation to be a rigorous description of the multiplexed CE floorplanning problem, it is apparent that a heuristic is required to obtain good solutions to problem instances of realistic size.

In the following sections we discuss four key parts of our floorplanning problem formulation: subsystem orientation, serpentine topology size, relative subsystem positions, and well placement.

Subsystem Orientation. Each subsystem can have one of eight possible orientations (Fig. 11-4). Different orientations contribute to design compaction. We define a vector $\boldsymbol{Z} = (z_1, z_2, \ldots, z_{|\mathcal{N}|})$ which contains the subsystem orientation labels $z_i \in \{1, \ldots, 8\}$ where i is a unique element of $\mathcal{N}$. Each z_i defines the orientation of subsystem i and allows for the deterministic calculation of that subsystem's port locations. For example, the buffer port location B_i for orientation $z_i = 4$ is $B_i = (B^x, B^y) = (x_i + w_i - \omega - PAD, y_i + h_i)$. The port locations are a function of the subsystem's location x_i and y_i, width w_i and height h_i, as well as the feature spacing PAD and channel width ω. However, recall that rSIM has no knowledge of subsystem orientation. Thus, Eq. (11.3)

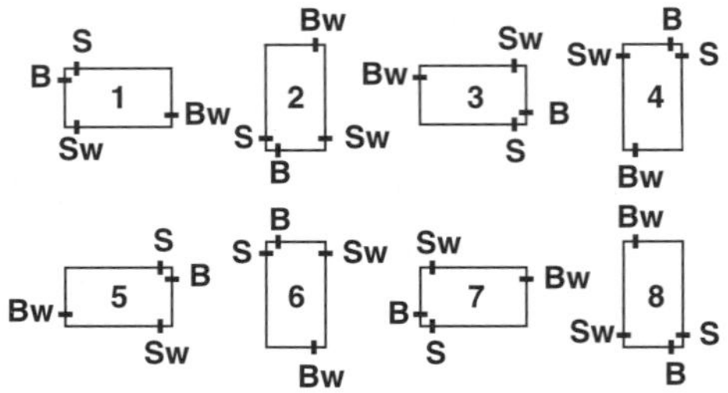

Figure 11-4. Eight possible subsystem orientations.

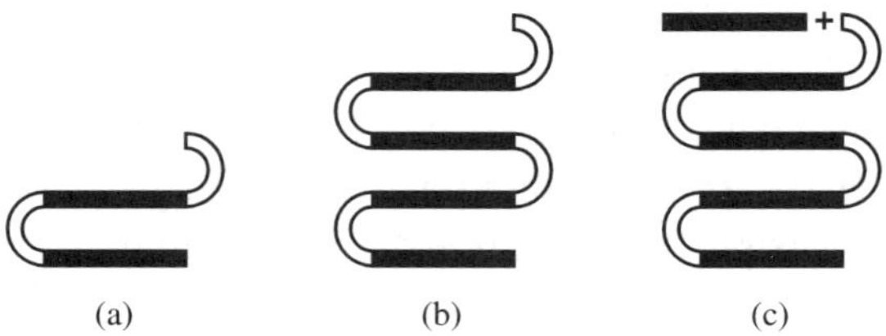

Figure 11-5. (a) One serpentine unit ($\tau = 1$). (b) Two serpentine units ($\tau = 2$). (c) Completing the topology with an initial straight section.

must be applied to translate the general subsystem dimensions, X_i and Y_i, used in the simulator to the actual subsystem dimensions w_i and h_i.

$$w_i = \begin{cases} X_i & \text{if } z_i \text{ even,} \\ Y_i & \text{otherwise.} \end{cases} \qquad h_i = \begin{cases} X_i & \text{if } z_i \text{ odd,} \\ Y_i & \text{otherwise.} \end{cases} \tag{11.3}$$

Serpentine Topology Size. The types list $\mathcal{T}$ can be reduced to a scalar value τ indicating the number of serpentine units within a subsystem topology (Fig. 11-5). Figs. 11-5(a) and 11-5(b) show a topology with $\tau = 1$ and $\tau = 2$ respectively. The topology is completed by adding an initial straight section to the serpentine (Fig. 11-5(c)). We define a vector $\boldsymbol{S} = (\tau_1, \tau_2, \ldots, \tau_{|\mathcal{N}|})$ which contains the number of serpentine units $\tau_i \in \{1, \ldots, \tau^{UB}\}$ for each subsystem, $i \in \mathcal{N}$. Here, τ_i can range from 1 to a user specified upper bound, τ^{UB}. Our experience indicates that a conservative value for τ^{UB} is 15, which corresponds to a serpentine composed of 61 individual channel sections where the number of sections $= 4 \cdot \tau_i + 1$.

Relative Subsystem Positions. We encode the constraints to prevent subsystems from overlapping using a VLSI floorplan representation known as the Sequence Pair (SP) [18]. SP encodes the $\{left, right, above, below\}$ spatial relations between objects in the plane into a concise structure.

This structure is readily searchable using standard heuristic methods such as Simulated Annealing (SA) [21].

A SP is composed of two $|\mathcal{N}|$-tuples $\Gamma = (\Gamma_+, \Gamma_-)$, where Γ_+ and Γ_- are ordered lists of the subsystem labels in $\mathcal{N}$. A SP maps to $\frac{|\mathcal{N}|(|\mathcal{N}|-1)}{2}$ linear math programming constraints as shown in Eq. (11.4).

$$\begin{aligned} x_i + w_i \le x_j &\leftarrow (\ldots i \ldots j \ldots\ ,\ \ldots i \ldots j \ldots) \\ y_i + h_i \le y_j &\leftarrow (\ldots j \ldots i \ldots\ ,\ \ldots i \ldots j \ldots) \end{aligned} \tag{11.4}$$

Here, i and j are unique subsystem labels in $\mathcal{N}$. If i appears before j in both Γ_+ and Γ_-, then subsystem i is left of j. Likewise, if i appears after j in Γ_+ and before j in Γ_-, then subsystem i is below j. The SP representation has several attractive features. First, the solution space, although large, is finite i.e. $\Psi \propto (|\mathcal{N}|!)^2$. Second, the neighborhood of a particular arrangement $\mathcal{A}$ is readily constructible through simple perturbations of Γ. Furthermore, every evaluation of Γ results in a feasible planar arrangement, which is not true for all non-slicing floorplan representations (e.g. Corner Block List presented by [22]). Finally, SP is a general floorplan representation, which means that optimal solutions are not excluded due to assumptions about floorplan structure.

Well Placement. Confining the fluid I/O wells to the edges of the chip facilitates fluid transport on and off the chip. We have developed a graph-based approach that places a well on the chip edge closest to its associated port.

For a given arrangement $\mathcal{A}$ of subsystems, we construct an undirected, edge-weighted planar graph $G_{\mathcal{A}} =< \mathcal{V}, \mathcal{E} >$ by noting that a compacted arrangement of subsystems is essentially a connected planar graph (Fig. 11-6). The graph we create is similar to a *placement graph* used in VLSI circuit design [15].

We construct the graph by first generating a set of preliminary nodes by taking the Cartesian product of the bounding box dimensions, subsystem locations and port locations Eq. (11.5).

$$\begin{aligned} nodes =&\{0, W, x_i, S_i^x, Sw_i^x, B_i^x, Bw_i^x; \forall i \in \mathcal{N}\} \\ &\times \{0, H, y_i, S_i^y, Sw_i^y, B_i^y, Bw_i^y; \forall i \in \mathcal{N}\} \end{aligned} \tag{11.5}$$

A set of preliminary edges are formed by connecting each node with its nearest neighbor in the x and y directions. The graph's vertex set $\mathcal{V}$ and edge set $\mathcal{E}$ are generated by removing all the nodes and associated edges that reside within subsystem boundaries. Finally, four supernodes, v^L, v^T, v^R, and v^B are added to the vertex set. The supernodes are connected to the left-most, top-most, right-most, and bottom-most nodes of $G_{\mathcal{A}}$ as shown in Fig. 11-6(b).

The weights assigned to edges within the interior of $G_{\mathcal{A}}$ are simply the distances in the x and y directions between connected adjacent nodes. The

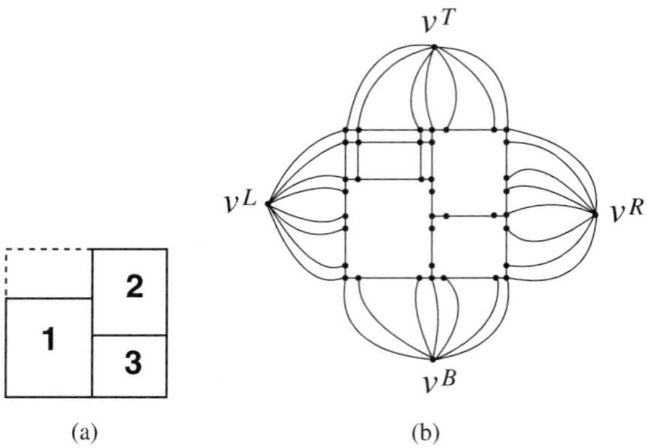

Figure 11-6. (a) Compacted arrangement of subsystems. (b) Corresponding well placement graph.

edges connecting supernodes are weighted based on the maximum bounding box dimension. This prevents short-circuit paths through the graph during the well placement procedure. Each well is placed on one of the four sides of $\mathcal{A}$ as follows:

1 Construct $G_{\mathcal{A}}$ and initialize sets: $left = right = top = bottom = \emptyset$.

2 Select a port vertex $p \in v_p$, where $v_p \subset \mathcal{V}$ is the set of ports.

3 Determine the shortest path between p and supernodes v^L, v^T, v^R, v^B and store distance in set $\mathcal{P}$, where $\mathcal{P}$ is the set of shortest paths between ports and wells.

4 Assign p to $left$ if p connects to v^L, to top if p connects to v^T, to $right$ if p connects to v^R or to $bottom$ if p connects to v^B.

5 Order wells in $left$ or $right$ based on the y-coordinate of its port.

6 Order wells in top or $bottom$ based on the x-coordinate of its port.

7 Obtain a total routing length estimate, ϕ, by summing the entries in $\mathcal{P}$. (note that $|\mathcal{P}| = |\mathcal{W}| = 4 \cdot |\mathcal{N}|$)

In step 3, we apply a shortest path algorithm to efficiently search $G_{\mathcal{A}}$. Steps 5 and 6 can be performed using an efficient sorting method.

Our solution methodology allows us to avoid two fundamental problems that would result from a more simplistic formulation. The first possible problem is illustrated in Fig. 11-7(a). Here wells are placed using a standard rectilinear distance metric. However, the shortest rectilinear distance between a port and

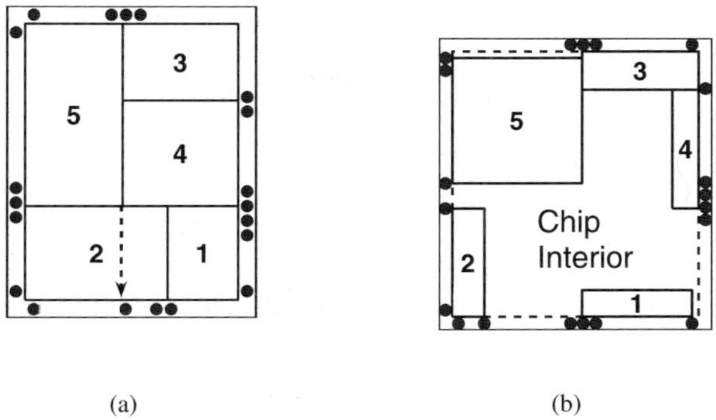

Figure 11-7. (a) Poor well placement based on rectilinear model (dotted arrow). (b) Compacted arrangement pulled apart by weighted-sum objective function.

a well, indicated by the dotted arrow, may pass through a subsystem and is therefore a poor estimate of the true distance between a port and the chip edge. The distance estimates produced using $G_{\mathfrak{A}}$ are far more accurate because they account for the fact that routes must go around subsystems.

A second problem can arise if a common VLSI floorplanning objective function of the form: $\alpha_1 \cdot A + \alpha_2 \cdot WL$ is used, where A is chip area, WL is wire length and α_1 and α_2 are weighting factors. This objective works well for typical circuit design problems because circuit elements are highly interconnected within the chip interior and a wide variety of α_1 and α_2 values will produce satisfactory results. Since we connect single ports within the chip's interior to single wells on the chip's edges, over-weighting α_2 *pulls-apart* a compact arrangement and results in a large amount of unused space within the chip interior (Fig. 11-7(b)). Since we generally do not know good values for α_1 and α_2 *a priori*, we first compact the subsystems and lock their positions. Then we place the wells, thereby decoupling the continuous variables of the problem while retaining the global nature of the combinatoric variables, Γ, $\boldsymbol{Z}$ and $\boldsymbol{S}$.

3.3 Floorplan Solution Method

Figure 11-8 is a flowchart illustrating our simultaneous design and layout approach. The main idea is to use a probabilistic search heuristic such as SA to deal with the combinatorial aspects of the problem and an efficient gradient based method for the remaining continuous-space problem. We have chosen this hybrid approach because SA has been shown to be an effective search method for difficult combinatorial problems, but is generally inferior to gradient-based methods on well posed continuous-space problems [23].

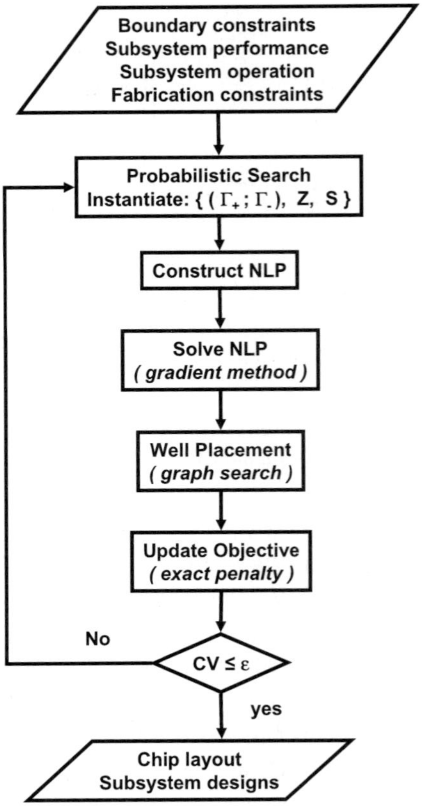

Figure 11-8. Flowchart for the floorplanning algorithm.

In our approach, the combinatorial states Γ, $\boldsymbol{Z}$ and $\boldsymbol{S}$ are instantiated by the SA algorithm resulting in a NLP. A new NLP is dynamically constructed for each new problem instance. This approach is conceptually similar to floorplanning methods meant to handle soft blocks (i.e. circuit modules with variable dimensions) where the resulting instantiated problem is an LP [24] or a convex program [25]. In our case, we are not only concerned with floorplanning, but also with the performance of each subsystem on the chip, where subsystem performance is a strong function of available chip area. The resulting NLP is non-convex, which results in convergence issues and increased computational complexity that are not present in linear or convex formulations. We combat these problems using an intuitive initialization procedure and a penalty function constraint-relaxation approach.

Stage 1: Instantiation. The algorithm in Fig. 11-8 begins by obtaining the relevant subsystem and chip design specifications. The search heuristic proposes a sequence pair, $(\Gamma_+;\Gamma_-)$, an instance of the block orientation vector $\boldsymbol{Z} = (z_1, \ldots, z_{|\mathcal{N}|})$ and an instance of the subsystem topology vector $\boldsymbol{S} = (\tau_1, \ldots, \tau_{|\mathcal{N}|})$.

For example: Suppose that we want to design a chip containing 5 subsystems. The SA instantiates a SP of $\Gamma = (< 5, 3, 4, 2, 1 >, < 2, 5, 1, 4, 2 >)$, a block orientation vector $\boldsymbol{Z} = (8, 2, 5, 5, 1)$ and a topology size vector $\boldsymbol{S} = (4, 5, 6, 4, 7)$. The relative position of each subsystem is defined by Γ, however, the dimensions of each subsystem and the resulting compact arrangement have not yet been determined. The instantiated values are then passed to the remaining stages in the procedure.

Stage 2: NLP Initialization. We have developed a tailored initialization procedure to aid in the convergence of our NLP. Our initialization procedure generates an upper bound on the chip design area by minimizing the size of each individual subsystem to produce a set of fixed-sized rectangles. The $|\mathcal{N}|$ fixed-size rectangles are generated by minimizing $(X_i + Y_i)$ for each subsystem subject to rSIM and the MSA constraint set using a standard sequential quadratic programming method. The z_i's from Stage 1 along with Eq. (11.3) are used to translate X_i and Y_i to the appropriate subsystem dimensions, w_i and h_i. When the dimensions of each subsystem are known, the (x, y) positions of each subsystem are determined by evaluating the Γ from Stage 1 using an efficient SP decoding algorithm [26]. The dimensions and positions are provided as initial values to the compaction NLP described in the following section.

Stage 3: NLP Construction. The minimum chip area (MCA) NLP is dynamically constructed using Γ, $\boldsymbol{Z}$ and $\boldsymbol{S}$ by simultaneously invoking the simulator rSIM and MSA constraint set for every subsystem. We use automatic differentiation [27] to generate the gradients $(\nabla E_i, \nabla \hat{R}_i, \nabla C_i)$ with respect to X_i, Y_i and V_i. Eq. (11.3) is used to map the general X_i and Y_i dimensions used in the simulator to the corresponding w_i and h_i for each subsystem. Eq. (11.4) is used to map Γ to a set of relative position constraints. Here we show the complete formulation of MCA where the objective is to minimize the design Decoding Γ for our 5 subsystem example results in 4 horizontal constraints and 6 vertical constraints. Notice that all of the nonlinear constraints in MCA are embedded within the simulator.

width ε^R and height ε^T.

$$
\begin{aligned}
\min \quad & \varepsilon^R + \varepsilon^T \\
\text{s.t.} \quad & [E_i, \hat{R}_i, C_i, \nabla E_i, \nabla \hat{R}_i, \nabla C_i] = \text{rSIM}(X_i, Y_i, V_i, \boldsymbol{\tau}_i, props_i) \\
& w_i = \text{mod}(z_i, 2) \cdot X_i + \neg\, \text{mod}(z_i, 2) \cdot Y_i \\
& h_i = \neg\, \text{mod}(z_i, 2) \cdot X_i + \text{mod}(z_i, 2) \cdot Y_i \\
& x_i + w_i \leq x_j \leftarrow (\Gamma_+; \Gamma_-) \\
& y_i + h_i \leq y_j \leftarrow (\Gamma_+; \Gamma_-) \\
& \varepsilon^R \geq x_i + w_i \\
& \varepsilon^T \geq y_i + h_i \\
& \hat{R}_i \geq \hat{R}_{spec} \\
& C_i \geq C_{min} \\
& E_i \leq E_{max} \\
& V_i \leq V_{max} \\
& x_i, y_i, w_i, h_i, X_i, Y_i, V_i, E_i \geq 0 \quad \forall i \in \mathcal{N} \qquad \text{(MCA)}
\end{aligned}
$$

The NLP is solved using a standard general reduced gradient method[1] starting from the initial values generated in Stage 2. The solution of the NLP results in an optimally compacted design for a given Γ, **Z** and **S**. This compacted design is then passed to the well placement stage.

Stage 4: Well Placement. Each subsystem is associated with 4 fluid I/O wells that must be placed on the edges of the chip. For our 5 subsystem example, each of the 20 fluid I/O wells are assigned to the *left*, *right*, *top*, or *bottom* sets by applying the well placement algorithm described previously to the compacted design obtained in Stage 3. In addition, the approximate total routing cost ϕ is also obtained.

Since wells occupy space on the chip, the design dimensions ε^R and ε^T must be updated. The width of the design is updated as shown in Eq. ((11.6)).

$$
\varepsilon^R = \varepsilon^R + \begin{cases} 2 \cdot (d + 2 \cdot PAD + \frac{\omega}{2}) & \text{if } left \wedge right \neq \emptyset, \\ d + 3 \cdot PAD + \omega & \text{elseif } left \veebar right \neq \emptyset \\ 2 \cdot (PAD + \frac{\omega}{2}) & \text{otherwise} \end{cases} \tag{11.6}
$$

Here, d is the well diameter, PAD is the feature spacing and ω is the channel width. Eq. ((11.6)) has three conditions: the first is applied when wells exist on both the left and right sides of a design, the second when wells occupy only the left or right sides of a design and the third when wells occupy neither the left nor right sides of a design. The design height is updated similarly, replacing *right* with *top* and *left* with *bottom*. In each case, ε^R or ε^T is adjusted to satisfy the well placement and feature spacing requirements.

Stage 5: Objective and Boundary Constraints. Once the approximate total routing cost ϕ and the updated ε^R and ε^T have been obtained, they are used to construct the objective function that is evaluated by the SA. The only remaining task is to enforce the boundary constraints that are necessary to confine a design to the available chip area.

We have chosen not to incorporate the boundary constraints in the compaction NLP discussed in Stage 3, and instead implement them using an *exact penalty* method [23]. This approach substantially reduces the likelihood that the NLP will return with an infeasible result. Therefore, we can evaluate the progress of our algorithm using information that would otherwise not exist.

Equation (11.7) shows the objective function evaluated by the SA.

$$\begin{aligned} \min \quad & \varepsilon^R + \varepsilon^T + \phi + \mathrm{P}(\varepsilon^R, \varepsilon^T) \\ & \mathrm{P}(\varepsilon^R, \varepsilon^T) = \rho \cdot (\max(\varepsilon^R - W, 0) + \max(\varepsilon^T - H, 0)) \\ & \rho = \max(W, H) \end{aligned} \tag{11.7}$$

The first, second and third terms in Eq. (11.7) are the updated chip dimensions and the total routing length estimate. The fourth term, $\mathrm{P}(\varepsilon^R, \varepsilon^T)$, is the penalty term that enforces the boundary constraints. A penalty is incurred in the objective when either ε^R is greater than the available chip width W or when ε^T is greater than the available chip height H. The penalty term is weighted by a scalar penalty parameter ρ which we define as the largest possible chip dimension. An attractive feature of the exact penalty method is that for a sufficiently large (finite) value of ρ, the optimal solution of the penalized problem will equal the solution of the original constrained problem. Practically speaking, this means that if ρ is initially set high enough, the boundary constraints will be met at the optimal solution. This is in contrast to other penalty methods where ρ is iteratively increased to reduce constraint violation. Our experience indicates that $\rho = \max(W, H)$ is large enough to ensure good convergence. Any non-smoothness created by the penalty function is of little consequence since Eq. (11.7) is evaluated by SA which does not require gradient information.

4. ROUTING OF AUXILIARY MICROFLUIDIC CHANNELS

The goal of the routing stage is to take a compacted arrangement generated by our floorplanning algorithm and to determine the placement or precise positions of subsystems and auxiliary channels that connect ports to wells. The auxiliary channels are kept as short and straight as possible to reduce fabrication costs and to allow for more convenient fluid loading and dispensing. In addition, short straight channels reduce dispersion effects [9] and concentration non-uniformities [3] as well as the operating voltage required to drive the chip.

4.1 Routing Problem Description

We define our routing problem as follows: Given a compact arrangement $\mathcal{A}$ of subsystems, find a routing solution which consists of the set of planar paths $\mathcal{P}$ through $\mathcal{A}$ such that each port is connected to a single well and that the total length and number of bends in $\mathcal{P}$ is minimized.

The routing of auxiliary channels is similar to wire routing in VLSI circuit design, however, there are several complicating features. Most importantly, LoC devices are generally fabricated in a single layer to reduce fabrication costs and complexity [28]. This means that all auxiliary channels must be routed in a planar fashion and can not be routed above or below a subsystem. Furthermore, the assumptions that channels can *feed through* a subsystem or that ports may move along a subsystem edge [15], do not apply in our formulation. Additionally, auxiliary channels occupy significant space on the chip, and therefore, the exact placement of routes is critical to the quality of the overall design.

The single-layer routing problem has been investigated in the VLSI literature [15]. However, much of this work is not directly relevant to our problem because it either pertains to global, multi-terminal, or detailed routing in bounded regions (eg. channel routing). In global routing, the paths that wires must follow around obstacles on the chip are determined in a general way. A subsequent detailed routing stage places the wires precisely. Also, most VLSI routing algorithms construct multi-terminal routes, since single wires must often connect several terminals on the chip. Unlike channel-routing, which takes place in small bounded regions of the chip, we are interested in finding routing solutions through the whole chip. Finally, many general VLSI routing algorithms employ a sequential (rip-up and re-route) approach to construct a routing solution. This approach lacks feasibility guarantees and can become artificially over-constrained when poor routing choices are made early in the procedure.

We are currently developing a routing procedure that is specifically designed to solve our two-terminal, single-layer, detailed routing problem. In addition, we directly address length and bend minimization. In our approach, we hope to exploit the network integrality property of the *minimum-cost network flow* (MCNF) formulation to find routing solutions in a simultaneous fashion using linear programming (LP). Large-scale MCNF problems can be solved using standard LP solvers or tailored algorithms [29]. Furthermore, LP can quickly provide rigorous feasibility guarantees. However, the bend-minimization constraints discussed in the following section break the MCNF structure, thus

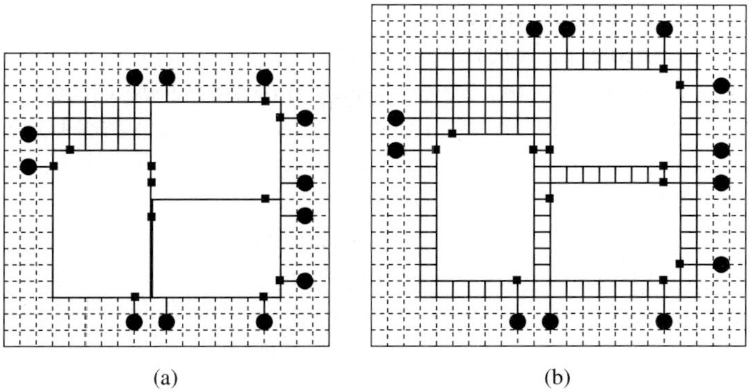

Figure 11-9. (a) Initially embedded arrangement. (b) Arrangement after two expansions. Black circles represent wells, black squares represent ports and solid lines represent available routing paths.

forcing us to use an integer programming (IP) approach. While many different IP formulations are possible to solve our routing problem, it is unclear which formulation will perform the best in all cases since there are non-trivial trade-offs between the number of constraints, the type of constraints and the number of variables.

We discuss our formulation, which is based loosely on the concept of MCNF, in the following section. We have successfully used this formulation to obtain good solutions for several small, compact test cases and are currently exploring algorithmic improvements that will allow us to deal with larger test cases.

4.2 Routing Problem Formulation

Our routing procedure is based on embedding a compacted arrangement of subsystems into a grid graph $G_{grid} =< \mathcal{V}, \mathcal{E} >$ as shown for a simple 3 subsystem example in Fig. 11-9(a). The black circles and squares in Fig. 11-9 represent wells and ports respectively and the solid grid-lines represent the complete set of all possible routing paths. The dashed lines are shown to illustrate valid positions for wells. Adjacent verticies, v_i and $v_j \in \mathcal{V}$, are spaced such that $(v_i, v_j) = PAD + \omega, \quad \forall (v_i, v_j) \in \mathcal{E}$ which represents the center-to-center distance between neighboring channels. All paths in $\mathcal{P}$ are implicitly feasible with respect to PAD and channel width ω. Every edge in G_{grid} is therefore of equal length and every vertex is equally spaced. G_{grid} is searched for a routing solution using our IP formulation. The action that is taken when no routing solution is found is discussed later.

Integer Programming Formulation. Since our IP formulation is based on MCNF, we construct a directed network $G_{net} =< \mathcal{V}, \mathcal{E} >$, which is formed by converting each undirected edge of G_{grid} into a forward directed edge and a backward directed edge. We arbitrarily assume that each port belongs to the set of sources, $s \subset \mathcal{V}$, and has a supply of 1, and that each well belongs to the set of sinks, $t \subset \mathcal{V}$, and has a demand of 1.

When an edge $(v_i, v_j) \in \mathcal{E}$ of G_{net} is occupied by a routing path, a binary variable $x_{(v_i,v_j)}$ representing the *flow* along that edge is equal to 1. Otherwise, if the edge is unoccupied, the flow is equal to 0. We then invoke continuity, planarity, and cycle elimination constraints for each $v_i \in \mathcal{V}$ to enforce feasible paths through G_{net} between ports and wells.

$$\sum_{(v_i,v_j)\in\mathcal{E}} x_{(v_i,v_j)} - \sum_{(v_j,v_i)\in\mathcal{E}} x_{(v_j,v_i)} = \begin{cases} 1 & \text{if } v_i \in s \\ -1 & \text{if } v_i \in t \\ 0 & \text{otherwise} \end{cases} \tag{11.8}$$

Equation (11.8) states that the flow into a node must equal the flow out of that node. A feasible solution consists of an unbroken path from a particular port in s to a particular well in t. Eq. (11.9) is used to enforce routing planarity.

$$\sum_{(v_i,v_j)\in\mathcal{E}} x_{(v_i,v_j)} + \sum_{(v_j,v_i)\in\mathcal{E}} x_{(v_j,v_i)} \leq 2 \tag{11.9}$$

Equation (11.9) states that the degree of every node in G_{net} must not be greater than 2. However, coupled with Eq. (11.8), it states that there can be at most one flow into and out of a given node. Eq. (11.10) is used to eliminate the existence of simple cycles between two adjacent nodes in G_{net}.

$$x_{(v_i,v_j)} + x_{(v_j,v_i)} \leq 1 \tag{11.10}$$

Recall that every undirected edge of G_{grid} is represented by a pair of directed edges in G_{net}. Eq. (11.10) is used to enforce that cycling does not occur for a pair of edges (v_i, v_j) and (v_j, v_i). At the optimum, Eq. (11.10) is not required since we are searching for minimum cost (shortest) paths. We include it to prevent cycles from existing if the search is terminated prematurely.

In general, there will exist a degenerate set of minimum-length paths between a particular port and a particular well. We can determine the straightest path between a port and well by essentially counting up the number of bends in each possible path and then choosing the path with the fewest bends. Bend-counting constraints are derived by examining connected vertex triplets v_i, v_j, and v_k. using logic propositions $\mathrm{P}_{(v_i,v_j)}$, $\mathrm{P}_{(v_j,v_k)}$ and $\mathrm{P}_{(v_i,v_k)}$ to indicate if a bend exists between v_i and v_k. [29, 30].

$$(\mathrm{P}_{(v_i,v_j)} \wedge \mathrm{P}_{(v_j,v_k)} \rightarrow \mathrm{P}_{(v_i,v_k)}) \vee (\mathrm{P}_{(v_j,v_i)} \wedge \mathrm{P}_{(v_k,v_j)} \rightarrow \mathrm{P}_{(v_i,v_k)}) \tag{11.11}$$

In Eq. (11.11), $P_{(v_i,v_k)}$ will be true if there is a flow along path (v_i, v_j, v_k) or (v_k, v_j, v_i). An equivalent math programming constraint, shown in Eq. (11.12), can be derived from Eq. ((11.11)).

$$\begin{aligned} x_{(v_i,v_j)} + x_{(v_j,v_k)} + x_{(v_j,v_i)} + x_{(v_k,v_j)} - 1 \leq y_{(v_i,v_k)} \\ \text{where, } y_{(v_i,v_k)} \geq 0 \end{aligned} \tag{11.12}$$

In Eq. (11.12), $y_{(v_i,v_k)}$ will equal 1 when a path through (v_i, v_j, v_k) or (v_k, v_j, v_i) exists, otherwise $y_{(v_i,v_k)}$ equals 0.

We have created a general method for deriving bend counting equations like the one shown in Eq. (11.12) based on the structure of the adjacency matrix $\boldsymbol{A}$ corresponding to G_{net}. $\boldsymbol{A}$ is a square symmetric matrix that defines the complete connectivity of G_{net}. We decompose $\boldsymbol{A}$ into a horizontal connectivity adjacency matrix $\boldsymbol{A}^{hor}$, and a vertical connectivity adjacency matrix $\boldsymbol{A}^{ver}$, such that $\boldsymbol{A} = \boldsymbol{A}^{hor} + \boldsymbol{A}^{ver}$.

Our algorithm for generating the set of bend counting constraints for a given G_{net} is shown below.

1 Initialize: $i = 1$, $\boldsymbol{n} = \boldsymbol{m} = \emptyset$

2 Scan row i of $\boldsymbol{A}^{hor}$ and $\boldsymbol{A}^{ver}$ for non-zero entries.

3 Store the indices of the non-zero entries from $\boldsymbol{A}^{hor}$ in $\boldsymbol{n}$ and from $\boldsymbol{A}^{ver}$ in $\boldsymbol{m}$.

4 Construct the equations: $x_{(v_i,v_n)} + x_{(v_i,v_m)} + x_{(v_n,v_i)} + x_{(v_m,v_i)} - 1 \leq y_{(v_n,v_m)} \quad \forall v_n \in \boldsymbol{n},\ \forall v_m \in \boldsymbol{m}$.

5 Increment: $i = i + 1$, set: $\boldsymbol{n} = \boldsymbol{m} = \emptyset$, and goto step 2. Stop when all rows of $\boldsymbol{A}^{hor}$ and $\boldsymbol{A}^{ver}$ have been scanned.

The constraints shown in step 4 are added to Eqs. (11.8)–(11.10) to generate an IP that is capable of minimizing both channel length and the number of bends.

Equation (11.13) shows the objective function that minimizes channel length and reduces the number of bends in a routing solution.

$$\min \sum_{(v_i,v_j)\in\mathcal{E}} x_{(v_i,v_j)} + \sum_{(v_j,v_i)\in\mathcal{E}} x_{(v_j,v_i)} + \sum_{(v_j,v_k)\in\mathcal{B}} y_{(v_j,v_k)} \tag{11.13}$$

The first two terms in Eq. (11.13) determine the shortest path and the final term is a summation over the set of all bends $\mathcal{B}$ that exist for a particular routing solution $\mathcal{P}$. The third term effectively penalizes routes that contain bends.

4.3 Routing Solution Method

Figure 11-10 shows a flowchart of our procedure for routing compacted arrangements. First, an initial arrangement is legalized so that it can be embedded in

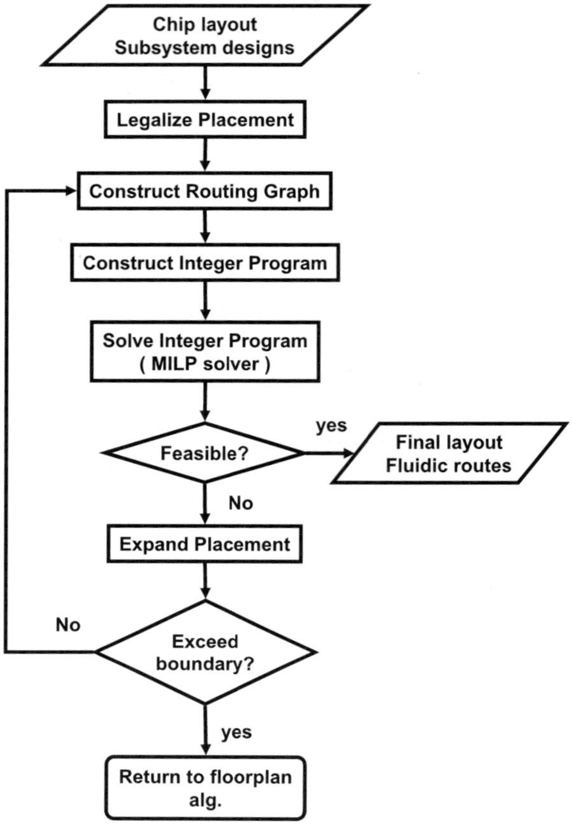

Figure 11-10. Flowchart for the routing algorithm.

G_{grid}. Next we construct G_{net} from G_{grid} and use it to generate the associated IP. The IP is solved using a standard commercial solver[2]. If the IP is feasible, then the IP solver searches for an optimal routing. If an optimal routing or a routing that meets a user specified tolerance is found, then the algorithm terminates successfully. If the IP is infeasible, the design is expanded to open up new routing paths and is searched again. If the expanded design exceeds the available chip area, the floorplan algorithm must be queried for a new arrangement that may have better routing characteristics. Two of the key steps of our routing procedure are discussed in the following sections.

Floorplan Legalization. Before an arrangement can be *placed* or embedded into G_{grid}, it must properly conform to the minimum allowable feature spacing requirement. If a subsystem's dimensions are not integer multiples of $PAD + \omega$, it cannot be properly embedded into the routing grid. Our floorplan algorithm does not guarantee that an arrangement will be embeddable. To account for this, we expand each subsystem in the x and y directions to the nearest integer multiple of $PAD + \omega$. For example, w_i is updated as follows: $w_i = \lceil w_i/(PAD + \omega) \rceil \cdot (PAD + \omega)$. The subsystem height h_i is treated in a similar fashion. The current Γ value of the floorplan is used to maintain the relative position of each subsystem. This procedure has a negligible effect on the performance and operation of each subsystem since the size-perturbation represents a relatively tiny fraction of the overall subsystem's size (typically $< 1.0\%$).

The wells on the chip's edges are updated in a similar fashion so that they too conform to valid grid points. As with other features on the chip, well edges are constrained to be no less than PAD apart. After legalization, port, well and subsystem locations conform to verticies in G_{grid}. Therefore, when a routing solution is found, it represents the exact location of auxiliary channels on the chip.

Floorplan Expansion. For a given G_{net}, the solution to the IP formulation will either yield a feasible routing solution, or it will declare the current routing problem infeasible. If the IP is infeasible, then we are guaranteed that no routing solution exists for the current instance of G_{net}.

We have developed a procedure that iteratively augments G_{grid} by expanding the arrangement to increase the number of possible routing paths in the grid. Each expansion adds one additional routing path between each subsystem. Fig. 11-9(b) shows a compacted arrangement that has undergone its first expansion.

An arrangement is expanded by first conceptually expanding each subsystem's width and height by $\gamma \cdot (PAD + \omega)$. We define γ as the number of times the arrangement is expanded. The arrangement's Γ is used to maintain the relative positions of each subsystem. Finally, each subsystem is shifted right and up from the origin by $\gamma \cdot (PAD + \omega)$. After each expansion, the arrangement is re-placed or re-embedded and a new G_{net} is generated. Expansions continue until either a feasible routing solution is obtained or until the placed design has become too large. Typically, arrangements require fewer than 5 expansions to achieve a feasible routing solution.

5. MULTIPLEX SYNTHESIS RESULTS

Here we discuss some of the key features and algorithmic behavior of the floorplanning and routing algorithms. The results are encouraging with respect to both computation time and solution quality, since our implementations have not yet been optimized. However, these results also indicate important areas for improvement and future study.

The result of our floorplanning algorithm for our 5 subsystem example is shown in Fig. 11-11(a). This result was obtained in approximately 5 minutes

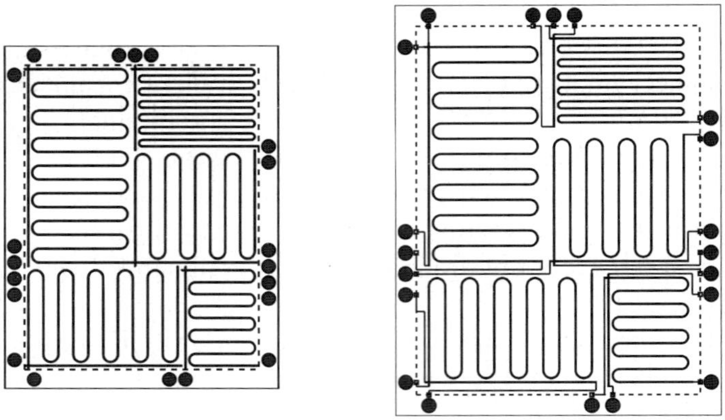

(a) (b)

Figure 11-11. Five-subsystem example: (a) Compacted arrangement (b) Final design (routed).

of CPU time[3]. At this point, the design is compact and all constraints are completely satisfied. The result of the routing algorithm for our 5 subsystem example is shown in Fig. 11-11(b). This design was routed in approximately 3 minutes of CPU time[3]. At this point, our design could be fabricated.

5.1 Floorplan Results

Figure 11-12(a) shows a placed and compacted design including well positions for a multiplexed chip containing 10 subsystems. The completed design shown in Fig. 11-12(b), was generated in 2 hours on a standard PC[3] with equal time allocated to the floorplanning and routing stages. While we are pleased with these results, significantly more time was required to generate the 10 subsystem design than was required for the 5 subsystem design. The compaction NLP presents the most significant computational bottleneck during the floorplanning stage.

It appears that 80% - 90% of the computation time used by our floorplanning algorithm is spent solving the compaction NLP. Fig. 11-13 shows the average change in objective value (*dotted line*) and the average function evaluation time (*dash & dotted line*) needed to solve the compaction NLP versus the number of subsystems in a given design. We created our test cases by first choosing typical physical property values, operating conditions and performance specifications for each subsystem. Each data point represents the average of 20 randomly generated instances. One standard deviation about the mean is also shown. As expected, the standard deviations of both time and objective value increase as

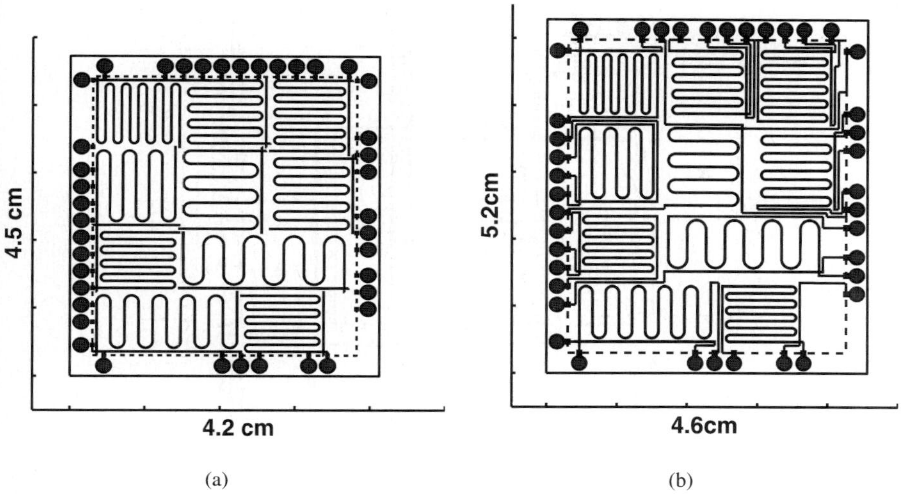

Figure 11-12. Ten-subsystem example: (a) Compacted arrangement (b) Final routed design.

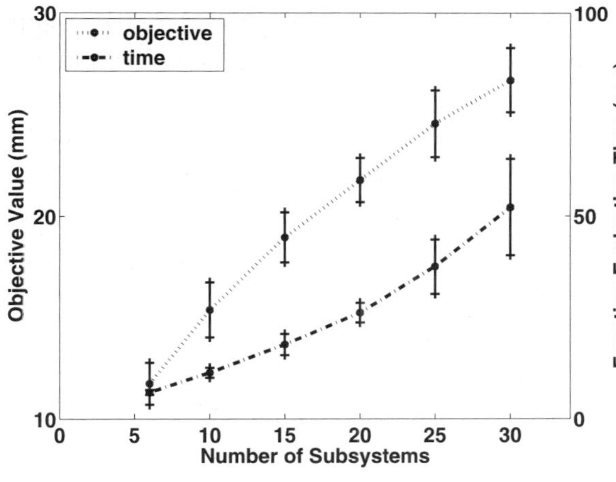

Figure 11-13. Scaling of NLP compaction subproblem in the floorplanning algorithm per size of design instance. Tests were performed on a standard PC (Intel Pentium 3 CPU: 1GHz CPU, 1Gb RAM).

the number subsystems increase. This is because the solution space grows in proportion to $(|\mathcal{N}|!)^2 \cdot 8^{|\mathcal{N}|}$.

Although the worst-case time complexity of typical NLP solvers is exponential, it appears that for the size problems we are currently investigating,

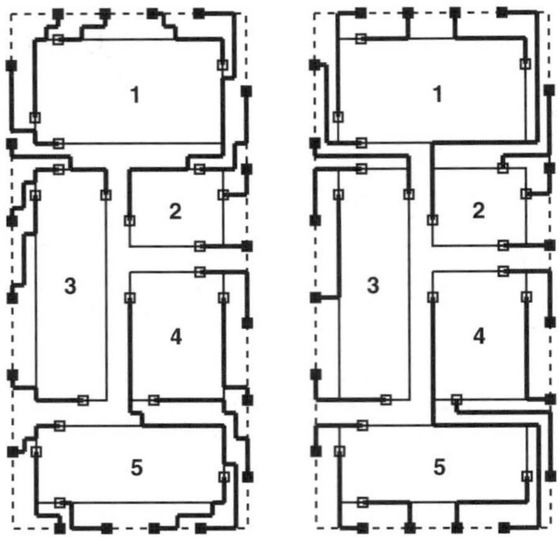

Figure 11-14. Example routing: (a) Minimum length only. (b) Minimum length and number of bends.

the average-case time complexity is not yet dominated by exponential behavior. This can be attributed to the success of our initialization procedure, which helps the NLP solver converge to locally optimal solutions in reasonable time. While the compaction NLP appears to scale acceptably, the SA must evaluate the NLP more often as designs become larger-scale to achieve satisfactory results. We have produced multiplexed serpentine designs containing up to 30 subsystems in under one day. As far as we know, these designs are significantly larger than any multiplexed serpentine designs presented in the literature. However, it is apparent that as problem size continues to increase, the computation time required by our algorithm will become prohibitive. We are currently investigating methods that will allow us to solve larger scale problems within a reasonable amount of time.

5.2 Microchannel Routing Results

Figure 11-14(a) shows a minimum length routing for a simple 5 subsystem floorplan. The thick lines are the auxiliary channels connecting ports to wells. This floorplan has been expanded two times (i.e. $\gamma = 2$). This means that there are at least three possible routing paths between each subsystem (each subsystem edge and one path in between). Notice that since there are many degenerate equal-length paths between ports and wells, the routing solution contains many bends. Fig. 11-14(b) shows a routing solution for the same example when both the channel length and the number of bends have been minimized. Typically,

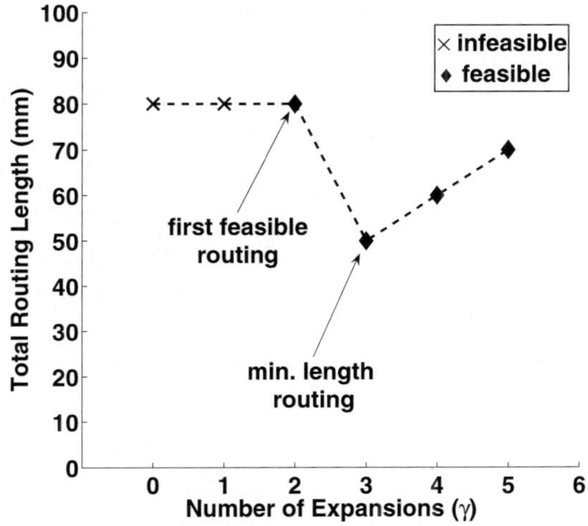

Figure 11-15. Total routing length vs. number expansions.

both the total channel length and the port-well assignment remain unaltered between the minimum length routing and the minimum length and bend routing (although this behavior is not guaranteed). This is because bend minimization represents only a small fraction of the overall objective value. In general, it does not cause re-routing to occur.

While bend minimization helps to reduce the number of non-unique solutions, it increases the optimality gap and makes the problem more computationally difficult. In our experiments, a standard IP solver[2] can typically prove a proposed routing grid infeasible in less than 30 seconds. Therefore the decision to expand the grid is made efficiently. When bend minimization is added to our problem, obtaining a proven optimal solution often becomes computationally prohibitive. Furthermore, no routing computation-time correlation can be drawn based on the number of subsystems in an arrangement since the routing problem scales with the number of grid edges and not the number of subsystems. Currently, we cope with these issues by allowing routing solutions to have a small relative optimality gap. Despite these complications, we have obtained reasonable quality solutions is under 1 hour of CPU time[3] for 10 subsystem designs.

Figure 11-15 shows the typical operation of our routing algorithm for the 5 subsystem example shown in Fig. 11-11(b). It illustrates the trade-off between minimum channel length and minimum design area. Fig. 11-15 depicts how iteratively expanding a design influences the total auxiliary channel length. For expansion numbers of $\gamma = 0$ and $\gamma = 1$, no feasible routing can be found. A feasible routing is obtained at $\gamma = 2$.

Normally, we would terminate the algorithm after optimizing the first feasible routing. However, if the design is expanded a third, fourth and fifth time, we notice that we achieve what appears to be a globally minimal length routing at $\gamma = 3$. This occurs because for low γ's, the routing grid is highly constrained. Therefore, the first feasible routing solution will often contain long channels. As γ increases, more possible routing paths are created and shorter routes become available. Eventually, continuing to increase γ will result in longer channels simply because the design is now larger and the channels must traverse a longer distance to connect ports to wells.

The minimum length solution at $\gamma = 3$ is also significant because this solution most closely approaches the solution obtained by the well placement algorithm described earlier in the Well Placement section. Recall that in the well placement algorithm, route continuity is enforced, but route planarity is not enforced. The well placement algorithm is in fact a lower bound relaxation of our IP routing formulation. We are currently investigating how to utilize this information to help us more quickly discover optimal or near optimal routing solutions.

6. CONCLUSION AND FUTURE WORK

We have presented a design automation methodology that is capable of generating full-custom multiplexed LoC's for CE applications in only minutes to hours. We have adapted and combined SoC circuit design techniques with optimal design methods to perform LoC design and layout. We are currently investigating methods for parallelizing our floorplanning algorithm to handle larger problem instances. We are also investigating ways to reduce the number of edges in routing grids while still maintaining a high level of connectivity. We believe that our experience with multiplexed design provides us with a tool set that can be extended to handle multi-function chips that contain many complex chemical operations.

ACKNOWLEDGMENTS

This research effort is sponsored by the Defense Advanced Research Projects Agency (DARPA) and U. S. Air Force Research Laboratory under agreement number F30602-01-2-0987 and by the National Science Foundation (NSF) under award CCR-0325344. The authors would like to thank members of the SYNBIOSYS group at Carnegie Mellon University.

Notes

1. ARKI Consulting and Development., CONOPT., http://www.conopt.com., Nov. 2004
2. CPLEX 9.0., ILOG Inc., http://www.ilog.com/products/cplex, Nov. 2004
3. Intel Pentium 4 CPU: 2GHz CPU, 1Gb RAM

REFERENCES

[1] D.R. Reyes, D. Iossifidis, A. Auroux, and A. Manz. Micro Total Analysis Systems. 1. Introduction, theory and technology. *Anal. Chem.*, 74:2623–2634, 2002.

[2] A.B. Kahng, I. Măndoiu, S. Reda, X. Xu, and A.Z. Zelikovsky. Evaluation of placement techniques for DNA probe array layout. In proceedings of *International Conference on Computer-Aided Design (ICCAD)*, pages 262–269, 2003.

[3] B. Mohammadi, I. Molho, and J.G. Santiago. Design of minimal dispersion fluidic channels in a CAD-free framework. In proceedings of *Center for Turbulence Research, Proceedings of the Summer Program 2000*, pages 49–62, 2000.

[4] A.J. Pfeiffer, J.D. Siirola, and S. Hauan. Optimal design of microscale separation systems using distributed agents. In proceedings of *Foundations of Computer-Aided Process Design (FOCAPD)*, pages 381–384, 2004.

[5] F. Su and K. Chakrabarty. Architectural-level synthesis of digital microfluidics-based biochips. In proceedings of *International Conference on Computer Aided Design (ICCAD)*, pages 223–228. 2004.

[6] S.V. Ermakov, S.C. Jacobson, and J.M. Ramsey. Computer simulations of electrokinetic injection techniques in microfluidic devices. *Anal. Chem.*, 72:3512–3517, 2000.

[7] M.G. Khaledi, editor. *High-Performance Capillary Electrophoresis*. John Wiley & Sons, 1998.

[8] R.S. Magargle, J.F. Hoburg, and T. Mukherjee. Microfluidic injector models based on artificial neural networks. in: *Design automation methods and tools for microfluidics-based biochips*, eds. K. Chakrabarty and J. Zeng, Springer, Norwell, MA, 2006.

[9] Y. Wang, Q. Lin, and T. Mukherjee. System-oriented dispersion models of general-shaped electrophoresis microchannels. *Lab-on-a-chip*, 4:453–463, 2004.

[10] Y. Wang, Q. Lin, and T. Mukherjee. A model for Joule heating-induced dispersion in microchip electrophoresis. *Lab-on-a-Chip*, 4:625–631, 2004.

[11] C.T. Culbertson, S.C. Jacobson, and M.J. Ramsey. Dispersion sources for compact geometries on microchips. *Anal. Chem.*, 70:3781–3789, 1998.

[12] A.J. Pfeiffer, T. Mukherjee, and S. Hauan. Design and optimization of compact microscale electrophoretic separation systems. *Ind. Eng. Chem. Res.*, 43:3539–3553, 2004.

[13] Y. Wang, Q. Lin, and T. Mukherjee. Composable modeling and simulation of electrokinetic Lab-on-a-Chips. in: *Design automation methods and tools for microfluidics-based biochips*, eds. K. Chakrabarty and J. Zeng, Springer, Norwell, MA, 2006.

[14] C.A. Emrich, H. Tian, I.L. Medintz, and R.A. Mathies. Microfabricated 384-lane capillary array electrophoresis bioanalyzer for ultrahigh-throughput genetic analysis. *Anal. Chem.*, 74:5076–5083, 2002.

[15] M. Sarrafzadeh and C.K. Wong. *An Introduction to VLSI Physical Design*. McGraw-Hill Inc., 1996.

[16] A.J. Pfeiffer, T. Mukherjee, and S. Hauan. Simultaneous design and placement of multiplexed chemical processing systems on microchips. In proceedings of*International Conference on Computer-Aided Design (ICCAD)*, pages 229–236. 2004.

[17] S. Lee and I.E. Grossmann. New algorithms for nonlinear generalized disjunctive programming. *Comp. Chem. Eng.*, 24(9–10):2125–2141, 2000.

[18] H. Murata, K. Fujiyoshi, S. Nakatake, and Y. Kajitani. VLSI module placement based on rectangle-packing by the Sequence Pair. *IEEE Trans. on CAD*, 15(12):1518–1524, 1996.

[19] H. Onodera, Y. Taniguchi, and K. Tamaru. Branch-and-bound placement for building block layout. In proceedings of *Design Automation Conference (DAC)*, pages 433–439, 1991.

[20] S. Sutanthavibul, E. Shragowitz, and J.B. Rosen. An analytical approach to floorplan design and optimization. *IEEE Trans. on CAD*, 10(6):761–769, 1991.

[21] S. Kirkpatrick, C.D. Galett, and M.P Vecchi. Optimization by simulated annealing. *Science*, 220, 4598:671–680, 1983.

[22] X. Hong, G. Huang, Y. Cai, J. Gu, S. Dong, C-K. Cheng, and J. Gu. Corner block list: an effective and efficient topological representation of non-slicing floorplan. In proceedings of *International Conference on Computer-Aided Design (ICCAD)*, pages 8–12. 2000.

[23] P.Y. Papalambros and D.J. Wilde. *Principles of Optimal Design: Modeling and Computation*. Cambridge University Press, 2000.

[24] J.G. Kim and Y.D. Kim. A linear programming-based algorithms for floorplanning in VLSI design. *IEEE Trans. on CAD*, 22(5):584–592, 2003.

[25] H. Murata and E.S. Kuh. Sequence-pair based placement method for hard/soft/preplaced modules. In proceedings of *International Symposium on Physical Design (ISPD)*, pages 167–172, 1996.

[26] X. Tang, R. Tian, and D.F. Wong. Fast evaluation of Sequence Pair in block placement by longest common subsequence computation. In proceedings of *Design Automation and Test in Europe (DATE)*, pages 106–111. 2000.

[27] INRIA Tropics Research Team. Tapenade. http://www.inria.fr, Nov. 2004.

[28] LioniX. LioniX MEMS foundry services. http://www.lionixbv.nl, November 2004.

[29] H.P. Williams. *Model Building in Mathematical Programming*. John Wiley and Sons, 4 edition, 1999.

[30] G. Nam, K. Sakallah, and R. Rutenbar. A boolean satisfiability-based incremental rerouting approach with application to FPGAs. In proceedings of *Design Automation and Test in Europe (DATE)*, pages 560–565, 2001.

Chapter 12

MODELING AND CONTROLLING PARALLEL TASKS IN DROPLET-BASED MICROFLUIDIC SYSTEMS

Karl F. Böhringer
University of Washington

Abstract: This paper presents general, hardware-independent models and algorithms to automate the operation of droplet-based microfluidic systems. In these systems, discrete liquid volumes of typically less than 1μl are transported across a planar array by dielectrophoretic or electrowetting effects for biochemical analysis. Unlike in systems based on continuous flow through channels, valves, and pumps, the droplet paths can be reconfigured on demand and even in real time. We develop algorithms that generate efficient sequences of control signals for moving one or many droplets from start to goal positions, subject to constraints such as specific features and obstacles on the array surface or limitations in the control circuitry. In addition, an approach towards automatic mapping of a biochemical analysis task onto a droplet-based microfluidic system is investigated. Achieving optimality in these algorithms can be prohibitive for large-scale configurations because of the high asymptotic complexity of coordinating multiple moving droplets. Instead, our algorithms achieve a compromise between high run-time efficiency and a more limited, non-global optimality in the generated control sequences.

Key words: Droplet-based microfluidic system, digital microfluidic system, parallel manipulation, lab on a chip (LOC)

1. INTRODUCTION

Advances in microfabrication and microelectromechanical systems (MEMS) over the past decades have lead to a rapidly expanding collection of techniques to build systems for the handling and analyzing of very small

K. Chakrabarty and J. Zeng (eds.),
Design Automation Methods and Tools for Microfluidics-Based Biochips, 301–327.

quantities of liquids (see, e.g., [1, 2]). These microfluidic systems typically consist of sub-millimeter scale components such as channels, valves, pumps, and reservoirs, as well as application-specific sensors and actuators. Microfluidic devices hold great promise, for example for novel fast, low-cost, portable, and disposable diagnostic tools. Applications include the massively parallel testing of new drugs, the on-site, real-time detection of toxins and pathogens, and PCR (polymerase chain reaction) for DNA sequence analysis. They usually operate with continuous flows of liquids, in analogy to traditional macro-scale laboratory set-ups, and can integrate all functionality into a complete *lab-on-a-chip (LOC)* or *bio-system-on-a-chip (bioSOC)*.

More recently, an alternative LOC approach has gained momentum using individual droplets, with volumes usually in the sub-microliter range. In these *droplet-based microfluidic systems,* droplets are generated, transported, merged, analyzed, and disposed on planar arrays of addressable cells; therefore they are also sometimes called *discrete* or *digital microfluidic systems*, and conveniently abbreviated *DMFS*. This architecture for microfluidic systems is attractive because of (a) greater flexibility – analyte handling may be reconfigured simply by re-programming rather than by changing the physical layout of the microfluidic components; (b) high droplet speeds – reportedly up to 25cm/s [3, 4]; (c) no dilution and cross-contamination due to diffusion and shear-flow; and (d) the possibility for massively parallel operation.

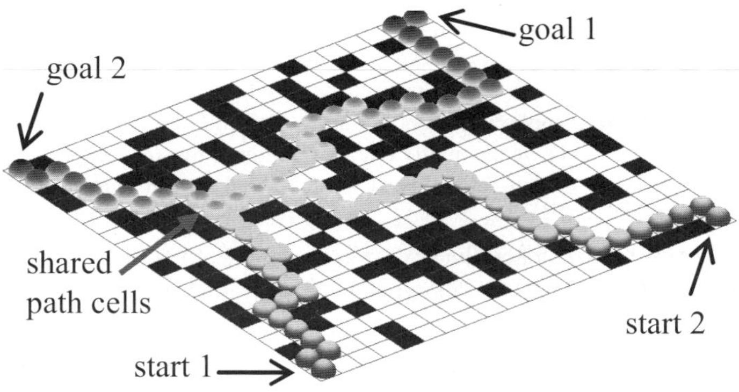

Figure 12-1. Two droplets moving in parallel on a DMFS consisting of a 20×20 array with obstacle cells (marked black). The droplets start from cells (1,1) and (20,1) and move to cells (20,20) and (1,20), respectively. Change in droplet color indicates the elapsed time. The droplets share cells (5,13) and (5,14) on their path but their coordinated schedule prevents any conflicts.

The DMFS approach assumes that it is advantageous to shift complexity from microfluidic hardware to control software. Therefore, for a DMFS to live up to its promise, it must be accompanied by a complementary set of software tools such that its usage can be largely automated. This includes software that helps the user to map a biochemical analysis protocol onto a given DMFS; as a specific subproblem, algorithms that automatically plan and schedule routes for simultaneous droplet motion are required. Fig. 12-1 shows a schematic example where two droplets move in parallel across a DMFS while circumnavigating numerous obstacles. Developing the formalisms, models, and control strategies for such automated droplet manipulation tasks is the goal of this paper.

Processing large numbers of discrete droplets simultaneously on an integrated microchip indicates a similarity to electronic digital circuits, giving rise to *microfluidic circuits* [5, 6]. This analogy also suggests that algorithms for layout, routing, and scheduling of droplet paths in a DMFS are computationally expensive, i.e., NP-hard.

Thus, this paper is organized as follows. Section 2 reviews background material on DMFS hardware, and discusses related work in control algorithms. Section 3 introduces a formal DMFS model and problem specification. Section 4 presents algorithms for coordinating parallel droplet motion on a DMFS, and investigates trade-offs between run-time efficiency and optimality. Section 5 extends these algorithms to allow for changes of the droplet type during DMFS operation, and develops an approach to automatically transform a laboratory protocol into a sequence of DMFS tasks. Section 6 concludes the paper with a summary and an outlook on future work.

2. RELATED WORK

Transferring a laboratory task such as DNA analysis, clinical diagnostics, or detection and manipulation of bio-molecules into a lab-on-a-chip system is a complex endeavor that can involve multiple challenges: the design of microfluidic hardware including sensing and actuation mechanisms for liquid analytes; the use of specialized techniques and materials such as modification and functionalization of surfaces with monolayers or antibodies; and the development of algorithms for layout and control of massively parallel microfluidic circuits.

Lab-on-a-chip design and manufacture has become an extensive research area with dedicated conferences (e.g., [7]) and journals (e.g., [8, 9]). This paper, however, focuses on the software aspects, and assumes a device model that abstracts away from details of the physical implementation. Here, we discuss very briefly important aspects of droplet-based microfluidics that are relevant to motivate and justify our modeling assumptions.

2.1 Droplet Transport Techniques

The most successful conventional droplet based system in the life sciences is arguably the fluorescence activated cell sorter (FACS) [10-12], a machine that can sort droplets containing single cells at rates well above 100kHz. It generates charged droplets, analyzes them in free flight via a laser fluorescence detection system, and sorts them accordingly via a modulated electrostatic field. Lab-on-a-chip FACS systems exist but so far work at much lower processing rates [13-15].

In micro-scale lab-on-a-chip systems, droplets can be moved across a planar surface effectively with a variety of techniques, including electric fields (e.g., [3, 16-19]), surface acoustic waves (e.g., [20-22]), thermocapillary and Marangoni effects (e.g., [23, 24]), electrochemical surface modulation (e.g., [25]), conformational changes in molecular surface layers (e.g., [26]), or gradients in surface chemistry (e.g., [27, 28]) and texture (e.g., [29, 30]). For this paper, droplet transport with high speed, accuracy, and full software control is essential, making electric fields the most suitable approach; hence we briefly discuss the two main techniques in this realm, dielectrophoresis and electrowetting [31, 32].

2.1.1 Dielectrophoresis

In dielectrophoresis (DEP), neutrally charged objects are first polarized by an electric field, and then experience a net force due to the field. This force can only be non-zero if a field gradient exists, i.e., the positively and negatively polarized regions of the object occupy areas of different field strengths. If the object has stronger polarization than the surrounding medium then it is pulled towards the areas of higher field strength (this is called positive DEP), but if the surrounding medium has higher polarization, then the object is pushed towards areas of lower field strength (negative DEP). DEP can be considered the electrostatic analogy of induced magnetism. Common examples for DEP are charged clothes that attract (neutral) lint particles. More information on dielectrophoresis can be found, e.g., at [33]. A DMFS system employing DEP with more than ten thousand array elements was demonstrated in [34].

2.1.2 Electrowetting

Electrowetting on dielectric (EWOD) exploits the decrease of contact angle that an aqueous droplet on a dielectric surface experiences when exposed to an electric field. If the field is localized at only one side of the droplet, then the difference in contact angle causes a pressure differential in the droplet, which drives it towards the region of higher field strength. Electrowetting and its applications in microfluidics have been investigated by several groups, including [3, 16, 17, 35, 36].

2.2 Droplet Transport Planning and Scheduling

Finding the optimal plan to generate, store, move, merge, split, and dispose multiple droplets on a droplet-based microfluidic system combines general path planning and scheduling with the more application-specific task of analyte droplet handling. Various researchers have studied parts of the overall problem and have shown important results on algorithmic solutions and their computational complexity.

One possible approach to this problem can be taken when the paths of the droplets are considered given a priori. This assumption leads to a scheduling problem, where the array cells en route are the limited resource that must be shared among different droplets. Griffith and Akella [37] show a solution with standard optimization tools guided by some user input, building on related work in coordinating multiple articulate robots [38, 39]. Many more references to related work in the areas of robot motion planning, flexible manufacturing systems, queuing theory and networking are also given in [37].

A related technique was used by Ding, Zhang, et al. [5, 6, 40] who attack the problem from the VLSI design perspective. As in [38, 39], this approach leads to an integer programming formulation. Both groups show NP-hardness of the scheduling problem even for fixed droplet routes.

VLSI circuit routing techniques could also be employed, which address the path planning problem but do not apply directly to the inherently two-dimensional layout of the droplet-based microfluidic platform.

This paper takes a different approach, by permitting the droplet paths to be chosen freely (except for constraints defined by the microfluidics hardware). Each droplet is interpreted as a point robot moving in a discrete two-dimensional configuration space. Under this assumption, path planning of the droplets becomes a motion planning problem with multiple moving robots. Erdmann and Lozano-Pérez showed in 1987 that this problem is NP-hard, but presented an algorithm that may find a good solution in polynomial time [41]. Their approach assigns priorities to each robot (droplet) and generates paths successively, starting with the highest priority robot. Lower priority robots consider higher priority robots as time-varying obstacles that must be avoided. The algorithm is not complete, and generated solutions depend on the priority ranking of the robots and may not be optimal.

In [42], this author described the problem as a graph search, and suggested search techniques such as A*. Even though this brute-force approach, unlike the other work mentioned above, guarantees optimality and completeness, it is not practical for larger scale problems because of its computational complexity, which is exponential in the number of moving droplets. Reference [43] introduced a formal problem definition and showed initial results with a more efficient approach based on Erdmann's algorithm [41].

3. DMFS FORMAL HARDWARE SPECIFICATION

Let us briefly review the most important physical properties and design parameters of a droplet-based microfluidic system (DMFS). Motivated by these characteristics, we can then develop an abstract DMFS model that captures the essential operational features without depending on specific implementation details.

3.1 DMFS Design Specifications

- Layout: Typically, a DMFS consists of a planar, rectangular array A with $m \times n$ cells (but, e.g., an arrangement of hexagonal cells would also be possible).
- Control circuitry: Various addressing schemes are possible to activate individual cells in a DMFS. In different physical implementations of DMFS, we can distinguish, for instance, individually addressable electrodes for each cell (e.g., [36]), or simpler row/column addressing (e.g., [44, 45]). For the latter, entire rows and columns are activated, and the droplet is attracted to a neighboring cell $A(x,y)$ only if it lies at the intersection of active column x and row y.
- Parallelism: The DMFS controller may be capable of simultaneous activation of more than one cell, which will allow simultaneous motion of multiple droplets. The total number of addressable cells may be limited by a number significantly smaller than $m \times n$.
- Location of cells with special functions: Droplet generators, reservoirs, cells for merging and splitting of droplets, sensors, waste, etc. may

require dedicated cells with special embedded hardware. These cells may not be available when planning a droplet path across the array.

These specifications provide a physical framework within which a DMFS can operate. Based on this framework, we can establish a formal description of the problem of controlling droplets in a DMFS. Once a sufficiently general DMFS model exists, we can investigate algorithmic solutions at an abstract level without worrying about the varying details of specific hardware implementations.

3.2 Abstract DMFS Specification

A droplet-based microfluidic system is specified by the droplets on the DMFS array, the DMFS hardware itself, and the task to be performed.

3.2.1 Droplets

Droplets are described by their type $\boldsymbol{T}$ and their volume $\boldsymbol{V}$. We are assuming here that all droplets in the DMFS have the same volume, except when two droplets have been merged. Therefore, we require that a merge operation is always immediately followed by a split operation that restores the original droplet volumes.

The droplet type $\boldsymbol{T}$ is a subset of all elementary droplet types, which we describe in general as a set $\mathrm{T} = \{T_1, T_2, T_3, \ldots\}$; thus, $\boldsymbol{T}$ is an element of the power set of T, $\boldsymbol{T} \in \Pi(\mathrm{T})$. For example, if T_1 represents "deionized water", T_2 "methanol", and T_3 "isopropanol", then a droplet of type $\boldsymbol{T} = \{T_1, T_3\}$ describes a mixture of DI water and IPA. Note that this convention provides a simple representation of mixed droplets, but does not keep track of sample concentrations. If needed, different concentrations could be represented as different elementary types.

3.2.2 DMFS Arrays and Tasks

The DMFS consists of an array A of $m \times n$ cells. Each cell in the array is either empty or occupied, which we represent by specifying its droplet type $\boldsymbol{T}$. Thus, the DMFS can be described by $A(x,y) = \boldsymbol{T}_{x,y}$ for $(x,y) \in \{1\ldots m\} \times \{1\ldots n\}$ and $\boldsymbol{T}_{x,y} \subseteq$ Ṭ. As a special case, $\boldsymbol{T}_{x,y} = \varnothing$ indicates an empty cell. We call $A \in \Pi(\text{Ṭ})^{m \times n}$ *the state of the DMFS.*

The location of a droplet can be specified by the pair $(x,y) \in \{1\ldots m\} \times \{1\ldots n\} = \boldsymbol{C}$; thus, $\boldsymbol{C}$ is the configuration space [46] of a single droplet, and $\boldsymbol{C}^d$ is the configuration space of d droplets, which we also call *the droplet placement of the DMFS.*

Time is assumed to be a discrete counter $t \in \{0, 1, 2, \ldots\}$, i.e., transitions in the array occur in integer time steps from t to $t+\Delta t$, where $\Delta t = 1$ unless noted otherwise. We write A_t to refer to the state of the array at a specific time t.

At this point, we can already outline the definition of a DMFS task: given a start state $A_s \in \Pi\,(\mathrm{T})^{m\times n}$ and a goal state $A_g \in \Pi\,(\mathrm{T})^{m\times n}$, we need to find a timed sequence of valid transitions that results in the desired droplet motions from A_s to A_g. Various kinds of transitions exist; they include simple droplet transport from cell to cell, but also droplet generation, disposal, merging, and splitting. In addition to motion, droplets may also be modified by operations on cells that change their type. All these operations are chosen from the following list of valid droplet transitions, which are usually associated to specific cells or groups of cells on the array:

- Droplet generation: For $(x,y) \in \boldsymbol{C}$ and some $\boldsymbol{T} \in \Pi\,(\mathrm{T})$, a droplet is generated at coordinate (x,y) if $A(x,y) = \varnothing$ at time t and $A(x,y) = \boldsymbol{T}$ at time $t+\Delta t$.
- Disposing: Definition analogous to droplet generation.
- Moving: Let (x,y) and $(x',y') \in \boldsymbol{C}$ and $|x-x'| + |y-y'| = 1$ (i.e., $A(x,y)$ and $A(x',y')$ are directly adjacent). At time t, $A(x,y) = \boldsymbol{T}$ and $A(x',y') = \varnothing$ and at time $t+\Delta t$, $A(x,y) = \varnothing$ and $A(x',y') = \boldsymbol{T}$.
- Merging: Let (x,y), (x',y'), and $(x'',y'') \in \boldsymbol{C}$ such that (x',y') and (x'',y'') are directly adjacent to (x,y) but not adjacent to each other. At time t, $A(x,y) = \varnothing$, $A(x',y') = \boldsymbol{T}_1$, and $A(x'',y'') = \boldsymbol{T}_2$, and at time $t+\Delta t$, $A(x,y) = \boldsymbol{T}_1 \cup \boldsymbol{T}_2$ and $A(x',y') = A(x'',y'') = \varnothing$, where $\boldsymbol{T}_1 \cup \boldsymbol{T}_2$ is the droplet type that results in merging droplet types $\boldsymbol{T}_1$ and $\boldsymbol{T}_2$.
- Splitting: Definition analogous to merging.
- Checking: For $(x,y) \in \boldsymbol{C}$, we require that a droplet remains at $A(x,y)$ from time t to time $t+\Delta t$. This allows, for example, sensing operations to be performed that neither change the location nor the type of the droplet (e.g., fluorescence detection).
- Changing: For $(x,y) \in \boldsymbol{C}$, we define a function $f\colon \Pi\,(\mathrm{T}) \rightarrow \Pi\,(\mathrm{T})$ such that $A(x,y) = \boldsymbol{T}_1$ at time t and $A(x,y) = \boldsymbol{T}_2$ at time $t+\Delta t$, and $f(\boldsymbol{T}_1) = \boldsymbol{T}_2$. This allows transition operations that modify the droplet type but not its location (e.g., heating/cooling for PCR).
- Blocking: For $(x,y) \in \boldsymbol{C}$, we define a set of forbidden droplet types $\boldsymbol{\Phi}_{x,y} \subseteq \mathrm{T}$ that are not allowed on $A(x,y)$. In particular, if $\boldsymbol{\Phi}_{x,y} \equiv \mathrm{T}$ then $A(x,y)$ is blocked for all droplets.

Finally, valid placement and motion of droplets on the array is subject to constraints:

- Placement: To avoid accidental merging of droplets, at least one empty cell is required between two occupied cells at all times, i.e., for any (x,y) and $(x',y') \in \boldsymbol{C}$ with $A(x,y) \neq \varnothing$ and $A(x',y') \neq \varnothing$, $|x - x'| > 1$ or $|y - y'| > 1$.

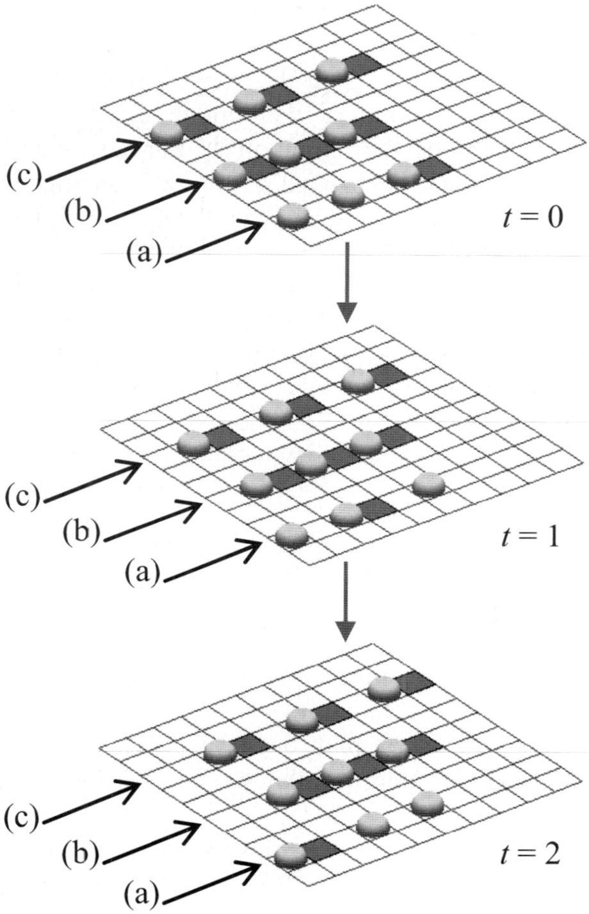

Figure 12-2. Parallel droplet transitions: Droplets (blue) and their activated neighbor cells (red squares) are shown at the instant when motion is commencing. The transitions in rows (a) and (c) are valid, but invalid in row (b) because during these transitions, two of the droplets have more than one activated neighbor cell, which could lead to unintentional splitting or merging.

- Parallel transitions: The previous constraint on placements must in particular also hold during transitions, i.e., for all pairs of droplet placements across the transition interval $[t,t+\Delta t]$ (see Fig. 12-2), except when merging or splitting is intended.

4. DROPLET PATH PLANNING

This section focuses on a central task in the control of DMFS: generating efficient paths for multiple droplets that move from a given start configuration A_s to a desired goal configuration A_g. For now, we require that the types of the droplets remain unchanged during the transition from A_s to A_g (this constraint will be removed in Section 5). We will first give a simple, complete algorithm based on A* search, but find that its computational complexity is very high (exponential in the number of droplets). We then present a more efficient algorithm for the *DMFS motion planning problem* that trades off completeness for faster execution times, while maintaining some "local" optimality guarantees.

4.1 Basic A* Search

This approach maintains a graph data structure to keep track of the droplet locations in the DMFS array. At any given time *t*, the state of the DMFS is described by A_t and identified with a node in this graph. A transition between two states A_t and $A_{t+\Delta t}$ defines a directed edge; this transition must conform with the conditions set forth in Section 3.2 above. Finding an optimal control strategy to transform start state A_s into goal state A_g then becomes a standard graph search problem: the shortest path between nodes A_s and A_g can be determined, e.g., using the A* algorithm known from artificial intelligence programming [47].

The A* algorithm outlined below maintains two lists of states, *Open* and *Closed*, which keep track of nodes that still need to be explored, and nodes that have already been processed, respectively. For each node, we maintain its predecessor *p*, the cost incurred *g* (i.e., number of transitions from A_s), the cost remaining *h* (i.e., number of transitions to A_g), and the total cost $f=g+h$. As has been widely discussed in the literature, *h*, which is not known in advance, can be estimated with an "admissible" heuristic function. The Manhattan metric provides such an admissible cost estimate, i.e., if droplet *i* at time *t* is at $(x_{t,i}, y_{t,i})$ and its goal is $(x_{g,i}, y_{g,i})$ then $h(t)$ can be estimated as $\Sigma_i |x_{g,i} - x_{t,i}| + |y_{g,i} - y_{t,i}|$.

```
Algorithm 1: A* for droplet path planning
Input: start state A_s, goal state A_g
Output: shortest path from A_s to A_g

Open ← { A_s };
Closed ← ∅;

while Open ≠ ∅ begin
  o ← pop state with smallest f from Open;
  Q ← list of all valid motion transitions from o;            // Line 5
  for each q in Q begin
    q.g ← o.g + 1;                                 // q is one step beyond o
    q.h ← distance estimate from q to A_g;
    q.f ← q.g + q.h;
    q.p ← o;                           // keep track of path from A_s via o to q
    if q = A_g, return q;                          // goal found, success
    if not (∃ q' ∈ Open such that q' = q and q'.f < q.f)
    and not (∃ q' ∈ Closed such that q' = q and q'.f < q.f)
    then add q to Open;                    // found new state q to be explored
  end
  add o to Closed;                               // finished exploring node o
end
return ∅;                                       // search exhausted, failure
```

Figure 12-3 shows a simple example where two droplets swap their position while avoiding an obstacle. The A* algorithm is guaranteed to always find an optimal solution if one exists, and indicate failure otherwise. However, the downside of this approach is its high asymptotic complexity. Suppose the number of droplets is d. In the simplest case, all are of the same type $\boldsymbol{T}_0$. Then the number of different placements of droplets on the array is $\binom{mn}{d}$, which for modest numbers $m=n=10$ and $d=10$ yields more than 1.7×10^{13} possibilities. If all droplets are of distinct type $\boldsymbol{T}_1 \ldots \boldsymbol{T}_d$, this number increases by $d!$ (to $\approx 6.3\times10^{19}$). One might hope that in practice, most of these choices need not be explored. However, at each step, d droplets offer up to $4d$ choices to be moved, assuming 4 neighbor cells per droplet. Thus, finding a strategy with s steps could mean checking up to $(4^d)^s$ choices or risk missing the solution, resulting again in astronomical numbers even for $s<10$.

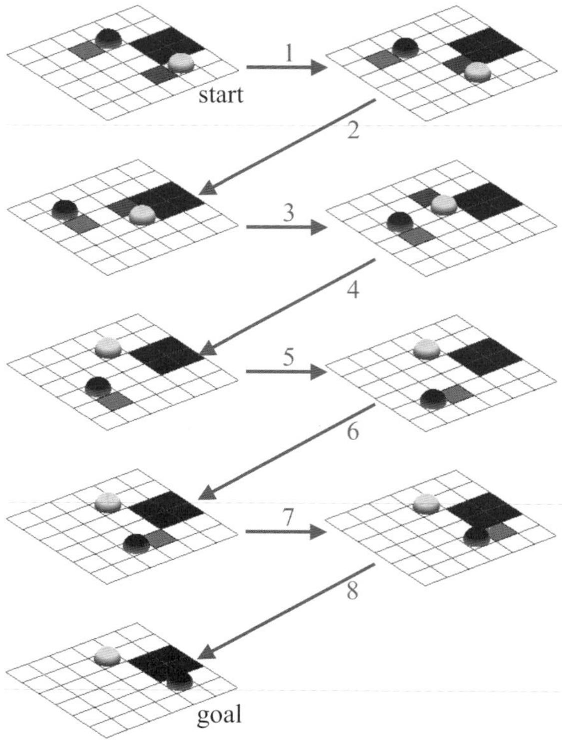

Figure 12-3. Two droplets moving simultaneously on a 6×6 DMFS array while avoiding an obstacle (black cells). The two droplets start at cells (5,2) and (4,5) and trade their places in 8 parallel transitions. The activated neighbor cells for their next transitions are also shown.

We conclude that the search graph explored with the A* algorithm has O((*mn*)!) nodes and a branching factor of O(4*d*), leading to a run-time complexity exponential in *d*, which is prohibitive for any non-trivial array size with more than a few droplets.

4.2 Prioritized A* Search

The discussion above has shown that droplet path planning for DMFS has two main aspects: generating efficient droplet path plans, and finding efficient algorithms to generate these plans. Erdmann and Lozano-Pérez's [41] NP-hardness results for coordinating multiple moving objects indicate that compromises need to be made to obtain practical solutions, and completeness or optimality in motion plans has to be traded off with efficiency in plan generation. They propose to impose a priority order on the moving objects, and sequentially find "locally" optimal solutions. In our case, the order can be assigned at random, or based on application-specific guidelines (e.g., water may have lower priority than droplets containing expensive or volatile compounds):

Algorithm 2: Prioritized A* for droplet path planning
Input: start state A_s, goal state A_g, priority order for droplets
Output: path from A_s to A_g

$S \leftarrow \varnothing$; // partial prioritized solution
for all droplets *i* in decreasing priority order **begin**
call Algorithm 1 to determine an optimal path for droplet *i* while considering all droplets with higher priority as moving obstacles and ignoring all droplets with lower priority;
if solution for droplet *i* exists
then add solution to *S*;
else return $\varnothing$; // failure
end
return *S*; // success

Figure 12-1 was generated using this algorithm. It eliminates the exponential complexity in *d*, where *d* is the number of droplets in the DMFS. Instead, the prioritized algorithm is linear in *d*. As a trade-off, (a) it is no longer complete: existing solutions may be missed; and (b) the solution may not be "globally" optimal: while each droplet *i* finds a "locally" optimal path among the moving droplets of higher priority, the complete solution will in general depend on the priority order and not be "globally" optimal. Thus, as was pointed out in [41], selecting the priority order can greatly influence the final solution. For instance, if a short path is important for a specific droplet type, then it should receive high priority. However, total run time is dominated likely by low priority droplets, since they may take convoluted paths to circumnavigate all higher priority droplets. A good heuristic for assigning priorities will take these points into account, as well as other, application-specific factors. For example, droplets whose type appears frequently on the DMFS could be assigned lower priorities than rare droplet types, because it is likely that one of the abundant droplets is already close to a desired destination.

4.3 Parallel Droplet Motion

The algorithms given so far are able to generate plans with simultaneous motion of multiple droplets. Beside the physical limitations to parallelism discussed with Fig. 12-2, the DMFS control hardware may impose additional constraints. For example, [44, 45] describe a DMFS with simpler row/column addressing, where a droplet moves to a neighboring cell $A(x,y)$ only if it lies at the intersection of activated column x and row y. Such conditions are encoded in line 5 of Algorithm 1:

" $Q \leftarrow$ list of all valid motion transitions from o; "

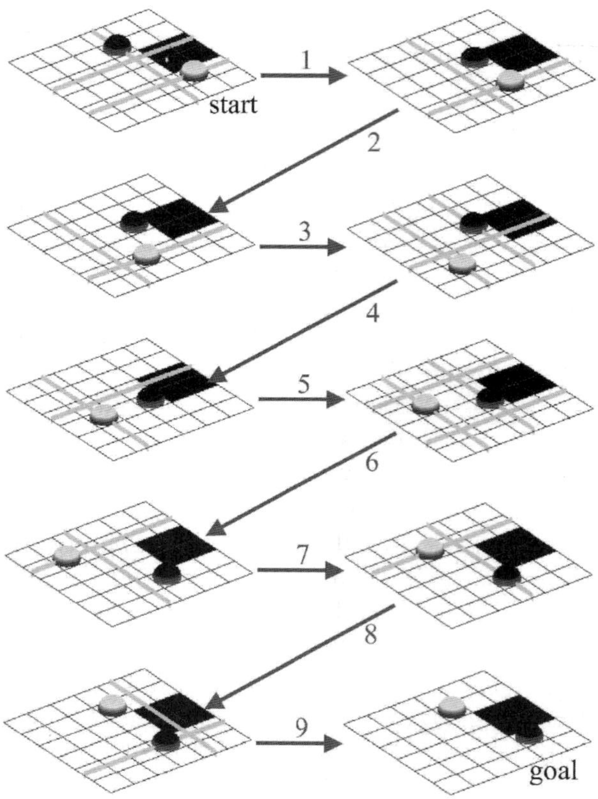

Figure 12-4. Two droplets trading places as in Fig. 12-3, but here droplets move only to neighbor cells whose row and column has been activated (indicated by a green line). An optimal strategy now requires 9 steps. Note that even though parallel droplet motion occurs in several steps, transitions 1 and 4 in Fig. 12-3 would not be possible with this addressing scheme.

Generation of this list of transitions must be implemented depending on the hardware specifications. Fig. 12-4 shows an optimal solution for the same start and goal states as in Fig. 12-3 but with this more limited row/column addressing scheme.

4.4 Duplicate Droplet Types

An important special case occurs when multiple droplets in the DMFS have the same droplet type. This is a likely scenario in practice, especially in DMFS with large numbers of droplets. In this case, there is no unique mapping between droplets in A_s and A_g (or with any intermediate state A_t). This complicates the calculation of the cost estimate h, but can also provide for more efficient plans by choosing opportune droplets that are closest to their respective goals.

Suppose we are given two sets of d droplet placements, S_1 and $S_2 \in \boldsymbol{C}^d$, and all droplets have the same type $\boldsymbol{T}$. We can find the minimum cost match between S_1 and S_2 efficiently (in analogy to bipartite graph matching [48]) by a greedy algorithm that sequentially matches up coordinate pairs with minimal Manhattan distance until all coordinates are paired up. This pairing leads to a monotone underestimate of the actual cost, and can thus be used as an admissible estimate for h. With this addition, both the basic and the prioritized A* algorithm for droplet path planning can efficiently handle inputs with duplicate droplet types.

5. DMFS TASK PLANNING

The previous section addressed the *DMFS motion planning problem*. However, transitions of the droplet types (due to mixing or other processing as discussed in Section 3) are essential parts of DMFS operation. Thus, this section extends the previously introduced algorithms to the *general DMFS planning problem*, which allows all remaining droplet transitions listed in Section 3.2, including merging, splitting, and changing of droplet type. This ultimately leads to the much broader question of how to transform a general laboratory protocol into a specific sequence of commands that can be executed on a DMFS.

5.1 Basic Graph Search

A straightforward algorithm to solve the general DMFS planning problem can be derived from Algorithm 1, where we can again modify line 5 to allow the complete set of transitions listed in Section 3.2, including in particular also changes of droplet type. However, this causes some immediate problems: (a) the number of possible transitions from each state (i.e., the branching factor in the search graph) becomes very large; (b) it is difficult to find an admissible heuristic for the A* algorithm, causing it to degenerate into breadth-first-search; in combination, this would result in very inefficient searches. These problems would apply equally to a modified Algorithm 2.

5.2 DMFS Task Protocols

To develop a more useful algorithm, it is important to keep in mind that the tasks to be executed here typically are laboratory protocols. Thus, it is reasonable to assume that the user (e.g., a chemical engineer or a researcher in molecular biology) has carefully worked out the individual steps in this protocol, and identified the intermediate products that are being generated during its execution. With this additional input, we can find efficient algorithms to perform these tasks on a DMFS, while leaving the task design to a knowledgeable human operator.

We now introduce a very simple *DMFS task language*; the user of a DMFS specifies the tasks to be executed in this language, based on the laboratory protocol for the process of interest. Our Algorithms 3 and 4 then interpret this task description and translate it into actual DMFS commands.

DMFS Task Language

// Textual description of DMFS protocol

// $x \in \{1\ldots m\}$; $y \in \{1\ldots n\}$; Δx, $\Delta y \in \{0, 1, \ldots\}$; $t \in \{1, 2, \ldots\}$

// $\boldsymbol{T} \in \Pi(\mathrm{T}\ )$; $f: \Pi(\mathrm{T}\ ) \rightarrow \Pi(\mathrm{T}\ )$

// *id* is an arbitrary textual identifier for a cell

in *x y* ***T*** *id* [**time** *t*]
out *x y* ***T*** *id* [**time** *t*]
waste *x y id*
mergesplit *x y* Δx Δy ***T*** *id* [**time** *t*]
check *x y* ***T*** *id* [**time** *t*]
change *x y f id* [**time** *t*]
block *x y* ***T*** *id*
connect *from-id to-id*

The statements in this language correspond to the DMFS array transitions listed in Section 3.2 with the following additional explanations:

x and y are the cell coordinates in the array. In general, we assume that transitions happen on a *single* array cell, except merge/split operations, which may require larger cells (specified by Δx and Δy such that $x+\Delta x \leq m$ and $y+\Delta y \leq n$).

- A droplet of type $\boldsymbol{T_d}$ is allowed on a cell with specified type $\boldsymbol{T}$ only if $\boldsymbol{T_d} \subseteq \boldsymbol{T}$.
- **in**, **out**, **mergesplit**, **check**, and **change** have an optional argument **time** t with default value $t = 1$ that specifies the time required for the transition.
- "**waste** *x y id*" is a short form for "**out** *x y* T *id*" which implies that the droplet type does not matter because the droplet will be discarded.
- "**block** *x y* ***T*** *id*" prohibits any droplet of type $\boldsymbol{T_d} \subseteq \boldsymbol{T}$.
- "**connect** *from-id to-id*" implies a single droplet moving between the two specified cells.

Note that identifiers need not be unique. However, if multiple transitions have the same identifier, then they belong to the same *cell group* and must describe the same transition. For example, we can write "**in** 1 1 H_2O DI-input" and "**in** 3 1 H_2O DI-input" to specify two cells (1,1) and (3,1) that provide a supply of DI water. Thus if we write "**connect** DI-input mix" then our algorithm will choose one of the DI water inputs to route a droplet to the cell with identifier "mix".

The following Fig. 12-5 gives a sample DMFS task input. Four input droplets of three different initial droplet types go through a sequence of merges, splits, type transitions, and checks, before finally reaching an output or waste cell. The user specifies these steps and their locations on the DMFS array. Our algorithms automatically generate the order of these operations, the selection of specific cells from cell groups, and the exact droplet paths and schedule.

In large DMFS with many moving droplets and many **in**, **out**, **mergesplit**, **check**, and **change** cells, choosing the locations where droplets are processed should also be automated. A greedy algorithm and simulated annealing are discussed in [49] to attack this NP-hard layout problem.

While this list of statements may look tedious, it is simply a textual description of a graph in which every node represents a transition (**in**, **out**, **waste**, **mergesplit**, **check**, **change**) at a specific location on the DMFS, and every edge corresponds to a droplet motion (**connect**). We call this directed graph, which specifies the flow of the droplets through the DMFS, the *task graph*. It gives a more intuitive representation of the DMFS task to be executed and will be discussed in the following section.

in 0 0 {R} *red*
in 0 2 {G} *green*
in 0 4 {B} *blue*
in 0 6 {B} *blue*
mergesplit 4 2 0 0 *magenta*
  connect *red magenta*
  connect *blue magenta*
mergesplit 4 5 0 0 *cyan*
  connect *green cyan*
  connect *blue cyan*
mergesplit 4 8 0 0 *pale*
  connect *magenta pale*
  connect *cyan pale*
change 14 1 {R,B}→{M) *modify*
  connect *magenta modify*
change 14 3 {G,B}→{C} *change*
  connect *cyan change*
change 14 5 {R,G,B}→{W} *process*
  connect *pale process*
check 10 2 **all** *sensor*
check 10 5 **all** *sensor*
  connect *modify sensor*
  connect *change sensor*
out 20 0 **all** *out*
out 20 2 **all** *out*
out 20 4 **all** *out*
  connect *process out*
  connect *pale out*
waste 20 10 *trash*
  connect *sensor trash*
  connect *sensor trash*

Figure 12-5. Sample DMFS tasks. There are three cell groups: *blue*, *sensor*, and *out*, consisting of multiple cells with the same identifier and the same transitions (**in**, **check**, **out**, respectively). The keyword **all** indicates the entire set of droplet types **T**. Note that only two out of the three **out** cells will be used.

5.3 DMFS Planning Algorithm

The final part of this paper is dedicated towards translating a DMFS task description, given in the language from the previous subsection, into a sequence of commands that can be executed on the array. This algorithm will do the following: (1) Generate the task graph from the textual input. (2) Identify initial transitions (typically, **in** nodes) that do not have any incoming edges. (3) Assign levels to all nodes in the task graph according to their precedence relationships such that transitions on the same level can be executed in parallel.

Algorithm 3: Task Graph Generation
Input: DMFS task description *tasks*
Output: task graph G with level assignments

parse *tasks* and generate the corresponding task graph G;
$old \leftarrow \varnothing$;
$new \leftarrow$ all nodes in G;
$current \leftarrow$ all nodes in G that do not have predecessors;
$i \leftarrow 0$;
while $current \neq \varnothing$ **begin**
mark all nodes in *current* with level i;
add all nodes in *current* to *old*;
$current \leftarrow$ all nodes in *new* that have only predecessors in *old*;
remove all nodes in *current* from *new*;
$i \leftarrow i + 1$;
end
if $new = \varnothing$
then return G (with level numbers); // success
else return $\varnothing$; // failure

If the directed graph is acyclic then this algorithm finds a level assignment with a minimum number of levels (which we call l), thus maximizing the potential for parallel execution of the transitions represented by its nodes and edges. Note that these level assignments merely reflect precedence relationships, not actual execution times: droplet transitions on a specific level and droplet motions between levels may have varying transition times (and the latter are not yet known). Thus, faster droplets may have to wait until slower droplets are finished on each level.

Algorithm 3 assumes that there are no resource conflicts between droplets on any given level, i.e., no two transitions require the same cell on the DMFS array. If this cannot be guaranteed during the specification of the DMFS task, then the algorithm must be modified to assign conflicting transitions to different levels. See, e.g., [50] for a comprehensive approach to dealing with such resource constraints.

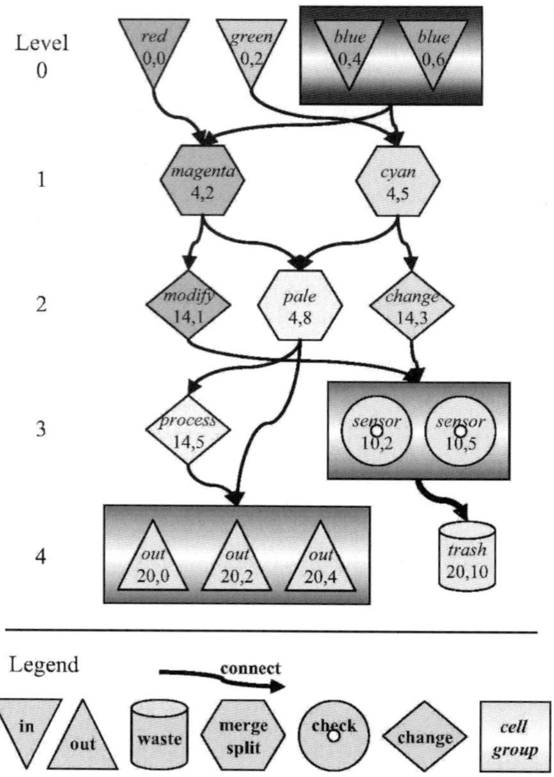

Figure 12-6. Task graph with level assignments generated from the task description in Fig. 12-5. Transitions on the same level can be executed in parallel. Note: (a) there are two droplets moving from the *sensor* cell group to *trash,* indicated by a double thickness arrow; (b) only two of the three output cells will be used.

Figure 12-6 shows the task graph generated by Algorithm 3 from the DMFS command input given in Fig. 12-5. From its level assignment, we can immediately generate array states A_{i-} and A_{i+} that correspond to each level $i \in \{0, \ldots, l\}$, such that A_{i-} and A_{i+} are the state of A immediately before and after the transitions of level i, respectively. Then, we can use Algorithm 2 to determine the droplet motions between arrays A_{i-1+} and A_{i-} for all $0 < i \leq l$:

<u>Algorithm 4: DMFS Planning</u>
<u>Input:</u> DMFS task description *tasks*
<u>Output:</u> task graph G and corresponding droplet motions S

```
G ← call Algorithm 3 with input tasks
if G = ∅ then return ∅;                         // no task graph exists, failure
S ← ∅;
for i ← {1… l} begin                  // l is the maximum level number of G
determine A_{i-1+} and A_{i–} , using G and S;
S_i ← call Algorithm 2 with start A_{i-1+} and goal A_{i–};
if S_i = ∅
then return ∅;                                 // no droplet path i exists, failure
else add S_i to S;                                       // droplet path i found
end
return G and S;                                                    // success
```

Table 12-1. Start, intermediate, and goal states generated from the task graph in Fig. 12-6. For each state A_i, $i \in \{0, \dots, 4\}$, droplet placements and their respective types are shown before and after the transition on level i. Three or four droplets move simultaneously during the four transitions from $A_{i\text{-}1+}$ to A_{i-} (for i>0), indicated by downward arrows.

DMFS Task States and Transitions					
State		Droplet Placements			
A_0		(0,0)	(0,2)	(0,4)	(0,6)
	+	{R}	{G}	{B}	{B}
		↓	↓	↓	↓
A_1		(4,2)	(4,5)	(4,2)	(4,5)
	–	{R}	{G}	{B}	{B}
	+	{R,B}	{G,B}	{R,B}	{G,B}
		↓	↓	↓	↓
A_2		(14,1)	(14,3)	(4,8)	(4,8)
	–	{R,B}	{G,B}	{R,B}	{G,B}
	+	{M}	{C}	{R,G,B}	{R,G,B}
		↓	↓	↓	•
A_3		(10,2)	(10,5)	(14,5)	(4,8)
	–	{M}	{C}	{R,G,B}	(R,G,B}
	+	{M}	{C}	{P}	{R,G,B}
		↓	↓	↓	↓
A_4		(20,10)	(20,10)	(20,4)	(20,2)
	–	{M}	{C}	{P}	{R,G,B}

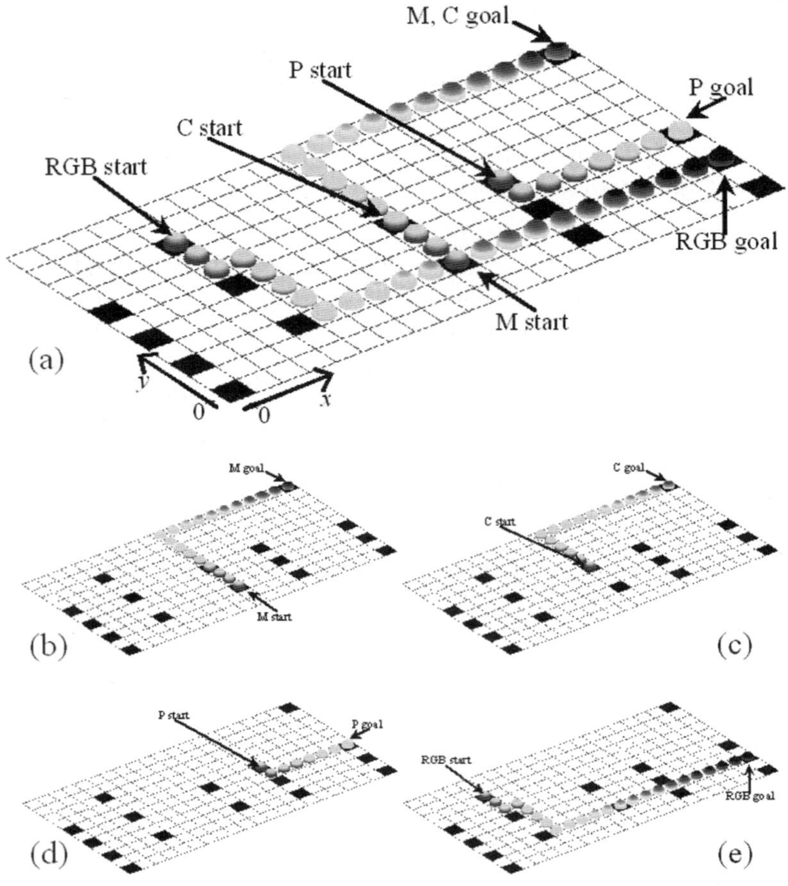

Figure 12-7. Simultaneous droplet motion during transition between states A_{3+} and A_{4-}. (a) shows all droplets, with change in color indicating progressing time. Cells with special functions are marked as black squares. (b), (c), (d), and (e) show individual droplet paths for the droplets of type {M}, {C}, {P}, and {R,G,B}, respectively. Note: In (b) and (c), droplet {C} follows the path of droplet {M} at a distance of 3 cells; in (e), the droplet circumnavigates the mergesplit cells at (4,2) and (4,5) but is allowed to pass over the sensor cell at (10,2).

Table 12-1 lists all the states and transitions generated by Algorithm 4 from the task graph in Fig. 12-6. Fig. 12-7 attempts to visualize parallel motion of multiple droplets on the DMFS for the transition from A_{3+} to A_{4-}.

Algorithm 3 is linear in the number of nodes and edges in the task graph. The complexity of Algorithm 4 is dominated by the calls to Algorithm 2, which occur $l - 1$ times total. These algorithms were implemented in Java. The total run times for the examples in this paper are in the millisecond range. The code is available upon request from the author.

6. CONCLUSION

This paper makes the following contributions: (1) A formal, hardware independent model of droplet-based microfluidic systems (DMFS). (2) Novel algorithms for motion and task planning with DMFS, leading to efficient (albeit not necessarily optimal or complete) solutions for coordinating large numbers of simultaneously moving droplets on a two-dimensional array. (3) An approach to automate the transition from general laboratory protocols to DMFS control command sequences. (4) Results using an implementation of these algorithms in Java.

The developed models and algorithms are "modular", such that results from the different sections are largely independent; e.g., DMFS *task* planning in Section 5 does not rely on a particular droplet *path* planning algorithm so some other algorithm could be readily substituted for prioritized A*. Similarly, the path planning algorithms from Section 4 could be applied to a different task planning algorithm.

Droplet manipulation based on electrowetting on arrays with up to hundred cells has been demonstrated by several groups (e.g., [3, 44, 51]), and an electrophoresis-based system with integrated CMOS addressing of tens of thousands of cells by [34]. The computational complexity for generating optimal droplet motion plans has been shown to be prohibitive even for much smaller systems. Thus, we have focused on finding an acceptable trade-off between efficiency and optimality.

A very different approach to this problem could be to limit droplet manipulation to a few standard, "pre-packaged" strategies. For example, on a 100×100 array, about 50 droplets could move in parallel across the array, followed by another wave of 50 droplets, etc., resembling a repetitive "peristaltic" motion [43]. However, in this case, the fundamental advantage of flexibility and reprogrammability in DMFS versus conventional (channel, valve, and pump based) microfluidic architectures is lost. In addition, the question still remains how to initially generate the "pre-packaged" strategies if they involve more complicated motion paths by many simultaneously moving droplets.

Other future work should explore the following directions:

1. Polynomial approximation algorithms exist for NP-hard problems (e.g., traveling salesman [52, 53]), which guarantee a tight limit on non-

optimality. If, e.g., a control strategy for a complex DMFS can be generated in polynomial time that is guaranteed to be at most twice as long as an optimal solution then this might be sufficient for most practical purposes.

2. While the (prioritized) A* algorithm has been effective in solving graph search problems, it is incomplete and worst-case exponential in the branching factor. More detailed benchmark tests could provide insights about scenarios where the algorithm fails to find solutions efficiently.
3. The optimal level number *l* produced by Algorithm 3 does not automatically imply maximal parallelism in droplet motion. Some nodes in the task graph can be assigned to a range of levels without affecting *l*, but varying level assignments may produce droplet motion plans with varying efficiency. For example, the droplet motion from (4,8) to (20,2) in Table 12-1 and Fig. 12-7(e) can be executed during transition $A_{2+} \rightarrow A_{3-}$ or during $A_{3+} \rightarrow A_{4-}$.
4. A related question is whether it is essential to allow parallel droplet motion in line 5 of Algorithm 1. An alternative approach would first generate plans without parallelism, and then post-process the generated plan to identify all droplet motions that could be executed in parallel.
5. More generally, it may be possible to improve the output of Algorithm 2 with some post-processing that locally improves the droplet motions.
6. The previous three points hint that our DMFS formalism could be developed much further. A general approach in this direction based on *state complexes* was given recently in [54], which presents efficient algorithms to detect and optimize parallelism.
7. As mentioned in Section 3.1, parallelism may be limited by the hardware controller to a number smaller than the total droplet count. This was not explicitly addressed in this paper, but could again appear as an additional constraint in line 5 of Algorithm 1.

ACKNOWLEDGMENT

The author thanks Srinivas Akella, Sankar Basu, Bruce R. Donald, Mike Erdmann, Rajinder Khosla, Eric Klavins, Xiaorong Xiong, and the anonymous reviewers for helpful insights and comments, Ji Hao Hoo and Tsung-Hao Suh for programming, Rohit Malhotra also for programming of an earlier software version, and Masayoshi Esashi, Hiroyuki Fujita and Osamu Tabata for their hospitality during a sabbatical visit at their laboratories.

Support for this project was provided in part by NSF SGER grant 0342632, NIH grant 1 P50 HG002360-01, and an invitational fellowship for research in Japan from the Japan Society for the Promotion of Science.

REFERENCES

1. Kovacs, G.T.A., *Micromachined Transducers Sourcebook*. 1998: McGraw-Hill.
2. Stone, H.A., A.D. Stroock, and A. Ajdari, *Engineering Flows in Small Devices: Microfluidics Toward a Lab-on-a-Chip*. Annual Review of Fluid Mechanics, 2004. **36**:381-411.
3. Moon, H., S.K. Cho, R.L. Garrell, and C.-J. Kim, *Low voltage electrowetting-on-dielectric*. Journal of Applied Physics, 2002. **92**(7):4080-4087.
4. Fair, R.B., V. Srinivasan, H. Ren, P. Paik, V.K. Pamula, and M.G. Pollack. *Electrowetting-based on-chip sample processing for integrated microfluidics*, in *IEEE International Electron Devices Meeting (IEDM)*. 2003.
5. Zhang, T., K. Chakrabarty, and R.B. Fair, *Integrated hierarchical design of microelectrofluidic systems using SystemC*. Microelectronics Journal, 2002. **33**:459-470.
6. Zhang, T., K. Chakrabarty, and R.B. Fair, *Design of Reconfigurable Composite Microsystems Based on Hardware/Software Codesign Principles*. IEEE Transactions on Computer-aided Design of Integrated Circuits and Systems, 2002. **21**(8):987-995.
7. *International Conference on Miniaturized Chemical and Biochemical Analysis Systems (microTAS)*. Annual.
8. *Sensors and Actuators B: Chemical*. Monthly, Elsevier.
9. *Lab on a Chip*. Monthly, Royal Society of Chemistry.
10. Shapiro, H.M., *Practical flow cytometry*. 1995, New York: Wiley.
11. Melamed, M.R., T. Lindmo, and M.L. Mendelsohn, *Flow cytometry and sorting*. 1990, New York: Wiley.
12. Crosland-Taylor, P.J., *A device for counting small particles suspended in a fluid through a tube*. Nature, 1953. **171**(4340):37-38.
13. Fu, A.Y., C. Spence, A. Scherer, F.H. Arnold, and S.R. Quake, *A microfabricated fluorescence-activated cell sorter*. Nature Biotechnology, 1999. **17**(11):1109-1111.
14. Krueger, J., K. Singh, A. O'Neill, C. Jackson, A. Morrison, and P. O'Brien, *Development of a microfluidic device for fluorescence activated cell sorting*. Journal of Micromechanics and Microengineering, 2002. **12**:486-494.
15. Tartagni, M., L. Altomare, R. Guerrieri, A. Fuchs, N. Manaresi, G. Medoro, and R. Thewes, *Microelectronic Chips for Molecular and Cell Biology*, in *Sensors Update*, H. Baltes, G.K. Fedder, and J.G. Korvink, Editors. 2004, Wiley-VCH. p. 156-200.
16. Beni, G. and M.A. Tenan, *Dynamics of electrowetting displays*. Applied Physics, 1981. **52**(10):6011-6015.
17. Pollack, M.G., R.B. Fair, and A.D. Shenderov, *Electrowetting-based actuation of liquid droplets for microfluidic applications*. Applied Physics Letters, 2000. **77**(11):1725-1726.
18. Jones, T.B., M. Gunji, M. Washizu, and M.J. Feldman, *Dielectrophoretic liquid actuation and nanodroplet formation*. Journal of Applied Physics, 2001. **89**(2):1441-1448.
19. Nanolytics, *www.nanolytics.com*.

20. Wixforth, A., *Verfahren und Vorrichtung zur Manipulation kleiner Flüssigkeitsmengen auf Oberflächen*, German Trademark and Patent Office. 2002, Advalytix AG, 85649 Brunnthal, DE: Germany.
21. Wixforth, A. and C. Gauer, *Mischvorrichtung und Mischverfahren für die Durchmischung kleiner Flüssigkeitsmengen*, in *European Patent Office*. 2004: European Union.
22. Wixforth, A., A. Rathgeber, C. Gauer, and J. Scriba, *Vorrichtung und Verfahren zur Vermessung kleiner Flüssigkeitsmengen und/oder deren Bewegung*, German Trademark and Patent Office. 2002, Advalytix AG, 80799 München, DE: Germany.
23. Kataoka, D.E. and S.M. Troian, *Patterning Liquid Flow at the Microscopic Scale.* Nature, 1999. **402**(6763):794-797.
24. Darhuber, A.A., J.P. Valentino, J.M. Davis, S.M. Troian, and S. Wagner, *Microfluidic actuation by modulation of surface stresses.* Applied Physics Letters, 2003. **82**(4):657-659.
25. Gallardo, B.S., V.K. Gupta, F.D. Eagerton, L.I. Jong, V.S. Craig, R.R. Shah, and N.L. Abbott, *Electrochemical principles for active control of liquids on submillimeter scales.* Science, 1999. **283**(5398):57-60.
26. Lahann, J., S. Mitragotri, T.-N. Tran, H. Kaido, J. Sundaram, I.S. Choi, S. Hoffer, G.A. So-morjai, and R. Langer, *A Reversibly Switching Surface.* Science, 2003. **299**(5605):371-374.
27. Chaudhury, M.K. and G.M. Whitesides, *How to make water run uphill?* Science, 1992. **256**(5063):1539-1541.
28. Daniel, S., S. Sircar, J. Gliem, and M.K. Chaudhury, *Ratcheting Motion of Liquid Drops on Gradient Surfaces.* Langmuir, 2004. **20**(10):4085-4092.
29. Sandre, O., L. Gorre-Talini, A. Adjari, J. Prost, and P. Silberzan, *Moving droplets on asymmetrically structured surfaces.* Physical Review E, 1999. **60**(3):2964-2972.
30. Shastry, A., M. Case, and K.F. Böhringer. *Engineering Surface Texture to Manipulate Droplets in Microfluidic Systems*, in *IEEE Conference on Micro Electro Mechanical Systems (MEMS)*. 2005. Miami Beach, FL.
31. Jones, T.B., J.D. Fowler, Y.S. Chang, and C.-J. Kim, *Frequency-Based Relationship of Electrowetting and Dielectrophoretic Liquid Microactuation.* Langmuir, 2003. **19**(18):7646-7651.
32. Zheng, J. and T. Korsmeyer, *Principles of droplet electrohydrodynamics for lab-on-a-chip.* Lab on a Chip, 2004. **4**:265-277.
33. Gascoyne, P.R.C., www.dielectrophoresis.org.
34. Fuchs, A., N. Manaresi, D. Freida, L. Altomare, C.L. Villiers, G. Medoro, A. Romani, I. Chartier, C. Bory, M. Tartagni, P.N. Marche, F. Chatelain, and R. Guerrie. *A Microelectronic Chip Opens New Fields in Rare Cell Population Analysis and Individual Cell Biology*, in *Micro Total Analysis Systems (MicroTAS)*. 2003. Squaw Valley, CA.
35. Paik, P., V.K. Pamula, and R.B. Fair, *Rapid droplet mixers for digital microfluidic systems.* Lab on a Chip, 2003. **4**:253-259.
36. Cho, S.K., H. Moon, and C.-J. Kim, *Creating, transporting, cutting, and merging liquid droplets by electrowetting-based actuation for digital microfluidic circuits.* Journal of Microelectromechanical Systems, 2003. **12**(1):70-80.
37. Griffith, E. and S. Akella. *Coordinating multiple droplets in planar array digital microfluidics systems*, in *Sixth Workshop on the Algorithmic Foundations of Robotics.* 2004. Utrecht, Zeist, The Netherlands.
38. Peng, J. and S. Akella. *Coordinating Multiple Robots with Kinodynamic Constraints along Specified Paths*, in *Workshop on the Algorithmic Foundations of Robotics (WAFR)*. 2002.

39. Akella, S. and S. Hutchinson. *Coordinating the Motions of Multiple Robots with Specified Trajectories*, in *IEEE International Conference on Robotics and Automation*. 2002. Washington D.C.
40. Ding, J., K. Chakrabarty, and R.B. Fair, *Scheduling of Microfluidic Operations for Reconfigurable Two-Dimensional Electrowetting Arrays*. IEEE Transactions on Computer-aided Design of Integrated Circuits and Systems, 2001. **20**(12):1463-1468.
41. Erdmann, M. and T. Lozano-Pérez, *On Multiple Moving Objects*. Algorithmica, 1987. **2**(4):477-521.
42. Böhringer, K.F. *Optimal Strategies for Moving Droplets in Digital Microfluidic Systems*, in *Seventh International Conference on Miniaturized Chemical and Biochemical Analysis Systems (MicroTAS'03)*. 2003. Squaw Valley, CA.
43. Böhringer, K.F. *Towards Optimal Strategies for Moving Droplets in Digital Microfluidic Systems*, in *IEEE International Conference on Robotics and Automation (ICRA)*. 2004. New Orleans, LA.
44. Fan, S.-K., P.-P.d. Guzman, and C.-J. Kim. *EWOD Driving of Droplet on NxM Grid Using Single Layer Electrode Patterns*, in *Solid-State Sensor, Actuator, and Microsystems Workshop*. 2002. Hilton Head Island, SC.
45. Fan, S.-K., C. Hashi, and C.-J. Kim. *Manipulation of multiple droplets on NxM grid by cross-reference EWOD driving scheme and pressure contact packaging*, in *IEEE International Conference on Microelectromechanical Systems*. 2003. Kyoto, Japan.
46. Lozano-Pérez, T., *Spatial planning: A configuration space approach*. IEEE Transactions on Computers, 1983. **C-32**(2):108-120.
47. Nilsson, N.J., *Principles of Artificial Intelligence*. 1982, Berlin Heidelberg New York: Springer Verlag.
48. Aho, A.V., J.E. Hopcroft, and J.D. Ullman, *Data Structures and Algorithms*. 2 ed. 1987, Reading, WA: Addison-Wesley.
49. Su, F. and K. Chakrabarty. *Design of fault-tolerant and dynamically-reconfigurable microfluidic biochips*, in *Design, Automation and Test in Europe (DATE)*. 2005.
50. Su, F. and K. Chakrabarty. *Architectural-level synthesis of digital microfluidics-based biochips*, in *IEEE International Conference on Computer Aided Design*. 2004.
51. Srinivasan, V., V.K. Pamula, M.G. Pollack, and R.B. Fair. *Clinical Diagnostics on Human Whole Blood, Plasma, Serum, Urin, Saliva, Sweat, and Tears on a Digital Microfluidic Platform*, in *Micro Total Analysis Systems (MicroTAS)*. 2003. Squaw Valley, CA.
52. Christofides, N. *Worst-case analysis of a new heuristic for the traveling salesman problem*, in *Symposium on New Directions and Recent Results in Algorithms and Complexity*. 1976. Orlando, FL: Academic Press.
53. Arora, S., *Polynomial Time Approximation Schemes for Euclidean Traveling Salesman and Other Geometric Problems*. Journal of the ACM, 1998. **45**(5):753-782.
54. Abrams, A. and R. Ghrist, *State Complexes for Metamorphic Robots*. The International Journal of Robotics Research, 2004. **23**(7-8):811-826.

Chapter 13

PERFORMANCE CHARACTERIZATION OF A RECONFIGURABLE PLANAR ARRAY DIGITAL MICROFLUIDIC SYSTEM

Eric J. Griffith
Data Visualization Group
Delft University of Technology
600 GA Delft, The Netherlands
E.J.Griffith@ewi.tudelft.nl

Srinivas Akella
Department of Computer Science
Rensselaer Polytechnic Institute
Troy, New York 12180, USA
sakella@cs.rpi.edu

Mark K. Goldberg
Department of Computer Science
Rensselaer Polytechnic Institute
Troy, New York 12180, USA
goldberg@cs.rpi.edu

Abstract: This chapter describes a computational approach to designing a digital microfluidic system (DMFS) that can be rapidly reconfigured for new biochemical analyses. Such a "lab-on-a-chip" system for biochemical analysis, based on electrowetting or dielectrophoresis, must coordinate the motions of discrete droplets or biological cells using a planar array of electrodes. We earlier introduced our layout-based system and demonstrated its flexibility through simulation, including the system's ability to perform multiple assays simultaneously. Since array layout design and droplet routing strategies are closely related in such a digital microfluidic system, our goal is to provide designers with algorithms that enable rapid simulation and control of these DMFS devices. In this chapter, we

K. Chakrabarty and J. Zeng (eds.),
Design Automation Methods and Tools for Microfluidics-Based Biochips, 329–356.

characterize the effects of variations in the basic array layout design, droplet routing control algorithms, and droplet spacing on system performance. We then consider DMFS arrays with hardware limited row-column addressing and develop a polynomial-time algorithm for coordinating droplet movement under such hardware limitations. To demonstrate the capabilities of our system, we describe example scenarios, including dilution control and minimalist layouts, in which our system can be successfully applied.

Keywords: Digital microfluidics, lab-on-a-chip, biochips, array layout, droplet routing, performance analysis, row-column addressing.

1. INTRODUCTION

Miniature biochemical analysis systems that use microfluidics technology have the potential to function as complete "lab-on-a-chip" systems. These systems offer a number of advantages, including reduced reagent requirements, size reduction, power reduction, increased throughput, and increased reliability. An important goal is to create reconfigurable and reprogrammable systems capable of handling a variety of biochemical analysis tasks.

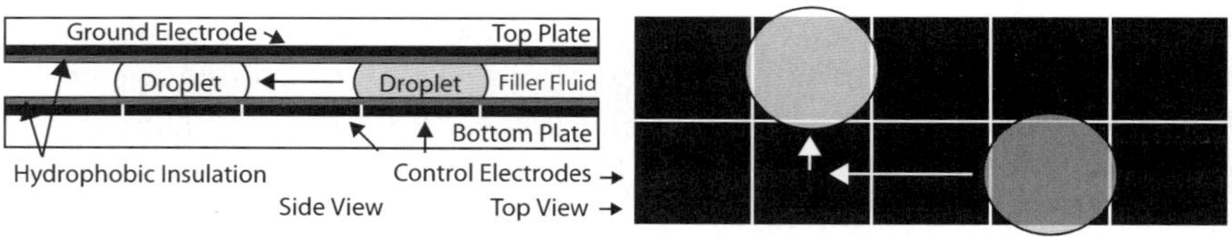

Figure 13-1. Droplets on an electrowetting array (side and top views). A droplet moves to a neighboring control electrode when the electrode is turned on. The electrode is turned off when the droplet completes its motion. Based on [29].

A promising new class of lab-on-a-chip systems are *digital microfluidic systems* (DMFS) that use phenomena such as electrowetting [31, 29, 8] and dielectrophoresis [22, 26]. Electrowetting-based microfluidic systems manipulate discrete droplets by modulating the interfacial tension of the droplets with a voltage [29]. Droplets have been moved at 12–25 cm/sec on planar arrays of 0.15 cm wide electrodes [14, 8]. Dielectrophoresis-based systems apply a spatially nonuniform electric field to actuate neutral charge particles [22, 26]. Arrays with 20 μm wide electrodes that manipulate biological cells have been demonstrated [16]. The ability to control individual droplets or biological cells on a planar array enables complex analysis operations to be performed in biochemical lab-on-a-chip systems (Figure 13-1). For example, they can be used to perform DNA polymerase chain reactions for DNA sequence analysis, to perform glucose assays, or to fuse biological cells with drug molecules. These systems have the potential to rapidly process hundreds or even thousands of samples on a single biochip. A key challenge in using digital microfluidic systems is developing computationally tractable algorithms to automate the simultaneous coordination of operations on a potentially large number of droplets or biological cells.

Our focus is the development of algorithms to automatically coordinate the transport and reaction operations on droplets or biological cells in a DMFS. We describe our approach in the context of droplet-based systems that use electrowetting; the same approach and algorithms may also be applied to dielectrophoresis-based systems that manipulate biological cells. The broad problem we are interested in is: *Given a chemical analysis graph describing the sequence in which chemicals should mix, coordinate the droplet operations on the DMFS array for a set of droplets so as to permit mixing with prescribed mix times while avoiding undesired contact between droplets.* Our approach to countering the complexity of this problem is to impose a virtual layout on the DMFS array and coordinate droplet operations by dynamically routing droplets to components in the layout. The layout permits us to abstract away from the underlying array hardware and provides additional structure that simplifies droplet coordination. We previously described this approach to creating a general-purpose DMFS (Griffith and Akella [18, 19]), which combines a semi-automated approach to array layout design using modular virtual components with algorithms for components to dynamically route the droplets. The resulting system has been simulated in software to perform analyses such as DNA polymerase chain reaction. The algorithms have been able to coordinate hundreds of droplets simultaneously and perform one or more chemical analyses in parallel.

In this chapter, we explore variations on the basic DMFS layout design and routing control for increased versatility and performance, and describe example scenarios in which our system can be applied. Since array layout design and droplet routing strategies are closely related in a reconfigurable DMFS, our goal is to provide designers with simulation tools for both rapid evaluation and real-time control of these DMFS devices. After summarizing our previous work in Section 3 to provide the background, we describe the effects on system performance of variations in design and control including different layout schemes, routing algorithms, and increased spacing between droplets in Section 4. We then develop a new approach to droplet coordination with limited row-column addressing in Section 5. We use a polynomial-time graph coloring algorithm to coordinate droplet movements under such hardware limitations. Finally, in Section 6, we outline two application scenarios involving droplet dilution control and minimal layouts to demonstrate the capabilities of our system.

2. RELATED WORK

Digital Microfluidic Systems: Digital microfluidic systems are a novel and emerging class of lab-on-a-chip systems. Most work in this area has focused on developing hardware to demonstrate the feasibility of this new technology. Pollack, Fair, and Shenderov [31] demonstrated rapid manipulation of discrete microdroplets by electrowetting-based actuation. Fair et al. [14] describe experiments on injection, dispensing, dilution, and mixing of samples in an electrowetting DMFS. Cho, Moon, and Kim [8] demonstrated creating, merging, splitting, and move operations using electrodes covered with dielectrics, and identified conditions under which these operations can be performed in an air environment. Fan, Hashi, and Kim [15] developed an orthogonal cross-reference grid of single layer electrodes to manipulate droplets with limited row-column addressing. Gong, Fan, and Kim [17] developed a portable digital microfluidics lab-on-chip platform using electrowetting. They use a time-multiplexed control scheme to control droplets with limited row-column addressing, where the number of steps is proportional to the number of array rows. Paik, Pamula, and Fair [29] studied the effects of droplet aspect ratios and mixing strategies on the rate of droplet mixing. Dielectrophoresis is another mechanism to actuate neutral charge particles and cells by applying a spatially nonuniform electric field [22, 26]. Jones et al. [22] demonstrated dielectrophoresis based liquid actuation and nanodroplet formation. Arrays with 20 μm wide electrodes that manipulate biological cells have been demonstrated [16].

More recently, work on DMFS has focused on applications. Srinivasan et al. [39] demonstrate the use of a DMFS as a biosensor for glucose, lactate, glutamate and pyruvate assays, and use it for clinical diagnostics on blood, plasma, serum, urine, saliva, sweat, and tears [40]. Pollack et al. [32] have demonstrated the use of electrowetting-based microfluidics for real-time polymerase chain reaction (PCR) applications. Wheeler et al. [46] demonstrate an electrowetting-based DMFS for analysis of proteins by matrix-assisted laser desorption/ionization mass spectrometry, for high-throughput proteomics applications.

Coordination of droplet operations and architectural design for DMFS, the topics most closely related to the current chapter, have been far less studied. In early work, Ding, Chakrabarty, and Fair [11] described an architectural design and optimization methodology for scheduling biochemical reactions using electrowetting arrays. They identified a basic set of droplet operations and used an integer programming formulation to minimize completion time. Droplet paths and areas on the array for storage, mixing and splitting operations are predefined by the user. Zhang, Chakrabarty, and Fair [47] describe hierarchical techniques for the modeling, design, performance evaluation, and optimization of microfluidic systems. They compared the performance of a continuous flow system and a droplet-based system and showed that the droplet-based system has a less complex design that provides higher throughput and processing capacity. Su and Chakrabarty [41] recently proposed architectural-level synthesis techniques for digital microfluidics-based biochips, and describe an integer programming formulation and heuristic techniques to schedule assay operations under resource constraints, prior to geometry-level synthesis. Our work is motivated by the above body of work, as well as the work of Böhringer [4, 5], who viewed each droplet in a DMFS as a simple robot that translates on an array and outlined an approach for moving droplets from start to goal locations, subject to droplet separation constraints, obstacles, and control circuitry limitations. He uses an A* search algorithm to generate optimal plans for droplets. To overcome the exponential complexity of this approach, he plans the droplet motions in prioritized order. However a DMFS must have additional capabilities, such as the ability to combine and split droplets as needed, sometimes with different mixing durations.

Multiple Robot Coordination: The coordination of droplets in a DMFS is closely related to multiple robot motion coordination, as pointed out above. Hopcroft, Schwartz, and Sharir [21] showed that even a simplified two-dimensional case of motion planning for multiple translating robots is PSPACE-hard. Erdmann and Lozano-Perez [13] developed a heuristic approach for planning the motions of multiple robots that orders robots by assigned priority and sequentially searches for collision-free paths; this approach was used by Böhringer [5]. Owing to the computational complexity of the multiple robot motion planning problem, recent efforts have focused on probabilistic approaches (Švestka and Overmars [44], Sanchez and Latombe [35]).

When the paths of the robots are specified, as in Ding, Chakrabarty, and Fair [11]'s DMFS model, a path coordination problem arises. Path coordination was first studied by O'Donnell and Lozano-Perez [28] for two robots. LaValle and Hutchinson addressed a similar problem in [24] where each robot was constrained to a C-space roadmap during its motion. Simeon, Leroy, and Laumond [37] coordinate over 100 car-like robots, where robots with intersecting paths are partitioned into smaller sets. Akella and Hutchinson [1] developed a mixed integer linear programming (MILP) formulation for the trajectory coordination of 20 robots by changing robot start times. Peng and Akella [30] developed an MILP formulation to coordinate many robots with simple double integrator dynamics along specified paths. Conflict resolution among multiple aircraft in a shared airspace (Tomlin, Pappas, and Sastry [43], Bicchi and Pallottino [3], Schouwenaars et al. [36]) is also closely related to multiple robot coordination.

Flexible Manufacturing Systems: Our approach to droplet coordination in a DMFS shares similarities with flexible manufacturing systems, where product assembly is like droplet mixing. One example is a reconfigurable, automated precision assembly system that uses cooperating, modular, robots [34]. Such systems have been modeled and analyzed using several techniques including Petri nets [10]. Of particular interest to flexible manufacturing systems is the issue of deadlock avoidance, which has been analyzed for certain classes of systems (Reveliotis, Lawley, and Ferreira [33], Lawley [25]).

Networking: We can view our DMFS as a network. This system differs from typical networking systems in nontrivial ways, including the fact that droplets cannot be dropped and that the system has multiple classes of nodes and operations. However techniques for network flow and rate control [42, 2] may be modified for a DMFS. Related research in networking includes work on hot-potato or deflection routing (Choudhury and Li [9], Busch, Herlihy, and Wattenhofer [7]) for different classes of networks, and work on rate control to ensure stability (Kelly, Maulloo, and Tan [23]).

3. SYSTEM OVERVIEW

In this section, we provide an overview of our system, previously described in [18, 19]. We create a general-purpose reconfigurable DMFS by first generating a *virtual layout* that logically partitions the array into virtual components that perform different functions, and then applying specialized algorithms for routing droplets to appropriate components. The layout is created by combining one or more modular *tiles* that each contain the same pattern of virtual components. Each *virtual component* is a logical grouping of cells that can perform one or more functions. A *cell* corresponds to an electrode of the array, and may have additional capabilities, such as the ability to optically sense droplets. We initially assume individual cells of the array are addressable by direct activation of individual electrodes. A droplet moves to a neighboring cell (electrode) when that electrode is activated; the electrode is turned off when the droplet completes its motion. We assume each droplet has a unit volume, except during mixing. Each mix operation is followed by a split operation, which is performed by simultaneously activating the two electrodes on either side of the droplet. Droplets are dynamically allocated to virtual components based on the operation (such as mixing or transport) to be performed on them. We adapt network routing algorithms to route the droplets to destination components in the layout. When the routing algorithms, provided with knowledge of the electrode addressing mechanism, are used as the software controller for a DMFS, the droplet motions can be downloaded to a microcontroller at each clock cycle. The microcontroller will activate the requested set of electrodes to enable droplet motion.

Our approach of imposing a layout on a digital microfluidic array to suit given chemical reactions is similar to programming a reconfigurable field programmable gate array (FPGA) [27]. However, unlike an FPGA, whose elements have distinct functions such as logic or routing, the interchangeable functionality of the DMFS cells permits instantaneous reconfigurations of the layout through just software changes. For example, a cell with a droplet transport function in one layout may be used for droplet mixing or sensing in another layout.

This DMFS is reconfigurable in several ways. In the simplest sense, it can be reconfigured to run a variety of analyses that require moving, mixing, and splitting of different types of droplets just by changing the types of the input droplets and their associated mixing operations. One or more of these reactions can also be run in parallel. This reconfigurability potentially requires no actual change of the layout, but just changes to inputs to the software. Second, the actual layout design itself can be modified by altering the number of tiles and their arrangement, the number of components in a tile and their arrangement, and the locations of the sources and the sinks. We can even partition a large array into multiple DMFS layouts. This type of reconfigurability offers control over the system performance, and supports a wider variety of biochemical analyses. Third, the system offers reconfigurability by the ability to introduce new component types such as droplet storage components or if supported by the array hardware, optical sensor components. This offers flexibility for tailoring to specific analysis needs and for future expansion. Finally, the system can easily incorporate changes to the droplet routing and scheduling algorithms to optimize performance.

3.1 Array Layout Design using Components

We partition the array into a set of "virtual" components, where each type of component performs a specific set of operations. This partitioning is enabled by the versatility of the array electrodes, which can perform droplet movement, merging, mixing, and splitting operations practically anywhere on the array. Each component controls droplets within its cells, and, by linking a sufficient set of components together, a DMFS can be created to perform one or more biochemical analyses. Figure 13-2 illustrates an example system comprised of six component types. These six virtual components (Figure 13-3) perform droplet transportation (street, connector, and intersection components) or droplet mixing, input, and output operations (work area, source, and sink components).

The Street Component: The street component is the general-purpose droplet transportation component. Streets are one-way to prevent two droplets from moving in opposite directions through the component.

The Connector Component: The connector component is a specialized version of a street component where a droplet only moves through a single cell. A droplet in a connector is adjacent to two components simultaneously.

The Intersection Component: The intersection components route droplets through the system, using the algorithms described in Section 3.2.

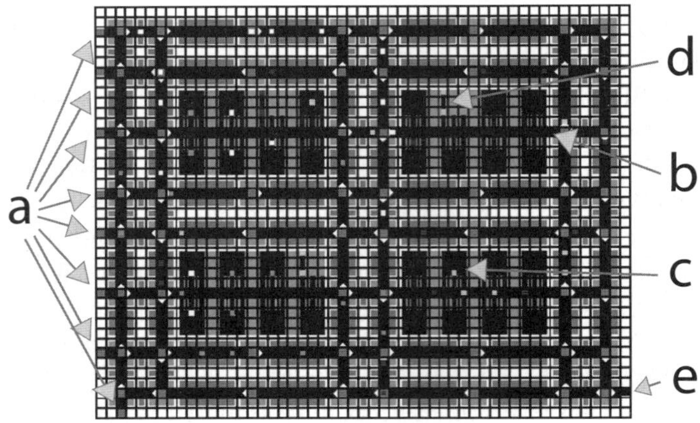

Figure 13-2. Array layout for the PCR analysis described in Section 3.3. Each cell of the array is represented by a square; arrowheads indicate valid droplet motion directions. On the left side of the array are **(a)** eight sources, which supply the input sample droplets to the system. There are **(b)** four work areas on the array, in which droplets are **(c)** mixed together and **(d)** split apart. In the lower right corner of the array is a **(e)** sink, which moves the droplets of the final products off the array.

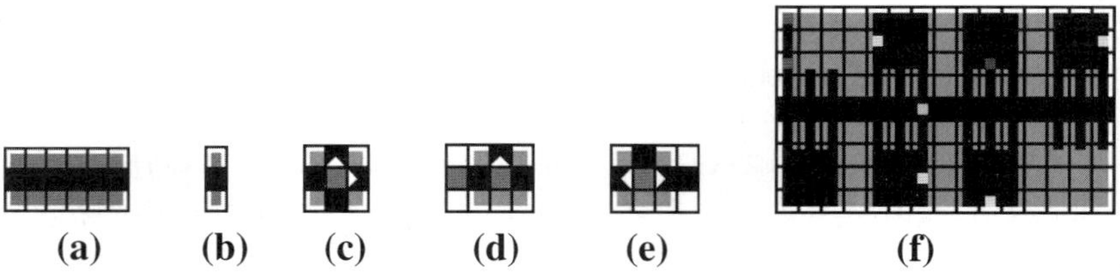

Figure 13-3. The components. **(a)** A street. **(b)** A connector component. **(c)** An intersection. **(d)** A source connected to an intersection. **(e)** A sink connected to an intersection. **(f)** An active work area, showing several mixing units with droplets (depicted as small squares).

The Work Area Component: The work area component is where mixing and splitting take place. Each work area has a transit area and multiple *mixing units*. Each mixing unit may function as a *mixer* and/or as a *splitter*. A work area can mix and split multiple droplets at the same time.

The Source Component: The source component represents an input point for droplets into the array.

The Sink Component: The sink component represents an output point for droplets from the array.

The layout is designed to have sufficient capacity to both transport droplets between components and to process droplets. We do this by first grouping one-way streets and intersections into two-way streets and rotaries (Figure 13-4). Then we couple this with a work area to form a pattern, shown in Figure 13-5, which can be tiled periodically to create the layout. The layout is completed with an alternating sequence of rotaries and streets along its upper and right edges. To generate the layout, the user must know the physical size of the array and specify the locations of sources and sinks. Our design can be expanded to accommodate new types of components for specific or general operations.

3.2 Droplet Destination Selection and Routing Algorithms

The core algorithms in our approach deal with deciding where to send droplets, and how to get them there. With these droplet destination selection and routing algorithms, we transform a set of interconnected components into a functional DMFS. The intersection components execute these algorithms to route droplets through the system.

Assigning a destination to a droplet depends on the droplet type and the available components. The droplet type determines whether it is to mix with another type of droplet in a work area or leave the array from a sink. An available work area is either one that has already had one of the two droplets for a mixing operation assigned and is requesting the other type, or one with free mixing units that can accept any type of droplet. Each available work area and sink adds itself to a (global) ordered list of components accepting droplets for operations. There is also a (global) ordered list of higher priority containing requests from work areas for specific droplet types required to complete a mix and split operation. Intersections assign work areas and sinks on a rotating basis, except when the second droplet in a mixing operation is being requested.

When a new droplet enters the system, or is created through a mixing operation, the droplet type determines the operation it is assigned. When the droplet enters an intersection, the intersection tries to find a destination component to send the droplet to by first checking the high priority list and then, if necessary, the low priority list. If any component is actively requesting that droplet type for its operation, the droplet is assigned to that component. Failing that, the

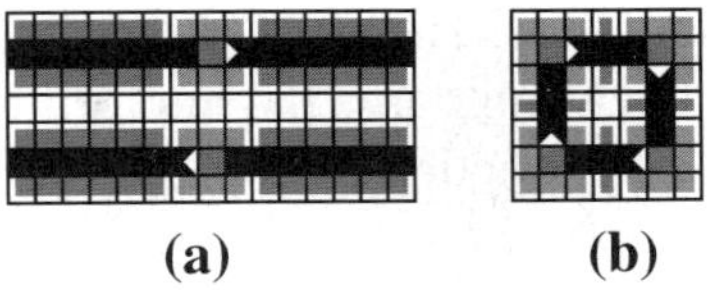

Figure 13-4. Simulating two-way transportation: **(a)** Two-way street **(b)** Rotary.

droplet is assigned to the first component that can accept droplets of its type. If no components are available to assign the droplet to, then the next intersection the droplet enters attempts to assign it a destination.

The droplet routing method we use can be viewed as a deflection routing variant [6] of the Open Shortest Path First (OSPF) network protocol [42]. When the system is initialized, each intersection uses Dijkstra's algorithm to compute a routing table, which maps the shortest legal path between the intersection and each component to a corresponding exit from which to leave the intersection.

At each clock cycle, the intersections are processed in a fixed order to select their droplet routing moves, as described in Section 3.3. Subsequently, synchronous motion of droplets is executed. If a droplet entering the intersection has no destination, then the intersection attempts to assign it one. If that fails, then the droplet is sent to a random, valid exit. For droplets with destinations, the intersection finds the destination component in its routing table and selects the exit that corresponds to the shortest path to the destination. If the droplet is able to move toward that exit, it does so. Otherwise, the intersection randomly chooses a valid exit for the droplet. If no viable exit is available, then the droplet waits.

3.3 A General-Purpose Digital Microfluidic System

We create a general-purpose DMFS by combining the component based layout design approach and droplet destination selection and routing algorithms. The basic layout is designed to handle a variety of analyses. Furthermore, the DMFS can be reconfigured by altering the number of mixing units in the work areas, the overall size of the layout, the locations of the sources and sinks, and the types of analyses it is to perform. The layout approach presented here can be extended to produce new layouts, and to incorporate new types of components into the system. To fully define the system, the user must specify additional parameters based on the chemical analyses to be performed, including the type of droplets introduced at each source, when and how often they are produced, the types of droplets to send to the sinks, and information about the various intermediate operations to perform on the droplets. A complete example 2×2 layout with eight sources and one sink can be seen in Figure 13-2.

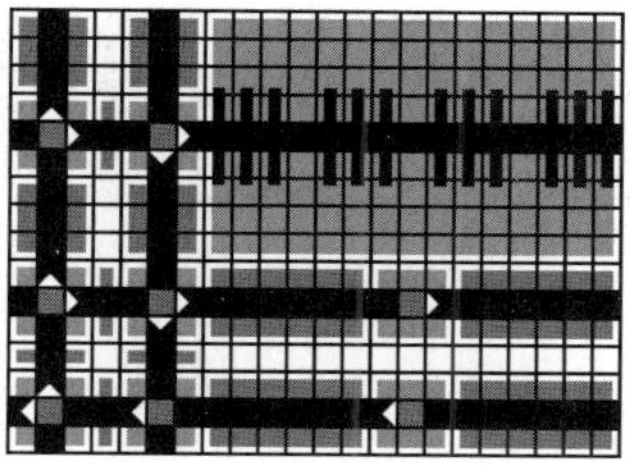

Figure 13-5. The pattern tile that is a modular building block for the layout.

DMFS Control. The above approach to DMFS organization yields a collection of communicating components organized into a network. Components may move droplets at will within themselves, but before moving droplets into cells bordering a neighboring component or into a neighboring component, they must consult the neighbor to ensure this would not result in two droplets being adjacent. Therefore, the system first processes the components serially at each clock cycle and then executes motion in parallel. The system does this by maintaining an ordered master list of components. At each clock cycle, each component in the list is instructed to attempt to move its droplets. When a particular component wishes to move a droplet into an array cell adjacent to or into a neighbor component, it first asks that component if the move will result in two droplets being adjacent. If it will, then it requests the neighbor component to attempt to move its droplets, and then it asks again if the move will result in two droplets being adjacent. If the move would still result in adjacent droplets, then it waits to move those droplets that would result in violations. A separate master list is kept containing the current location of all droplets and their desired location in the next clock cycle. As each component is processed, it updates the list of droplets to reflect the current and desired locations of each droplet within it. The set of consistent droplet movements can then be collected so motion can be performed in parallel.

System Stability. The behavior of a general-purpose system changes with the chemical analysis it performs. We define a DMFS to be *stable* if it does not get deadlocked after 10 million clock cycles of operation. We define a DMFS to be in deadlock if no droplet in the system is able to move. A system operating continuously may or may not be stable depending on its parameters, especially the input flow rate of droplets. In an unstable system, droplets enter the system faster than the system is able to process them, and a steady-state flow cannot be guaranteed [20]. In time, such a system will become heavily congested and finally become deadlocked. We identify stable systems by simulating them and checking at each clock cycle whether they are in a state where no droplet may move.

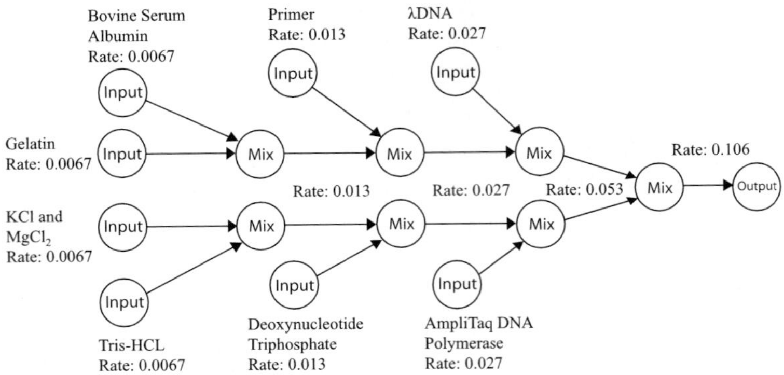

Figure 13-6. PCR analysis graph. Input nodes are labeled with the samples they introduce and the rate at which they introduce them, in droplets per cycle. Edges out of mix nodes are labeled with the droplet rate resulting from the operation.

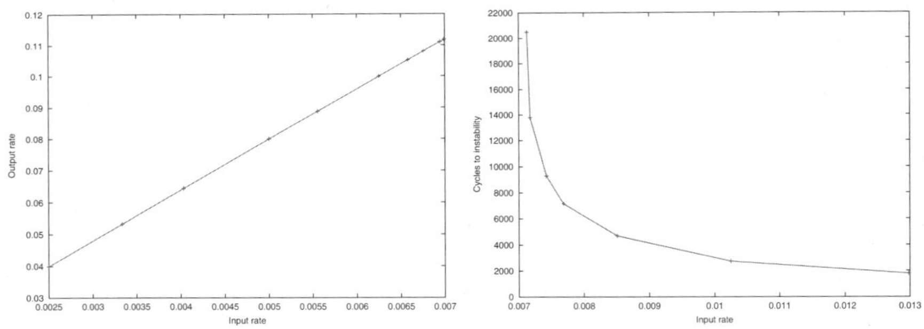

Figure 13-7. Simulation data for the PCR analysis illustrating (a) variation of droplet output rate with input rate in the stable range, and (b) number of cycles at which the system goes into deadlock, as input rate is increased in the unstable range. For this example, mixing time is 128 cycles, the number of mixing units per work area is 8, and the tiles are in a 2×2 pattern.

System Simulation. We have simulated several analyses, including one based on the DNA polymerase chain reaction (PCR) operations outlined in [11]. The analysis involves eight input droplet types and seven mixing operations. See Figure 13-6 for an analysis graph of the system. (Note that the PCR analysis requires heating steps. We assume that droplets may be routed off-chip for heating.) Immediately following each mixing operation, the resulting droplet is split into two droplets. The layout is set up with four work areas, eight sources, each introducing an input droplet type, and one sink to collect the final product (Figure 13-2). This layout with a 2×2 tile arrangment has 53×41 cells. The system has an average of 66 droplets on the array. Our simulation environment is the stand-alone C++ software that we have created for this application; this software may also be used in a controller for a DMFS chip. The routing computations for this array are performed at a rate of about 60,000–70,000 cycles a second on a 1.7 GHz Pentium-M laptop with 512 MB of RAM. This enables rapid simulation of the system to verify stability. For example, at this speed, we can simulate 1,000,000 cycles in approximately 15–20 seconds. Animations of the PCR analysis, as well as multiple analyses running in parallel, are available at `www.cs.rpi.edu/~sakella/microfluidics/`.

The simulation approach has provided insight into the behavior of the system. When the system is in its stable operating range, there is a linear relation between the input droplet rate and output droplet rate, since no droplets are accumulating on the array (Figure 13-7(a)). Once a critical input rate is exceeded, there is a rapid dropoff in the number of clock cycles at which deadlock occurs (Figure 13-7(b)). Here the "input rate" is the rate at which each of the four chemicals on the left of Figure 13-6 is introduced. The subsequent input chemicals are introduced at correspondingly higher multiples of the input rate. We have observed sharp variations in behavior when simulating systems that are on the borderline between stability and instability. Small changes in the input rate at which droplets enter the system can mean the difference between becoming deadlocked in 5,000 cycles, becoming deadlocked in 2,000,000 cycles, or running continuously for 10,000,000 cycles without deadlock.

4. VARIATIONS ON THE EXISTING SYSTEM

We now briefly describe our efforts to optimize the system performance. We experimented with a variety of modifications to the original system to gauge their effects on the stability of the system, and to determine which modifications allowed the system to be stable at the highest input rates.

4.1 Variations on the Layout Tile

We first experimented with altering the modular tile pattern used to create the layout (Figure 13-5). Our goal was to increase the percentage of space on

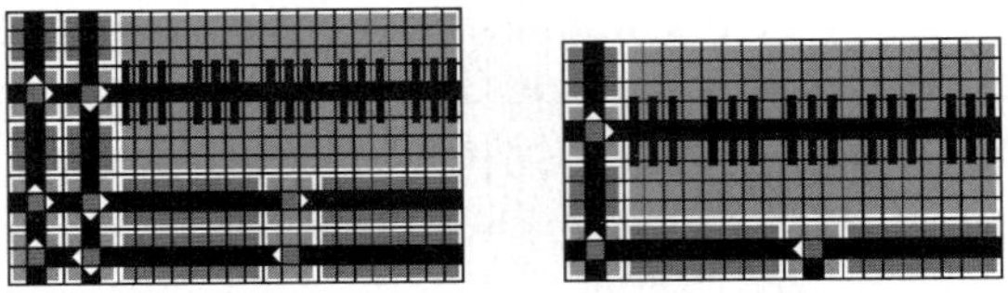

Figure 13-8. Tile variations: **(a)** With no connectors between streets. **(b)** With only one way streets.

the tile devoted to droplet mix and split operations. We created two alternative layouts, shown in Figure 13-8. The first tile removes the connector components between streets, and the second tile has only one horizontal and vertical street, rather than oppositely directed pairs of each.

These alternative tiles were not effective, however. In the tile without the connectors between the streets, rotaries become deadlocked whenever the situation in Figure 13-9 arises. Once one set of intersections has become deadlocked, the system usually ceases being able to operate soon after due to the resulting droplet traffic backup. The layout with only one way streets suffers from a diminished capacity for droplet traffic, which is exacerbated by droplets often needing to travel a greater distance to reach their destinations. The three layout designs are compared in Table 13-1.

Table 13-1. Comparison of the stability of three tile layout patterns with a 2×2 tile arrangement, for the PCR analysis. Input rate is measured in droplets per clock cycle.

Tile Layout	Mixing Units per Work Area	Highest Stable Rate (Approx.)
Default	8	0.0065
No Connectors	8	0.0040
One Way	8	0.0050
Default	10	0.0080
No Connectors	10	0.0030
One Way	10	0.0055
Default	12	0.0090
No Connectors	12	0.0040
One Way	12	0.0060

4.2 Variations in Routing Control

We also experimented with three changes to component behavior. The first change was to modify the droplet destination selection and routing algorithm to assign droplets to the closest available component instead of the original method of assigning them to components on a rotating basis. The second change was to

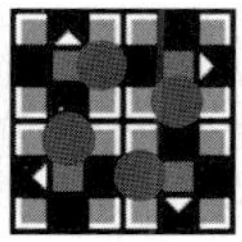

Figure 13-9. When droplets are in this particular configuration, they cannot move again. Attempting to advance any droplet would require activating the adjacent electrode, which is also diagonally adjacent to another droplet. This activation could result in unexpected droplet movement or mixing, and therefore is disallowed.

have half of the work areas on the array be right-to-left (i.e., droplets enter from the right and exit from the left side of the work area) instead of all work areas being left-to-right. The third change was varying the order in which components attempt to move their droplets. In the original implementation, the components were assigned an initial order, and they attempted to move their droplets in that order at each cycle. The order is generally sources and work areas first and then the remaining components; the order could vary a little at each cycle based on droplet movement dependencies. We instead compute a random permutation of the components at each clock cycle, and then the components try to move their droplets in that order, subject to droplet movement dependency variations.

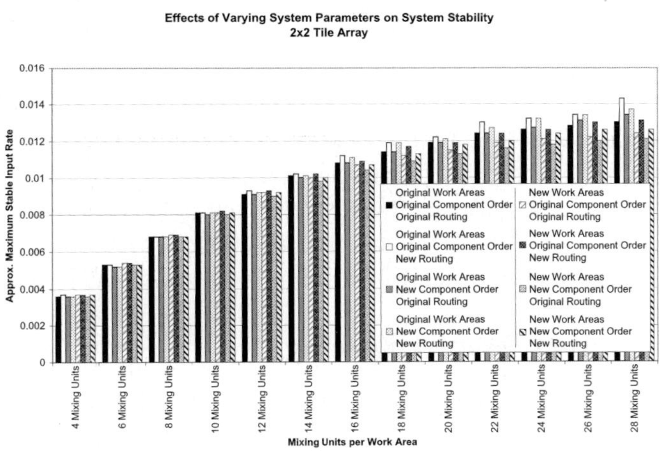

Figure 13-10. Chart depicting the effects of each of the three routing control variations on a 2×2 tile PCR simulation. Input rate is measured in droplets per clock cycle. Each bar in the graph corresponds to operating the system under a certain set of parameters. Parameters labeled as 'new' correspond to the new methods in Section 4.2. Parameters labeled as 'original' correspond to the original methods described in Section 3.

The effects of these variations are depicted in Figure 13-10. The best performance is obtained by using the new routing algorithm with the original work areas and fixed component order. In general, all combinations with the new routing algorithm performed better than their counterparts with the old routing

Table 13-2. Comparison of the maximum increase in stable rate due to different variations in routing control, for different values of mixing units per work area. Data is for a 2×2 tile layout simulation of the PCR analysis. For lower number of mixing units per work area, the maximum increase is achieved with new work areas and new routing, while for higher number of mixing units per work area, it is achieved with new routing and the original work areas and component order. Rate is measured in droplets per clock cycle.

Mixing Units per Work Area	Maximum % Increase in Stable Rate
4	2.778
6	1.887
8	1.471
10	1.235
12	2.198
14	0.990
16	3.704
18	4.386
20	2.521
22	4.839
24	4.762
26	4.688
28	10.0

algorithm. The opposite is true with the mixture of left-to-right work areas with right-to-left work areas versus just left-to-right work areas. Similarly, the new component order offers slightly inferior performance to the original component ordering. The other interesting characteristic is that the effects of the various changes are negligible with small arrays that can only operate at lower input rates, but, as the size of the array and thus its capacity for processing droplets increases, the effects of the changes become more pronounced (Table 13-2).

4.3 Increased Droplet Spacing

We earlier assumed that multiple droplets moving in a line could be moved in synchrony in the same direction with only a single empty array cell between droplets. However, this assumption requires a high degree of synchronization of electrode activation, and may make this type of movement hard to implement or even infeasible. We now assume that in addition to the requirement that droplets must have at least one empty array cell on all sides except when mixing is about to occur, that any droplets moving in the same direction simultaneously must have at least two empty cells between them to avoid undesired mixing or splitting (Figure 13-11). There should be at least three empty cells between droplets when there is a 90 degree bend in the path. This change has not significantly affected the performance of the system because it is rare, under stable conditions, for droplets to be moving in the same direction with only one empty array cell between them.

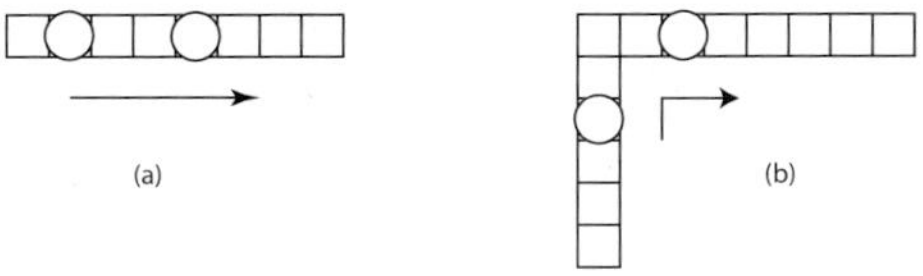

Figure 13-11. The minimum number of empty cells between two occupied cells to ensure that the droplets cannot combine or split inadvertently depends on the path shape. (a) When the two cells are on a straight line. (b) When the two cells are around a bend in the path.

4.4 Additional Enhancements

Although we have implicitly described all mixing operations as taking the same amount of time, the system accomodates mixing operations with differing durations based on the droplet types. There are other enhancements to the system that can be easily incorporated. We can add virtual storage components to the layout by treating one or more of the mixing units in a work area as storage units. Similarly, if some or all of the array cells have optical sensing capabilities, we can create sensing components for the layout, located in the work areas, for example, or even in the streets or intersections. These sensors can permit monitoring of reaction results based on droplet color.

5. LIMITED ROW-COLUMN ADDRESSING

We have so far assumed that every electrode on the 2D array can be individually addressed, so an arbitrary set of cells can be activated at each cycle. In a limited row-column addressing scheme, individual cells are not directly addressable. Only entire rows and columns can be activated and only electrodes at intersections of activated rows and columns will be turned on [15, 5, 17]. For example, Fan, Hashi, and Kim [15] developed a cross-referencing scheme by arranging two vertically separated electrode layers orthogonal to each other. While this simplifies the hardware and reduces fabrication and packaging costs, it provides less flexibility in moving several droplets in synchrony and complicates droplet control. The *interference graph* (Figure 13-12) represents potential conflicts between droplet movements. Here two vertices connected by an edge represent droplets that cannot be moved in the same clock cycle.

5.1 Modified Schemes for Limited Row-Column Addressing

The central issue with limited row-column addressing is how to serialize the previously synchronous motion of the droplets at each clock cycle. In direct

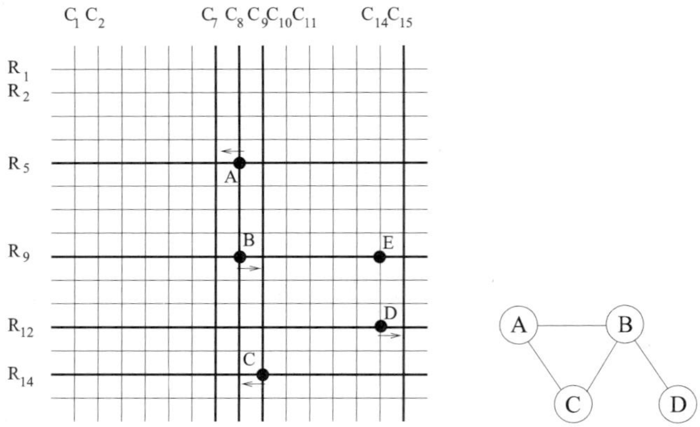

(a) (b)

Figure 13-12. A schematic illustration of droplet motion in an array with limited row-column addressing. (a) Each line represents a control wire connected to all electrodes in the corresponding row or column. Bold lines represent columns or rows to be activated. Droplet A is to be moved from the cell at (C_8, R_5) to (C_7, R_5), droplet B from (C_8, R_9) to (C_9, R_9), droplet C from (C_9, R_{14}) to (C_8, R_{14}), droplet D from (C_{14}, R_{12}) to (C_{15}, R_{12}), and droplet E is to remain stationary. (b) The interference graph indicates the conflicts for simultaneous droplet motion. Each vertex represents a droplet, and two vertices connected by an edge represent droplets that cannot be moved at the same time. Simultaneously activating the rows R_5, R_9, R_{14} and columns C_7, C_8, C_9 would not guarantee the desired motion for droplets A, B, and C. Moving droplets B and D simultaneously would also move droplet E. Instead, in one clock cycle, droplet A can be moved by activating R_5 and C_7 and droplet D by activating R_{12} and C_{15}, in the next clock cycle droplet C can be moved by activating R_{14} and C_8, and in the next clock cycle droplet B can be moved by activating R_9 and C_9.

addressing mode, the movements for all droplets are calculated at each clock cycle, and they are then executed in parallel. For clarity, we will refer to one clock cycle in direct addressing mode as a *virtual clock cycle*. For row-column addressing, the droplet movements are computed at the beginning of each virtual clock cycle and then the droplet movements are executed over one or more *real clock cycles*.

We have developed two schemes to perform limited row-column addressing for the DMFS. The first is a simple row-column addressing scheme where only one cell is addressed each cycle, by simultaneously activating both its row and column. Hence only one droplet is moved each real clock cycle. Moving any droplet by a planned move will not result in it being inadvertently adjacent to any other droplet either before or after the droplet's movement. This is because the planning of the droplet movements (Section 3.3) ensures that no motions are allowed for droplets that would move adjacent to either the starting or ending location of a droplet in a particular virtual cycle.

We next describe a more complex row-column addressing scheme where multiple cells may be addressed by simultaneously activating their rows and columns. In this scheme, multiple droplets may be moved at each clock cycle such that their activation does not cause other droplets to move inadvertently, and they do not inadvertently move next to another droplet. See Figure 13-12 for an example scenario.

5.2 A Graph Coloring Approach

We have developed a graph coloring approach to limited row-column addressing, to reduce the number of real clock cycles per virtual clock cycle by performing multiple droplet motions simultaneously. The results below are quite general and in fact apply to any array layout with a planar grid of electrodes. Scheduling an interference-free movement of the droplets may be modeled as a vertex coloring problem. It is known that the general vertex coloring problem is NP-complete (see [38]); furthermore it is NP-complete even on the class of 3-colorable graphs. The fastest algorithms for 3-colorable graphs are exponential [12]. We introduce a heuristic, greedy, polynomial-time algorithm for coloring the the interference graph (or equivalently, the transition graph introduced below). Note that this algorithm is not guaranteed to produce an optimal coloring.

To address the problem of scheduling the movements of the droplets, we define a *transition graph* $T(V, E)$. The input to such a graph consists of a set L of the current locations of the droplets and the set M of the droplets' movements that are to be performed in the current virtual clock cycle. Every movement is an ordered pair of coordinates $[(x_s, y_s); (x_d, y_d)]$, where the first term, (x_s, y_s) is the current (start) location of the droplet, and the second one, (x_d, y_d), is the next destination. Since all movements are either horizontal or vertical movements in the grid, the pair describing a movement satisfies the following condition:

$$\begin{array}{ll} |x_s - x_d| = 1 \text{ and } y_s = y_d, & \text{for a horizontal movement,} \\ |y_s - y_d| = 1 \text{ and } x_s = x_d, & \text{for a vertical movement.} \end{array}$$

In Figure 13-13 below, we present an example set of movements, including $[(2, 4); (3, 4)]$, a horizontal movement, and $[(7, 6), (7, 5)]$, a vertical movement.

The vertex set $V(T)$ of the transition graph T is the set of all movements that must be performed during a virtual clock cycle. The set $E(T)$ of edges of T consists of all pairs (u, v), $u, v \in V(T)$, such that the corresponding movements cannot be performed in the same real clock cycle of a given virtual clock cycle.

For an arbitrary graph G, a (legal) vertex coloring of the vertex set $V(G)$ is an assignment $F : V(G) \to C$, where C is a finite set called a *color set*, such that no two adjacent vertices are colored the same color. Usually, C is a set of non-negative integers $\{0, 1, 2, \cdots\}$. The *chromatic number* $\chi(G)$ is the smallest number of colors needed to legally color the vertices of G. In the context of the

The stability behavior of the system remains the same under these addressing schemes.

Table 13-3. Comparison of the efficiency of three addressing schemes for a 2×2 tile layout simulation of the PCR analysis.

Addressing Scheme	Virtual Cycles Completed	Real Clock Cycles Taken
Direct Addressing	1,000,000	1,000,000
Simple Row-Column	1,000,000	39,579,750
Coloring-based Row-Column	1,000,000	10,035,243

6. SYSTEM APPLICATION SCENARIOS

In this section, we discuss two scenarios that our system is capable of handling. The first scenario deals with adjusting the concentration levels of the droplets being used on the array. The second scenario describes an approach to use a minimal layout for glucose assays.

6.1 Dilution Control

Having the ability to dilute chemicals on chip is useful for improving the sensitivity and accuracy of bioanalyte detection [39]. Fair et al. [14] describe an interpolating serial dilution scheme. Each exponential dilution step mixes a unit volume chemical droplet with a unit volume buffer droplet to obtain two unit volume droplets of half the concentration. Each interpolation step combines unit volume droplets of concentrations C_1 and C_2 to obtain two droplets of concentration $(C_1 + C_2)/2$. In principle, a droplet with an arbitrary dilution level can be created through a sequence of interpolating and exponential dilution steps.

We have implemented an algorithm for automated droplet dilution control. We associate a concentration level with each droplet type the system is to process. If a droplet of a particular type and concentration is specified as an input to the system, and a mixing operation is specified that takes that droplet type but with a lower concentration as input, then the system will recognize that the input droplet needs to be diluted. A set of mixing operations to create the desired concentration is computed by applying Algorithm 2, which is based on a binary search strategy. To facilitate the dilution, two special droplet types are introduced. The first, a buffer droplet, has a concentration level of 0 and can be used to reduce the concentration of any droplet it mixes with by half. The second is a waste droplet; any unwanted, extra droplets produced by the dilution process that are to be discarded are designated as waste droplets. Once the set of mixing operations M has been computed, droplets of matching concentrations can be linked together in a mixing graph, by comparing the input and output concentrations of pairs of operations. See the example graph in Figure 13-15.

Algorithm 2 Droplet Dilution

```
Input: d_i, d_b // Input droplet type with known concentration
                // and the buffer droplet type
       c // Desired concentration level.
       tol // The tolerance within which concentrations
           // are considered equal
Output: M // Set of mixing operations {((d_j, d_k) → (d_jk^mix1, d_jk^mix2)}
          // that yield concentration c.
D ← {d_i, d_b} // Initializing D, set of droplets of varying
               // concentrations available for mixing
M ← ∅
range ← Concentration(d_i) − Concentration(d_b)
d_H ← d_i // d_H is upper bound for concentration
d_L ← d_b // d_L is lower bound for concentration
while range > tol do
  for all d_l, d_h ∈ D do
    if Concentration(d_l) < c and Concentration(d_h) > c then
      if Concentration(d_h) − Concentration(d_l) < range then
        range ← Concentration(d_h) − Concentration(d_l)
        d_H ← d_h
        d_L ← d_l
      end if
    end if
  end for
  m ← ((d_H, d_L) → (d_HL, d_w)) // d_w is identical to d_HL
                                 // but designated a waste droplet
  M ← M ⋃ m
  D ← D ⋃ d_HL
end while
return M
```

6.2 Minimalistic Layout for Glucose Assays

Experimentally demonstrated digital microfluidic systems range in size from small electrowetting arrays (for example, 5×5 cells [15]) to large dielectrophoresis arrays (for example, 320×320 cells [26]). The layouts we described above for our system are intermediate in size. We can also create a small layout of 11x17 cells (Figure 13-16), comparable in size to existing electrowetting-based

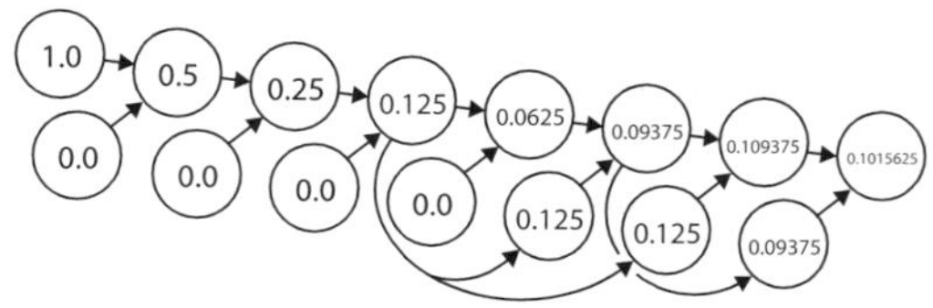

Figure 13-15. An example mixing graph for dilution control. The scheme assumes that droplets of a specified concentration level are given and that buffer droplets of 0% concentration are available. Any desired reduced concentration can be achieved; our approach is to identify the intermediate droplet concentrations through a binary search strategy. Here, the concentration is reduced to approximately 10% of its original level.

arrays [17]. These small layouts are most appropriate for simple reactions that require only a small number of droplet types.

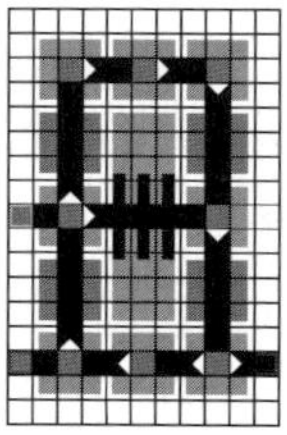

Figure 13-16. A 11x17 array layout for sample preparation for glucose assay.

Srinivasan, Pamula, and Fair [40] describe the use of a prototype DMFS for glucose assays in a variety of biological fluids. They mix sample droplets and reagent droplets in the system to dilute the sample. After splitting, one resulting droplet is discarded as waste and the other is sent to an on-chip concentration detection cell. We have successfully simulated the sample preparation phase of this glucose assay using the minimal 11x17 layout in Figure 13-16. Currently, we assume that the diluted samples are sent off-chip for glucose concentration sensing; an optical sensor component can be easily incorporated into the layout, in the work area or at the sink intersection. This glucose assay example, along with the PCR example, demonstrates that our system is highly scalable; it is able to operate successfully on a range of sizes consistent with current experimental systems.

7. CONCLUSION

Our approach to creating a general-purpose DMFS, previously described in [18, 19], consists of imposing a virtual layout of components on the planar array and coordinating the motions of droplets by developing decentralized routing algorithms. The system can perform real-time droplet manipulation, and can be easily used to act as a controller for a physical array. The same array can perform a variety of chemical analyses including the DNA polymerase chain reaction and glucose assays, and can even perform multiple analyses in parallel.

In this chapter, we enhanced the original system in a number of ways for greater versatility and performance. These included support for new layout schemes, routing algorithms, and increased spacing between droplets, and characterization of their effects on system performance. We found the system relatively stable to these variations, which implies the overall design is relatively robust. We then considered DMFS arrays with hardware limited row-column addressing and developed a polynomial-time graph coloring algorithm for the problem of droplet coordination under such hardware limitations. We demonstrated the capabilities of our system on example scenarios, including dilution control and minimalist layouts.

There are several directions for future work. Identifying the minimum number of steps to execute a set of droplet movements under limited row-column addressing is an open problem that we are working on using the graph coloring approach. The overall design of the components and the system allows for the introduction of new component types, such as droplet heater components, for example. Automatically generating the optimal layout for a given analysis requires methods for optimizing the number of tiles and their arrangement, as well as the locations of the sinks and sources on the array. Modeling the system as a network can potentially provide insights into changes to the array design and improve system performance. The design and control of dynamically reconfigurable layouts, where any part of the array may be reallocated for any desired operation, pose particularly interesting challenges. Developing layouts that can adapt to electrode failures is another direction that will lead to robust systems.

ACKNOWLEDGMENTS

Many thanks to Karl Böhringer for introducing us to this problem and providing encouragement and advice. This work was supported in part by NSF under Award No. IIS-0093233 and Award No. IIS-0541224.

REFERENCES

[1] S. Akella and S. Hutchinson. Coordinating the motions of multiple robots with specified trajectories. In *IEEE International Conference on Robotics and Automation*, pages 624–631, Washington, DC, May 2002.

[2] D. P. Bertsekas and R. G. Gallagher. *Data Networks*. Prentice-Hall, Englewood Cliffs, N.J., second edition, 1992.

[3] A. Bicchi and L. Pallottino. On optimal cooperative conflict resolution for air traffic management systems. *IEEE Transactions on Intelligent Transportation Systems*, 1(4):221–231, Dec. 2000.

[4] K.-F. Böhringer. Optimal strategies for moving droplets in digital microfluidic systems. In *Seventh International Conference on Micro Total Analysis Systems (MicroTAS '03)*, pages 591–594, Squaw Valley, CA, Oct. 2003.

[5] K. F. Böhringer. Towards optimal strategies for moving droplets in digital microfluidic systems. In *IEEE International Conference on Robotics and Automation*, New Orleans, LA, Apr. 2004.

[6] J. Brassil and R. Cruz. Nonuniform traffic in the Manhattan street network. In *IEEE International Conference on Communications (ICC '91)*, pages 1647–1651, June 1991.

[7] C. Busch, M. Herlihy, and R. Wattenhofer. Hard-potato routing. In *Proceedings of the 32nd Annual ACM Symposium on Theory of Computing (STOC 2000)*, pages 278–285, Portland, Oregon, May 2000.

[8] S. K. Cho, H. Moon, and C.-J. Kim. Creating, transporting, cutting, and merging liquid droplets by electrowetting-based actuation for digital microfluidic circuits. *Journal of Microelectromechanical Systems*, 12(1):70–80, Feb. 2003.

[9] A. K. Choudhury and V. O. K. Li. An approximate analysis of the performance of deflection routing in regular networks. *IEEE Journal on Selected Areas in Communications*, 11(8):1302–1316, Oct. 1993.

[10] A. A. Desrochers. *Modeling and Control of Automated Manufacturing Systems*. IEEE Computer Society, Washington, DC, 1990.

[11] J. Ding, K. Chakrabarty, and R. B. Fair. Scheduling of microfluidic operations for reconfigurable two-dimensional electrowetting arrays. *IEEE Transactions on Computer-Aided Design of Integrated Circuits and Systems*, 20(12):1463–1468, Dec. 2001.

[12] D. Eppstein. Improved algorithms for 3-coloring, 3-edge-coloring, and constraint satisfaction. In *Proc. 12th Symp. Discrete Algorithms*, pages 329–337. ACM and SIAM, January 2001.

[13] M. Erdmann and T. Lozano-Perez. On multiple moving objects. *Algorithmica*, 2(4):477–521, 1987.

[14] R. B. Fair, V. Srinivasan, H. Ren, P. Paik, V. Pamula, and M. G. Pollack. Electrowetting-based on-chip sample processing for integrated microfluidics. In *IEEE International Electron Devices Meeting (IEDM)*, 2003.

[15] S.-K. Fan, C. Hashi, and C.-J. Kim. Manipulation of multiple droplets on NxM grid by cross-reference EWOD driving scheme and pressure-contact packaging. In *IEEE Conference on MEMS*, pages 694–697, Kyoto, Japan, Jan. 2003.

[16] A. Fuchs, N. Manaresi, D. Freida, L. Altomare, C. L. Villiers, G. Medoro, A. Romani, I. Chartier, C. Bory, M. Tartagni, P. N. Marche, F. Chatelain, and R. Guerrieri. A microelectronic chip opens new fields in rare cell population analysis and individual cell biology. In *Seventh International Conference on Micro Total Analysis Systems (MicroTAS '03)*, pages 911–914, Squaw Valley, CA, Oct. 2003.

[17] J. Gong, S.-K. Fan, and C.-J. Kim. Portable digital microfluidics platform with active but disposable lab-on-chip. In *Tech. Digest of 17th IEEE International Conference on Micro Electro Mechanical Systems (MEMS'04)*, pages 355–358, Maastricht, The Netherlands, Jan. 2004.

[18] E. Griffith and S. Akella. Coordinating multiple droplets in planar array digital microfluidics systems. In M. Erdmann, D. Hsu, M. Overmars, and A. F. van der Stappen, editors, *Algorithmic Foundations of Robotics VI*, pages 219–234. Springer-Verlag, Berlin, 2005.

[19] E. J. Griffith and S. Akella. Coordinating multiple droplets in planar array digital microfluidic systems. *International Journal of Robotics Research*, 24(11):933–949, Nov. 2005.

[20] D. Gross and C. M. Harris. *Fundamentals of Queueing Theory*. Wiley, New York, third edition, 1998.

[21] J. E. Hopcroft, J. T. Schwartz, and M. Sharir. On the complexity of motion planning for multiple independent objects: PSPACE-hardness of the "warehouseman's problem". *International Journal of Robotics Research*, 3(4):76–88, 1984.

[22] T. B. Jones, M. Gunji, M. Washizu, and M. J. Feldman. Dielectrophoretic liquid actuation and nanodroplet formation. *Journal of Applied Physics*, 89:1441–1448, Jan. 2001.

[23] F. Kelly, A. Maulloo, and D. Tan. Rate control in communication networks: shadow prices, proportional fairness and stability. *Journal of the Operational Research Society*, 49:237–252, 1998.

[24] S. M. LaValle and S. A. Hutchinson. Optimal motion planning for multiple robots having independent goals. *IEEE Transactions on Robotics and Automation*, 14(6):912–925, Dec. 1998.

[25] M. A. Lawley. Deadlock avoidance for production systems with flexible routing. *IEEE Transactions on Robotics and Automation*, 15(3):497–509, June 1999.

[26] N. Manaresi, A. Romani, G. Medoro, L. Altomare, A. Leonardi, M. Tartagni, and R. Guerrieri. A CMOS chip for individual cell manipulation and detection. *IEEE Journal of Solid-State Circuits*, 38(12):2297–2305, Dec. 2003.

[27] C. Maxfield. *The Design Warrior's Guide to FPGAs: Devices, Tools, and Flows*. Elsevier, Burlington, MA, 2004.

[28] P. A. O'Donnell and T. Lozano-Perez. Deadlock-free and collision-free coordination of two robot manipulators. In *IEEE International Conference on Robotics and Automation*, pages 484–489, Scottsdale, AZ, May 1989.

[29] P. Paik, V. K. Pamula, and R. B. Fair. Rapid droplet mixers for digital microfluidic systems. *Lab on a Chip*, 3:253–259, 2003.

[30] J. Peng and S. Akella. Coordinating multiple robots with kinodynamic constraints along specified paths. *International Journal of Robotics Research*, 24(4):295–310, Apr. 2005.

[31] M. G. Pollack, R. B. Fair, and A. D. Shenderov. Electrowetting-based actuation of liquid droplets for microfluidic applications. *Applied Physics Letters*, 77:1725–1726, 2000.

[32] M. G. Pollack, P. Y. Paik, A. D. Shenderov, V. K. Pamula, F. S. Dietrich, and R. B. Fair. Investigation of electrowetting-based microfluidics for real-time PCR applications. In *Seventh International Conference on Miniaturized Chemical and Biochemical Analysis Systems (MicroTAS '03)*, pages 619–622, Squaw Valley, CA, Oct. 2003.

[33] S. A. Reveliotis, M. A. Lawley, and P. M. Ferreira. Polynomial-complexity deadlock avoidance policies for sequential resource allocation systems. *IEEE Transactions on Automatic Control*, 42(10):1344–1357, Oct. 1997.

[34] A. A. Rizzi, J. Gowdy, and R. L. Hollis. Distributed coordination in modular precision assembly systems. *International Journal of Robotics Research*, 20(10):819–838, Oct. 2001.

[35] G. Sanchez and J. Latombe. On delaying collision checking in PRM planning — application to multi-robot coordination. *International Journal of Robotics Research*, 21(1):5–26, Jan. 2002.

[36] T. Schouwenaars, B. De Moor, E. Feron, and J. How. Mixed integer programming for multi-vehicle path planning. In *European Control Conference 2001*, pages 2603–2608, Porto, Portugal, 2001.

[37] T. Simeon, S. Leroy, and J.-P. Laumond. Path coordination for multiple mobile robots: A resolution-complete algorithm. *IEEE Transactions on Robotics and Automation*, 18(1):42–49, Feb. 2002.

[38] S. S. Skiena. *The Algorithm Design Manual*. Springer-Verlag, New York, 1998.

[39] V. Srinivasan, V. Pamula, M. Pollack, and R. Fair. A digital microfluidic biosensor for multianalyte detection. In *IEEE 16th Annual International Conference on Micro Electro Mechanical Systems*, pages 327–330, 2003.

[40] V. Srinivasan, V. K. Pamula, and R. B. Fair. An integrated digital microfluidic lab-on-a-chip for clinical diagnostics on human physiological fluids. *Lab on a Chip*, 4:310–315, 2004.

[41] F. Su and K. Chakrabarty. Architectural-level synthesis of digital microfluidics-based biochips. In *Proc. IEEE International Conference on CAD*, pages 223–228, 2004.

[42] A. S. Tanenbaum. *Computer Networks*. Prentice Hall, Upper Saddle River, NJ, third edition, 1996.

[43] C. Tomlin, G. J. Pappas, and S. Sastry. Conflict resolution for air traffic management: A study in multi-agent hybrid systems. *IEEE Transactions on Automatic Control*, 43(4):509–521, Apr. 1998.

[44] P. Švestka and M. Overmars. Coordinated path planning for multiple robots. *Robotics and Autonomous Systems*, 23(3):125–152, Apr. 1998.

[45] D. B. West. *Introduction to Graph Theory*. Prentice Hall, Upper Saddle River, NJ, second edition, 2001.

[46] A. R. Wheeler, H. Moon, C.-J. C. Kim, J. A. Loo, and R. L. Garrell. Electrowetting-based microfluidics for analysis of peptides and proteins by matrix-assisted laser desorption/ionization mass spectrometry. *Analytical Chemistry*, 76(16):4833–4838, Aug. 2004.

[47] T. Zhang, K. Chakrabarty, and R. B. Fair. *Microelectrofluidic Systems: Modeling and Simulation*. CRC Press, Boca Raton, Florida, 2002.

Chapter 14

A PATTERN-MINING METHOD FOR HIGH-THROUGHPUT LAB-ON-A-CHIP DATA ANALYSIS

Sungroh Yoon*
Computer Systems Laboratory
Stanford University
Stanford, CA 94305, USA
sryoon@stanford.edu

Luca Benini
Department of Electrical Engineering and Computer Science (DEIS)
University of Bologna
40136 Bologna, Italy
lbenini@deis.unibo.it

Giovanni De Micheli
Integrated Systems Center
Swiss Federal Institute of Technology (EPFL)
CH-1015 Lausanne, Switzerland
giovanni.demicheli@epfl.ch

Abstract: Biochips are emerging as a useful tool for high-throughput acquisition of biological data and continue to grow in information quality and in discovering new applications. Recent advances include CMOS-based integrated biosensor arrays for deoxyribonucleic acid (DNA) expression analysis [35, 17], and active research is ongoing for the miniaturization and integration of protein microarrays [36, 19, 33], tissue microarrays (TMAs) [37, 8], and fluorescence-based multiplexed cytokine immunoassays [41]. The main advantages of microfluidic lab-on-a-chip include ease-of-use, speed of analysis, low sample and reagent

*To whom correspondence should be addressed.

K. Chakrabarty and J. Zeng (eds.),
Design Automation Methods and Tools for Microfluidics-Based Biochips, 357–400.

consumption, and high reproducibility due to standardization and automation. Without effective data analysis methods, however, the merit of acquiring massive data through biochips will be marginal. The high-dimensional nature of such data requires novel techniques that can cope with the curse of dimensionality better than conventional data analysis approaches. In this paper, we propose a pattern-mining method to analyze large-scale biological data obtained from high-throughput biochip experiments. In particular, when a data set is given as a matrix, our method can find patterns appearing in form of (possibly overlapping) submatrices of the input matrix. Our method exploits the techniques developed for the symbolic manipulation of Boolean functions. Leveraged by this approach, our method can find, given a data matrix, all patterns that satisfy specific input parameters. The authors tested our method with several large-scale biochip data and observed that the proposed method outperforms the alternatives in terms of efficiency and the number of patterns discovered.

Keywords: Biomedical transducers, biomedical signal analysis, bioinformatics, computer aided analysis, data management, logic design

1. INTRODUCTION

Interest in *in vivo* and *in vitro* applications of lab-on-a-chip, also called microfluidics-based biochips or bio-MEMS, is growing [15, 40]. The main advantages of this technology include ease-of-use, speed of analysis, low sample and reagent consumption and high reproducibility due to standardization and automation. Biochips has become one of the standard tools for high-throughput acquisition of biological data, as is evident from the recent advances in integrated biosensor arrays [35, 17], protein microarrays [36, 19, 33], tissue microarrays (TMAs) [37, 8], fluorescence-based multiplexed cytokine immunoassays [41].

However, the usefulness of this fascinating innovation may be limited without an effective means of analysis of the data obtained. In fact, technical breakthroughs in biotechnologies have already led to a rapid growth of biological data, both in size and complexity. For example, in recent years the rate at which the GenBank database (`http://www.ncbi.nlm.nih.gov/Genbank`) has grown exceeds the pace set by Moore's Law, as seen in Fig 14-1. Therefore, it is of utmost importance to have a fast and statistically robust data analysis tool that can lead to breakthrough improvements in quality and time-to-market, by providing the designers of high-throughput biochips with the necessary feedback for the next design iteration in timely manner.

Multiple new methods have been proposed to effectively analyze large-scale biological data obtained from high-throughput biotechnologies, despite the mature literature on traditional clinical data analysis. This is partly because the data acquired from biochips often exhibit different characteristics from traditional clinical data. For instance, as seen in Fig. 14-2, the number of variables

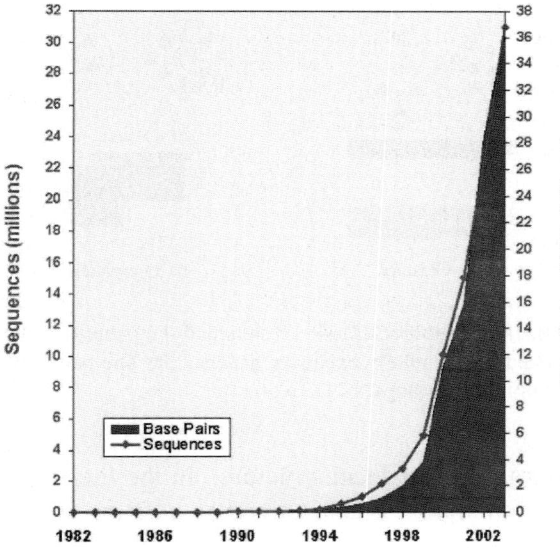

Figure 14-1. Growth of GenBank database. The growth rate exceeds the pace set by Moore's Law.

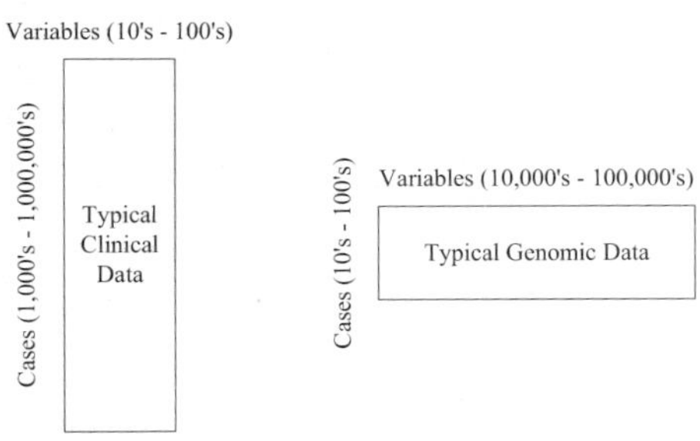

Figure 14-2. A major difference between classic clinical studies and genomic studies [20]. In contrast to clinical data, genomic data often results in a highly underdetermined system.

involved in a typical genomic study is far more than that of the observations, in contrast to a typical clinical study where there are normally more observations than variables [20]. Thus, in typical genomic studies we often encounter the curse of dimensionality and the problem of identifying a highly underdetermined system.

Among the methods that have been proposed to handle this challenge, one of the most natural and in fact effective approaches is to focus only on subsets of the entire data [2]. By performing simultaneous clustering of rows and columns in a data matrix, we can discover some local structure appearing in the form of overlapping submatrices of the matrix. In this paper, we use the

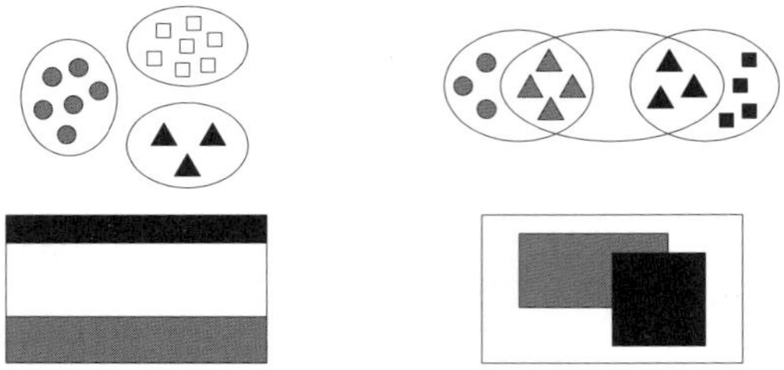

(a) Clusters (b) Our patterns

Figure 14-3. Comparison between classic clusters and the patterns our method can find. (a) Objects are partitioned into mutually exclusive groups. (b) The patterns in our definition are allowed to belong to multiple groups.

term *pattern* to refer to this local structure. In the literature, the local structure modeled by a submatrix is also termed *co-clusters*, *biclusters*, or *modules*. Fig. 14-3 informally compares conventional clustering with the pattern mining method described in this chapter. The interested reader is directed to [23] for a review.

We observe that many patterns defined in the literature possess a common property. Suppose that $\mathfrak{D}$ is a certain condition under which a pattern, P, is defined. Here we refer to the pattern P as *homogeneous* if any legitimate sub-pattern of P also satisfies the condition $\mathfrak{D}$. Examples of homogeneous patterns in the literature include *conserved gene expression motifs* (xMOTIFs) [29], *δ-valid kj-patterns* [7], *gene expression modules* (GEMS) [43], *order-preserving submatrices* (OPSMs) [3], *OP-Clusters* [21], and *δ-pClusters* [42], just to name a few.

Despite its relevance, the problem of homogeneous pattern mining is often computationally challenging. Let A be a matrix with row set R and column set C. The matrix A can be converted to a weighted bipartite graph $G = (V, E)$, where the vertex set $V = R \cup C$ and the edge set E consists of edge $\{i, j\}$ connecting row $i \in R$ and column $j \in C$ with weight a_{ij}. A submatrix of A then corresponds to a biclique in the graph G. To find not just any submatrix but a useful one, we need to consider individual elements of a submatrix, or equivalently the edge weights of a biclique. Moreover, in order to avoid redundancy, we usually focus on finding maximal submatrices. Therefore, the problem of discovering patterns with certain semantics is at least as hard as that of finding the maximum edge biclique in a bipartite graph, a problem known to be NP-complete [23, 30].

In this paper, we propose a novel pattern mining method that exploits the techniques commonly used for the symbolic manipulation of Boolean functions. The techniques have been reported useful to solve many practical instances of intractable problems [12, 5, 6, 25, 34]. In particular, we use the

Zero-suppressed Binary Decision Diagrams (ZBDDs) [26, 27] to implicitly represent and manipulate massive intermediate data occurring in the pattern mining process. Leveraged by this approach, our method can find, given a data matrix, all homogeneous patterns that satisfy specific input parameters. Especially, our method can find three types of homogeneous patterns, which are defined in such a way that they can serve as representative examples of the homogeneous patterns frequently encountered in the literature.

In our experiments, we first tested the proposed method with synthetic data sets to verify its validity. We then applied our method to some biological data in order to evaluate its applicability to actual biological data sets. We used gene expression data obtained from genome chip experiments [13, 22]. This type of data is one of the most large-scale biochip data available. We observed that our method outperforms the alternative methods that are designed to find the same patterns, not only in terms of efficiency but also with respect to the total number of patterns discovered. In particular, we confirmed that the use of ZBDDs can greatly enhance the scalability of our approach and enable us to apply it to large-scale data sets.

The remainder of this paper is organized as follows. In Section 2, we brief the reader on some biochip technologies in order to show the wide applicability of our method. Section 3 presents the formal definition of homogeneous patterns our method can find. In Sections 4 and 5, we explain at length the proposed method, which consists of essentially two stages. The first stage, which is detailed in Section 4, is to find special homogeneous patterns called *atomic patterns*. Section 5 presents the second stage of our method, which derives general (non-atomic) homogeneous patterns from the atomic patterns previously found. Section 6 provides our experimental results, followed by conclusions in Section 7.

2. BACKGROUND: DATA ACQUISITION BY HIGH-THROUGHPUT BIOCHIPS

After readout and preliminary data processing, biological data produced by high-throughput technologies are typically arranged in a matrix. Our method can analyze any type of biochip data, as long as the input data are represented as a matrix of real numbers. Here we present several examples, in order to provide an idea of the wide applicability of our approach. Some of these technologies have already been implemented into a biochip, whereas others are currently under active research for miniaturization and integration.

One of the most well-known and widely available is the DNA microarray technology [13, 22], which enables us to monitor the expression levels of a large number of genes simultaneously, providing a global view of gene expression information of the organism under study [2, 31, 20]. Depending upon the specific technology used, a DNA microarray data matrix reflects either absolute expression levels (*e.g.,* Affymetrix GeneChips [22]) or relative expression ratios (*e.g.,* cDNA microarrays [13]) of thousands of genes under hundreds of experimental conditions. Recently, CMOS-based integrated DNA microarrays have been reported [35, 17], and the scale of integration will continue to grow.

The tissue microarray (TMA) technique enables researchers to extract small cylinders of tissue from histological sections and arrange them in a matrix configuration on a recipient paraffin block such that hundreds can be analyzed simultaneously [8, 37]. TMA thus allows the rapid and cost effective validation of novel markers in multiple pathological tissue specimens.

The protein microarray is a crucial biomaterial for the rapid and high-throughput assay of many biological events where proteins are involved. In contrast to the DNA microarray, it has not been sufficiently established because of protein instability under the conventional dry conditions [19]. However, protein microarrays will eventually reveal vast amount of information essential to the understanding of gene functions and products.

Other examples include the fluorescence-based multiplexed cytokine immunoassays [41] and ligand chips [33]. In particular, using the cytokine chip, cytokine expression in breast cancer cells were examined and the chemokines associated with human cervical cancers were successfully identified [41].

3. DEFINITIONS AND OVERVIEW

Our method is a generalization of some homogeneous pattern mining techniques in the literature [29, 7, 43, 3, 21, 9, 44]. Thus, within a unified framework, our approach can find various types of homogeneous patterns. In particular, we focus on finding three specific types of homogeneous patterns in this paper. Their formal definitions are provided in Section 3.1. Some biological intuition behind these definitions is presented in Section 3.2. The problem statement and an overview of our approach will follow in Sections 3.3 and 3.4, respectively.

3.1 Definition of Homogeneous Patterns

Throughout the paper, we let A denote an input data matrix of real numbers with set of rows $R = \{1, 2, \ldots, n\}$ and set of columns $C = \{1, 2, \ldots, m\}$. That is, $A \in \mathbb{R}^{n \times m}$. We also denote the matrix A by pair (R, C). We first provide a formal definition of a homogeneous pattern.

DEFINITION 14.1 *Given $A = (R, C)$, an input matrix, and $\mathfrak{D}$, a certain condition defined on a matrix, let pair $P = (I, J)$ denote a submatrix of A, namely, $I \subseteq R$ and $J \subseteq C$. The submatrix P is called a* pattern *appearing in A under $\mathfrak{D}$, if P satisfies the condition $\mathfrak{D}$.*

EXAMPLE 14.2 *Figure 14-4(a) presents matrix* $A \in \mathbb{R}^{4\times4}$ *with* $R = C = \{1, 2, 3, 4\}$. *Let* $\mathfrak{D}$ *define a matrix in which the values on each row are constant. Fig. 14-4(b) shows* P_1, P_2, P_3, *some patterns appearing in the matrix A under the condition* $\mathfrak{D}$.

	1	2	3	4
1	1.0	1.0	1.0	1.0
2	1.0	1.0	1.0	3.0
3	2.0	2.0	3.0	2.0
4	3.0	3.0	2.0	3.0

(a) Input matrix A

	1	2	3
1	1.0	1.0	1.0
2	1.0	1.0	1.0

P_1

	1	2	4
1	1.0	1.0	1.0
3	2.0	2.0	2.0
4	3.0	3.0	3.0

P_2

	1	2
1	1.0	1.0
2	1.0	1.0
3	2.0	2.0
4	3.0	3.0

P_3

(b) Patterns P_1, P_2, and P_3

Figure 14-4. Example of an input matrix and some patterns appearing in the matrix. The condition $\mathfrak{D}$ here defines a matrix in which the values on each row are constant.

DEFINITION 14.3 *Let P be a pattern appearing in matrix A under condition* $\mathfrak{D}$. *The pattern P is called* homogeneous, *if any subset (or submatrix) of P is also a pattern appearing in the matrix A under the condition* $\mathfrak{D}$.

EXAMPLE 14.4 *In Fig. 14-4(b), it can be easily verified that any submatrix of* P_1, P_2, *and* P_3 *is another pattern appearing in the matrix A under the same condition* $\mathfrak{D}$, *since the values on each row of such a submatrix remain constant. Thus,* P_1, P_2, *and* P_3 *are all homogeneous patterns.*

We introduce three types of homogeneous patterns can be found by our pattern mining method proposed in this paper. Table 14-1 is for a quick lookup of related information. In what follows, the term *pattern* always means a homogeneous pattern, unless otherwise stated.

Table 14-1. Classification of homogeneous patterns.

Type	*Definition*	*Example*	*Related patterns in the literature*
1	14.6	Fig. 14-4, 14-5	xMOTIFs [29], δ-valid kj-pattern [7], GEMS [43]
2	14.9	Fig. 14-6	OPSM [3], OP-Cluster [21]
3	14.12	Fig. 14-7	δ-bicluster [9], δ-pCluster [42, 46], FLOC cluster [44]

Type 1 patterns.

DEFINITION 14.5 *For any set S on* $\mathbb{R}$, *the range of S, denoted by* RANGE(S), *is the difference between the largest and the smallest elements of S.*

DEFINITION 14.6 *Given matrix $A = (R, C)$ and threshold $\tau \geq 0$, a* Type 1 pattern *is a matrix denoted by (I, J) such that* (1) *$I \subseteq R$ and $J \subseteq C$; and* (2) *for each $i \in I$, $\text{RANGE}(\{a_{ij} | \forall j \in J\}) \leq \tau$.*

EXAMPLE 14.7 *Figure 14-5 presents an input matrix and some Type 1 patterns appearing in the matrix with respect to the parameter $\tau = 0.5$.*

	1	2	3	4
1	1.0	1.8	1.2	1.4
2	2.0	2.2	1.6	2.4
3	3.0	3.4	5.2	1.0
4	2.5	2.7	4.1	3.1

(a) Input matrix A

	1	2
2	2.0	2.2
3	3.0	3.4
4	2.5	2.7

P_1

	2	4
1	1.8	1.4
2	2.2	2.4
4	2.7	3.1

P_2

	1	3
1	1.0	1.2
2	2.0	1.6

P_3

	1	4
1	1.0	1.4
2	2.0	2.4

P_4

(b) Type 1 patterns P_1, P_2, P_3 and P_4

Figure 14-5. Type 1 patterns appearing in the input matrix A (the parameter $\tau = 0.5$).

Type 1 patterns are a representative example of the patterns that have a one-row-based or one-column-based definition. Examples include a pattern with constant values on rows, as seen in Fig. 14-4, or with constant values on columns. In the literature, patterns such as *xMOTIFs* [29], *δ-valid kj-patterns* [7] and *GEMS* [43] belong to this type.

The reader can easily verify that any Type 1 pattern is homogeneous. Furthermore, the following property holds for Type 1 patterns.

PROPERTY 1 *If (I_1, J_1) and (I_2, J_2) are both Type 1 patterns with respect to τ, then the pattern $(I_1 \cup I_2, J_1 \cap J_2)$ is also Type 1 with respect to τ.*

EXAMPLE 14.8 *The patterns shown in Fig. 14-4(b) satisfies Definition 14.6 with respect to $\tau = 0$, and thus P_1, P_2, and P_3 are all Type 1 patterns. Let $P_1 = (I_1, J_1)$, $P_2 = (I_2, J_2)$, and $P_3 = (I_3, J_3)$. Then, $I_3 = I_1 \cup I_2$ and $J_3 = J_1 \cap J_2$. Thus, Property 1 holds for these patterns.*

Type 2 patterns.

DEFINITION 14.9 *Given matrix $A = (R, C)$, let $J \subseteq C$ be a set of size $k \geq 2$ and let $\langle o_1, o_2, \ldots, o_k \rangle$ be a linear ordering of J. A* Type 2 pattern *is a matrix* *denoted by* (I, J) *such that* (1) $I \subseteq R$*; and* (2) *for each* $i \in I$, $a_{io_1} > a_{io_2} > \cdots > a_{io_k}$.

EXAMPLE 14.10 *Figure 14-6 presents a data matrix and Type 2 patterns appearing in it. The order of the values on each row is preserved. For example, for* $i \in I = \{1, 2\}$ *in* P_1, $a_{i1} > a_{i4} > a_{i2}$*; for* $i \in I = \{1, 2, 3\}$ *in* P_4, $a_{i3} > a_{i4} > a_{i2}$.

	1	2	3	4
1	3.0	1.0	4.0	2.0
2	4.0	1.0	3.0	2.0
3	2.0	1.0	4.0	3.0

(a) Input matrix A

	1	2	4
1	3.0	1.0	2.0
2	4.0	1.0	2.0

P_1

	2	3	4
1	1.0	4.0	2.0
2	1.0	3.0	2.0

P_2

	2	3	4
2	1.0	3.0	2.0
3	1.0	4.0	3.0

P_3

	2	3	4
1	1.0	4.0	2.0
2	1.0	3.0	2.0
3	1.0	4.0	3.0

P_4

(b) Type 2 patterns P_1, P_2, P_3 and P_4

Figure 14-6. An example of Type 2 patterns.

Type 2 patterns are a representative example of the patterns in which the order of the values (or some states defined by them) on a row or column is preserved for the other rows or columns as well. Examples in the literature include *OPSMs* [3] and *OP-Clusters* [21].

It can easily be verified that a Type 2 pattern is homogeneous. In addition, the following property holds for Type 2 patterns.

PROPERTY 2 *If both* (I_1, J_1) *and* (I_2, J_2) *are Type 2 patterns, then the pattern* $(I_1 \cup I_2, J_1 \cap J_2)$ *is also Type 2.*

EXAMPLE 14.11 *Property 2 holds for the patterns shown in Fig. 14-6(b). For example, let* $P_2 = (I_2, J_2)$, $P_3 = (I_3, J_3)$, *and* $P_4 = (I_4, J_4)$. *Then, it can be verified that* $I_4 = I_2 \cup I_3$ *and* $J_4 = J_2 \cap J_3$.

Type 3 patterns.

DEFINITION 14.12 *Given matrix* $A = (R, C)$ *and threshold* $\tau \geq 0$, *a* Type 3 pattern *is a matrix denoted by* $P = (I, J)$ *such that* (1) $I \subseteq R$ *and* $J \subseteq C$*; and* (2) *for any* 2×2 *submatrix* $\begin{bmatrix} e & f \\ g & h \end{bmatrix}$ *in* P, $|e - g - f + h| \leq \tau$.

EXAMPLE 14.13 *Figure 14-7 shows a data matrix and some Type 3 patterns appearing in the matrix with respect to the parameter $\tau = 1$.*

	1	2	3	4	5
1	6.0	9.0	2.0	5.0	4.0
2	2.0	3.0	4.0	3.0	6.0
3	2.0	2.0	3.0	6.0	7.0
4	3.0	4.0	5.0	2.0	6.0
5	4.0	7.0	3.0	1.0	4.0

(a) Input matrix A

	3	5
1	2.0	4.0
2	4.0	6.0
4	5.0	6.0
5	3.0	4.0

P_1

	1	2	3
2	2.0	3.0	4.0
3	2.0	2.0	3.0
4	3.0	4.0	5.0

P_2

	4	5
2	3.0	6.0
4	2.0	6.0
5	1.0	4.0

P_3

(b) Type 3 patterns P_1, P_2, and P_3

Figure 14-7. An example of Type 3 patterns (the parameter $\tau = 1$).

Type 3 patterns are to model a matrix in which the elements exhibit some coherent behavior. Examples include a matrix in which the value of the elements fluctuate in harmony and a matrix in which all elements have the same value. Type 3 patterns in Definition 14.12 are in essence equivalent to *δ-pClusters* [42] and closely related to *δ-biclusters*[1] [9] and *FLOC clusters* [44].

The reader can verify that Type 3 patterns are homogeneous. However, Properties 1 and 2 do not necessarily hold for Type 3 patterns. Also note that the same set of Type 3 patterns can be found from input A and the transpose of A. This is because two matrices $\begin{bmatrix} e & f \\ g & h \end{bmatrix}$ and $\begin{bmatrix} e & g \\ f & h \end{bmatrix}$ are indistinguishable in the definition since $|e - g - f + h| = |(e - g) - (f - h)| = |(e - f) - (g - h)|$.

3.2 Biology Behind the Definitions of Patterns

The three types of homogeneous patterns were defined in such a way that they can effectively capture important biological phenomena involved in various applications. For example, in gene co-regulation analysis, researchers are often interested in recognizing common fluctuations in the expression levels of multiple genes. Finding Type 2 and Type 3 patterns from gene expression data matrices may be useful in this application. Discovering Type 1 patterns can provide some biological insight for applications such as the task of marker gene identification, where we are interested in correlating the activity of one or more genes to specific subphenotypes and thus finding genes expressed only in some phenotypes. For more examples, the reader can refer to [23] as well as the references listed in Table 14-1.

3.3 Problem Statement

Given an input data matrix $A = (R, C)$, a specific definition $\mathfrak{D} \in \{$Definition 14.6, Definition 14.9, Definition 14.12$\}$, and the parameters specified in $\mathfrak{D}$, the problem of pattern mining is to find all maximal homogeneous patterns $P = (I, J)$ appearing in A under $\mathfrak{D}$. We search only maximal[2] patterns or those that are not contained by other patterns as a submatrix, since non-maximal patterns contain redundant information. Optionally, we can specify the minimum size of patterns in order not to generate too small patterns.

3.4 Overview of Our Approach

Our pattern mining algorithm consists of essentially two steps. The first step is to find special patterns called *atomic patterns*. The second step is to derive other general (non-atomic) patterns from the atomic patterns previously found. These two steps are detailed in Sections 4 and 5, respectively. Fig 14-8 provides a flowchart of our method, and Tables 14-2 and 14-3 list related information for a quick reference.

Table 14-2. Step 1 - finding atomic patterns.

Atomic pattern	*Definition*	*Algorithm*	*Example*
Type 1	Definition 14.14	Algorithm 1	Fig. 14-10
Type 2	Definition 14.16	Algorithm 2	Fig. 14-12
Type 3	Definition 14.18	Algorithm 3	Fig. 14-14

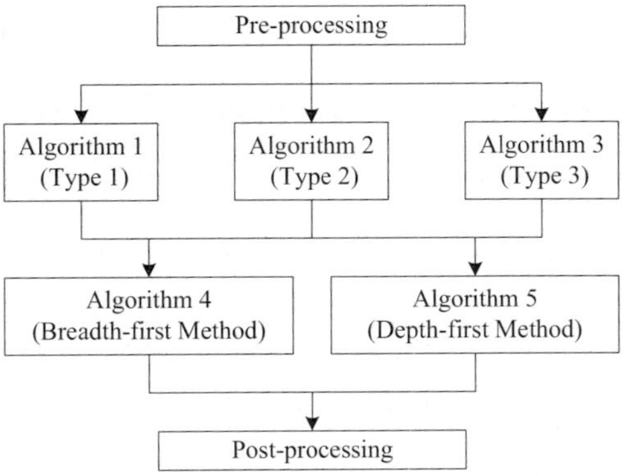

Figure 14-8. A flowchart of our method. The first step (Algorithms 1, 2, and 3) is to find atomic patterns. The second step (Algorithms 4 and 5) is to derive non-atomic patterns.

3.5 Notation

Table 14-4 lists some important notations that will be used throughout the paper, especially in Section 5.

Table 14-3. Step 2 - deriving non-atomic patterns from atomic patterns.

Method	*Details*	*Algorithm*	*Based upon*
Breadth-first	Section 5.3.0	Algorithm 4	Corollary 14.29
Depth-first	Section 5.3.0	Algorithm 5	Corollary 14.31

Table 14-4. Notations.

Notation	*Meaning*
$A = (R, C)$	Input matrix A with row set R and column set C; $A \in \mathbb{R}^{\lvert R\rvert \times \lvert C\rvert}$.
$P = (I, J)$	Pattern P with row set I and column set J; $I \subseteq R$ and $J \subseteq C$.
$\mathcal{J}$	Set of column index sets.
v	Vertex in the lattice graph (Algorithms 4 and 5).
$v.I$	Row index set associated with vertex v.
$v.\mathcal{J}$	Set of column index sets associated with vertex v.
$\mathfrak{J}$	Function $\mathfrak{J}$ (Definition 14.21).
$\mathfrak{J}(I)$	Image of set I under function $\mathfrak{J}$; essentially, set of column index sets.
$\mathfrak{J}_1, \mathfrak{J}_2, \mathfrak{J}_3$	Function $\mathfrak{J}$ with explicit type specification.

4. FINDING ATOMIC PATTERNS

Informally, an atomic pattern is represented by a matrix that has only one row (Type 1) or two rows (Types 2 and 3) but as many columns as possible. In this section, we provide the formal definition of atomic patterns and specific algorithms to find them.

4.1 Finding Type 1 Atomic Patterns

DEFINITION 14.14 *Given input matrix $A = (R, C)$ and threshold $\tau \geq 0$, a* Type 1 atomic pattern *for row $i \in R$ is a one-row matrix, denoted by pair $P = (\{i\}, J)$, that satisfies the following:* (1) *P is a Type 1 pattern on A; and* (2) *there is no J' such that $J' \supset J$ and $(\{i\}, J')$ is also a Type 1 pattern.*

The condition (2) in the above definition is not to generate those atomic patterns that are contained by others, since such patterns are redundant.

Algorithm 1 details our approach to find Type 1 atomic patterns of Definition 14.6. The key idea of this algorithm is simple: when the elements of a set S are sorted and arranged in the corresponding order, *range*(S) is simply the absolute difference between the first and the last elements of S. The worst-case complexity of the algorithm is polynomial in $|C|$, and the maximum number of atomic patterns found per row by Algorithm 1 is $(|C| - 1)$.

In **Lines 1–4**, the column indices are sorted in ascending order according to the value of the corresponding elements. The variables *begin* and *end* in

```
Algorithm 1: Finding Type 1 atomic patterns for one row
   input  : A = (R, C), a data matrix
   input  : i ∈ R, a row index
   input  : τ, a threshold
   output: J ⊆ C, ({i}, J) is a Type 1 atomic pattern
1  foreach j ∈ C do
2      D[j].val := a_ij;
3      D[j].ind := j;
4  Sort array D in ascending order with respect to the field val;
5  begin := 1;
6  end := 2;
7  while (end ≤ |C|) do
8      if (D[end].val − D[begin].val ≤ τ) then
9          end := end + 1;
10         if (end > |C|) then
11             Report J = {D[begin].ind, . . . , D[end − 1].ind};
12     else
13         Report J = {D[begin].ind, . . . , D[end − 1].ind};
14         repeat
15             begin := begin + 1;
16         until (begin = end) or
           (D[end].val − D[begin].val ≤ τ);
```

Figure 14-9. Algorithm 1.

Lines 5–6 are to point to the first and the last elements of the sub-array under consideration at some point. Inside the while loop in **Lines 7–16**, *J*, the column set of an atomic pattern, is generated as the variables *begin* and *end* are incremented. Note that multiple *J* can exist per row and overlap with each other. Since the array *D* is sorted, the algorithm only needs to compare in Line 8 the first element (*D*[*begin*]) and the last element (*D*[*end*]), in order to see if all the elements in the sub-array are similar. **In Lines 8–9**, the variable *end* is extended as long as $D[end].val - D[begin].val \leq \tau$. The algorithm reports *J* in **Line 11** or **Line 13**. **Lines 14–16** are to adjust the variable *begin* appropriately after one instance of *J* is found, because multiple overlapping instances of *J* can be found for each row.

EXAMPLE 14.15 *Figure 14-10(b) presents the Type 1 atomic patterns discovered by Algorithm 1 from the data matrix in Fig. 14-5(a), repeated here in Fig. 14-10(a) for convenience. The parameter used is* $\tau = 0.5$.

4.2 Finding Type 2 Atomic Patterns

DEFINITION 14.16 *Given input matrix* $A = (R, C)$, *a* Type 2 atomic pattern *for rows* $i, k \in R$ $(i \neq k)$ *is a two-row matrix, denoted by pair* $P = (\{i, k\}, J)$, *that satisfies the following:* (1) *P is a Type 2 pattern on A; and* (2) *there is no* J' *such that* $J' \supset J$ *and* $(\{i, k\}, J')$ *is also a Type 2 pattern.*

	1	*2*	*3*	*4*
1	1.0	1.8	1.2	1.4
2	2.0	2.2	1.6	2.4
3	3.0	3.4	5.2	1.0
4	2.5	2.7	4.1	3.1

(a)

Row $i \in R$	Column set $J \in 2^C$
1	$\{1,3,4\}$, $\{2,4\}$
2	$\{1,2,4\}$, $\{1,3\}$
3	$\{1,2\}$
4	$\{1,2\}$, $\{2,4\}$

(b)

Figure 14-10. (a) Input matrix *A*. (b) Type 1 atomic patterns found from *A* by Algorithm 1 ($\tau = 0.5$).

Our approach to find Type 2 atomic patterns is outlined in Algorithm 2. The main problem is to find a largest two-row matrix in which the order of the values on each row is preserved. The key is to exploit algorithms to solve the problem of finding *maximal common subsequences* (MCSs) of two sequences [11, 16].

Besides an input data matrix, Algorithm 2 takes an additional input parameter, min_J, to specify the minimum cardinality of the column set of an atomic pattern. This is to limit the total number of atomic patterns per pair of rows.

Algorithm 2: Finding Type 2 atomic patterns for pair of rows

input : $A = (R, C)$, a data matrix
input : $q, r \in R$, row indices ($q \neq r$)
input : $min_J \geq 0$, minimum column size
output: $J \subseteq C$, $(\{q, r\}, J)$ is an atomic Type 2 pattern

1 **foreach** $j \in C$ **do**
2 $\quad \hat{X}[j].val := a_{qj}$;
3 $\quad \hat{Y}[j].val := a_{rj}$;
4 $\quad \hat{X}[j].ind := \hat{Y}[j].ind := j$;
5 Sort arrays $\hat{X}$ and $\hat{Y}$ in descending order with respect to the field *val*;
6 Construct sequence $X = \langle x_1, x_2, \ldots, x_{|C|} \rangle$ such that $x_i = \hat{X}[i].ind$;
7 Construct sequence $Y = \langle y_1, y_2, \ldots, y_{|C|} \rangle$ such that $y_i = \hat{Y}[i].ind$;
8 Find $\mathcal{Z} = \{Z | Z$ is an MCS of X and Y, $|J| >= min_J\}$;
9 **foreach** $Z \in \mathcal{Z}$ **do**
10 $\quad$ Convert $Z = \langle z_1, z_2, \ldots \rangle$ to $J = \{z_1, z_2, \ldots\}$;
11 $\quad$ Return J;

Figure 14-11. Algorithm 2.

In **Lines 1–5**, the elements of each row are sorted with respect to their value, and the column indices are ordered accordingly. **Lines 6–7** are to convert arrays of column indices to sequences. In **Line 8**, an MCS-search algorithm is invoked. In **Lines 9-11**, each MCS found is converted to a set and returned.

The MCS problem has been extensively studied in the literature, and the typical solution relies on dynamic programming [11, 16]. The worst-case complexity of an algorithm to solve the MCS problem is polynomial in the length of sequences [11, 16]. Some details of an MCS algorithm can be found in the following example.

EXAMPLE 14.17 *Figure 14-12(c) presents the Type 2 atomic patterns discovered by Algorithm 2 from the data matrix in Fig. 14-6(a), repeated here in Fig. 14-12(a) for convenience. The parameter used is* $\min_J = 3$. *We can solve the MCS problem by modeling it as a sequence alignment problem [16]. In a sequence alignment problem, the scores for a match, a mismatch, and a space should first be assigned. For the MCS problem, the scores for a match, a mismatch, and a space are one, zero, and zero, respectively [16]. Fig. 14-12(b) shows the dynamic programming table for computing the MCS of two sequences* $X = \langle 3, 1, 4, 2\rangle$ *and* $Y = \langle 1, 3, 4, 2\rangle$*, derived from rows 1 and 2 of the input matrix A, respectively. We denote the entry in the i-th row and the j-th column by* $D[i, j]$*. We index the topmost row by* $i = 0$ *and use* $j = 0$ *to indicate the leftmost column. Let* x_i *and* y_j *denote the i-th and the j-th element of X and Y, respectively. Then, the optimal substructure of the MCS problem gives the following recursive formula [11]*

$$D[i, j] = \begin{cases} 0 & : \text{ if } i = 0 \text{ or } j = 0, \\ D[i-1, j-1] + 1 & : \text{ if } i, j > 0 \text{ and } x_i = y_j, \\ max(D[i-1, j], D[i, j-1]) & : \text{ if } i, j > 0 \text{ and } x_i \neq y_j, \end{cases}$$

In addition, we place a traceback pointer ($\nwarrow, \uparrow, \longleftarrow$) *in every entry* $D[i, j]$ *for* $i > 0$ *and* $j > 0$*, indicating where the value in the entry* $D[i, j]$ *originated (i.e.,* $D[i-1, j-1]$*,* $D[i-1, j]$*, or* $D[i, j-1]$*). Each MCS corresponds to a traceback path from the largest element in the table, and this path is obtained by following the traceback pointers, which are indicated by the bold arrows in Fig. 14-12(b). In this particular example, two MCS exist, namely,* $\langle 3, 4, 2\rangle$ *and* $\langle 1, 4, 2\rangle$*. More details on this procedure can be found in [11, 16].*

One possible improvement of Algorithm 2 would be to consider 'noisy ordering.' That is, we can devise an algorithm that can rearrange elements with similar values in such a way that a longer MCS can emerge. This heuristic will help to find atomic patterns with more columns, from which larger Type 2 patterns can potentially be derived.

4.3 Finding Type 3 Atomic Patterns

DEFINITION 14.18 *Given input matrix* $A = (R, C)$ *and threshold* $\tau \geq 0$*, a* Type 3 atomic pattern *for rows* $i, k \in R$ $(i \neq k)$ *is a two-row matrix, denoted by pair* $P = (\{i, k\}, J)$*, that satisfies the following:* (1) *P is a Type 3 pattern on A;* and (2) *there is no* J' *such that* $J' \supset J$ *and* $(\{i, k\}, J')$ *is also a Type 3 pattern.*

	1	2	3	4
1	3.0	1.0	4.0	2.0
2	4.0	1.0	3.0	2.0
3	2.0	1.0	4.0	3.0

(a)

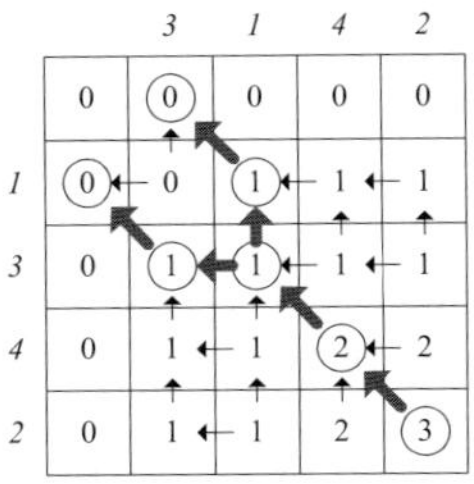

(b)

Row set $\{q, r\}$	Column set $J \in 2^C$
$\{1, 2\}$	$\{1, 2, 4\}$, $\{2, 3, 4\}$
$\{2, 3\}$	$\{2, 3, 4\}$
$\{1, 3\}$	$\{1, 2, 3\}$, $\{2, 3, 4\}$

(c)

Figure 14-12. (a) Input data. (b) Finding MCS for rows 1 and 2. (c) Type 2 atomic patterns found by Algorithm 2 ($min_J = 3$).

Algorithm 3 details our approach to find Type 3 atomic patterns defined in Definition 14.12. This algorithm is equivalent to Algorithm 1, except for Line 2. An informal explanation is as follows. Algorithm 1 is to find a Type 1 atomic pattern, or a one-row matrix in which the elements have similar values. Algorithm 3 is to find a two-row matrix in which any 2×2 submatrix $\begin{bmatrix} e & f \\ g & h \end{bmatrix}$ has similar values of $(e - g)$ and $(f - h)$, since $|e - g - f + h| = |(e - g) - (f - h)| \leq \tau$. Thus, we can use Algorithm 1 to find Type 3 atomic patterns, simply by subtracting the values in one row from the values in another and considering the result as a one-row matrix. This subtraction occurs in Line 2 of Algorithm 3. Some details helpful to understand our informal proof can be found in [42].

Algorithm 3: Finding Type 3 atomic patterns for pair of rows

```
   input : A = (R, C), a data matrix
   input : q, r ∈ R, row indices (q ≠ r)
   input : τ, a threshold
   output: J ⊆ C, ({q, r}, J) is an atomic Type 3 pattern
1  foreach j ∈ C do
2      D[j].val := a_qj − a_rj;
3      D[j].ind := j;
4  Sort array s in ascending order with respect to the field val;
5  begin := 1;
6  end := 2;
7  while (end ≤ |C|) do
8      if (D[end].val − D[begin].val ≤ τ) then
9          end := end + 1;
10         if (end > |C|) then
11             Report J = {D[begin].ind, ..., D[end − 1].ind};
12     else
13         Report J = {D[begin].ind, ..., D[end − 1].ind};
14         repeat
15             begin := begin + 1;
16         until (begin = end) or
           (D[end].val − D[begin].val ≤ τ);
```

Figure 14-13. Algorithm 3.

EXAMPLE 14.19 *Fig. 14-14(b) presents the Type 3 atomic patterns found by Algorithm 3 from the data in Fig. 14-7(a), repeated in Fig. 14-14(a). The parameter used is* $\tau = 1$.

	1	2	3	4	5
1	6.0	9.0	2.0	5.0	4.0
2	2.0	3.0	4.0	3.0	6.0
3	2.0	2.0	3.0	6.0	7.0
4	3.0	4.0	5.0	2.0	6.0
5	4.0	7.0	3.0	1.0	4.0

(a)

Row set $\{q, r\}$	Column set $J \in 2^C$
$\{1,2\}$	$\{3,5\}$
$\{1,3\}$	$\{3,4\}$
$\{1,4\}$	$\{1,4\}, \{3,5\}$
$\{1,5\}$	$\{1,2\}, \{3,5\}$
$\{2,3\}$	$\{1,5\}, \{1,2,3\}$
$\{2,4\}$	$\{1,2,3,5\}, \{4,5\}$
$\{2,5\}$	$\{3,4,5\}$
$\{3,4\}$	$\{1,2,3\}$
$\{3,5\}$	$\emptyset$
$\{4,5\}$	$\{3,4,5\}$

(b)

Figure 14-14. (a) Input matrix A. (b) Type 3 atomic patterns found from A by Algorithm 3 ($\tau = 1$).

5. OUR PATTERN MINING ALGORITHM

5.1 Overview

We can formulate the pattern mining problem in terms of a binary relation.

DEFINITION 14.20 *Given* $A = (R, C)$*, an input data matrix, and* $\mathfrak{D}$*, a specific definition of a pattern,* $\mathfrak{R}_{\mathfrak{D}}$ *is a binary relation on* $2^R \times 2^C$*:*

$$\mathfrak{R}_{\mathfrak{D}} = \{(I, J) | \textit{The pair } (I, J) \textit{ forms a pattern appearing in } A \textit{ under } \mathfrak{D}\}. \quad (14.1)$$

Under this definition, the objective of pattern mining is to find the elements of the relation $\mathfrak{R}_{\mathfrak{D}}$. We aim at finding only maximal patterns, as stated in Section 3.3.

Assume that we can find a function, denoted by $\mathfrak{I}$, that accepts as input $I \in 2^R$ and produces all maximal $J \in 2^C$ such that $(I, J) \in \mathfrak{R}_{\mathfrak{D}}$. Then, we may devise a naive algorithm that can provide all elements of $\mathfrak{R}_{\mathfrak{D}}$: First enumerate every $I \in 2^R$ and then feed it to the function $\mathfrak{I}$. Obviously, this approach is not feasible for a data matrix of non-trivial size since the powerset 2^R grows exponentially. Here we explain how to improve this idea of exploiting the function $\mathfrak{I}$ so that we can apply it to mining homogeneous patterns appearing in large-scale data matrices. Formally, the definition of $\mathfrak{I}$ is as follows.

DEFINITION 14.21 *Given matrix* $A = (R, C)$*,* $\mathfrak{I}$ *is a function that maps* $I \in 2^R$ *to the image* $\mathfrak{I}(I)$*, where*

$$\mathfrak{I}(I) = \{J \in 2^C | (I, J) \in \mathfrak{R}_{\mathfrak{D}} \textit{ and } \nexists J' \supset J \textit{ s.t. } (I, J') \in \mathfrak{R}_{\mathfrak{D}}\}.$$

In Section 5.2, we first explain how to define the function $\mathfrak{I}$ using the atomic patterns previously developed. In addition, we propose a novel technique to implement the function efficiently. The technique is based upon a data structure called *ZBDDs (zero-suppressed binary decision diagrams)* [27]. Section 5.3 then presents how to exploit the function $\mathfrak{I}$ to find homogeneous patterns, avoiding the exhaustive enumeration of $I \in 2^R$. We propose two algorithms: One uses a breadth-first approach and the other employs a depth-first approach. Finally, Section 5.4 provides remarks on algorithm complexity and other issues.

5.2 Representation and Implementation of the Function $\mathfrak{I}$

We first introduce the operator $\otimes$, which is essentially the pairwise intersection of two sets of subsets but does not contain redundant subsets.

DEFINITION 14.22 *Let $\mathcal{T}$ and $\mathcal{U}$ be two sets of subsets. Also let $Q = \{T \cap U | \forall T \in \mathcal{T}, \forall U \in \mathcal{U}\}$. Then, the binary operator $\otimes$ on $\mathcal{T}$ and $\mathcal{U}$ is defined as follows:*

$$\mathcal{T} \otimes \mathcal{U} = Q - \{Q | \exists Q' \in Q \text{ s.t. } Q' \supset Q\}. \tag{14.2}$$

THEOREM 14.23 *Let $\mathcal{T}, \mathcal{U}, \mathcal{W}$ be sets of sets. Then, $(\mathcal{T} \otimes \mathcal{U}) \otimes \mathcal{W} = \mathcal{T} \otimes (\mathcal{U} \otimes \mathcal{W})$.*

A proof of Theorem 14.23 is provided in Appendix 14.B. The associative law thus holds for the operator $\otimes$, and it is trivial to show that the commutative law, $\mathcal{T} \otimes \mathcal{U} = \mathcal{U} \otimes \mathcal{T}$, holds. Consequently, we can develop the following notation.

DEFINITION 14.24 *The pairwise intersection of the k sets of sets $\mathcal{T}_1, \mathcal{T}_2, \ldots, \mathcal{T}_k$ is denoted by*

$$\mathcal{T}_1 \otimes \mathcal{T}_2 \otimes \cdots \otimes \mathcal{T}_k = \bigotimes_{i=1}^{k} \mathcal{T}_i. \tag{14.3}$$

In addition, we define the operator COVER(S) for a set S in order to facilitate further explanation.

DEFINITION 14.25 *Given a set $S = \{s_1, s_2, \ldots, s_k\}$ with $k \geq 2$,* COVER(S) *is a minimum edge cover of K_k, the complete graph with k vertices, in which the set of vertices corresponds to S.*

EXAMPLE 14.26 $\{\{0, 1, 2\}, \{2, 3, 4\}\} \otimes \{\{0, 2\}, \{4, 5\}\} = \{\{0, 2\}, \{4\}\}$. *Let $S_1 = \{1, 2, 3, 4\}$ and $S_2 = \{10, 11, 12\}$. Then, a possible instance of* COVER(S_1) $= \{\{1, 3\}, \{2, 4\}\}$, *and an example of* COVER(S_2) $= \{\{10, 11\}, \{10, 12\}\}$.

Re-defining $\mathfrak{I}$ in terms of atomic patterns (APs). The image $\mathfrak{I}(I)$ defined in Definition 14.21 can be re-defined using atomic patterns by the following theorem (see Appendix 14.C for a proof).

THEOREM 14.27 *Let $\mathfrak{I}_1$, $\mathfrak{I}_2$, and $\mathfrak{I}_3$ denote the function $\mathfrak{I}$ for Types 1, 2, and 3, respectively. Given input data $A = (R, C)$, the image of $I \in 2^R$, or $\mathfrak{I}(I)$, can be represented as follows.*

- *When the set I has only one or two elements:*

$$\mathfrak{I}_1(\{r\}) = \{J|(\{r\}, J) \text{ is Type 1 AP for } r \in R\} \quad (14.4)$$

$$\mathfrak{I}_2(\{q, r\}) = \{J|(\{q, r\}, J) \text{ is Type 2 AP for } q, r \in R\} \quad (14.5)$$

$$\mathfrak{I}_3(\{q, r\}) = \{J|(\{q, r\}, J) \text{ is Type 3 AP for } q, r \in R\} \quad (14.6)$$

- *Otherwise:*

$$\mathfrak{I}_1(I) = \bigotimes_{\forall i \in I} \mathfrak{I}_1(\{i\}) \quad (14.7)$$

$$\mathfrak{I}_2(I) = \bigotimes_{\forall I' \in \text{COVER}(I)} \mathfrak{I}_2(I') \quad (14.8)$$

$$\mathfrak{I}_3(I) = \bigotimes_{\forall \{i,k\} \subseteq I} \mathfrak{I}_3(\{i, k\}) \quad (14.9)$$

To evaluate Equations (14.7),(14.8), and (14.9), we need to invoke the operator $\otimes$ at most $(|I| - 1)$, $(\lceil \frac{|I|}{2} \rceil - 1)$, and $\binom{|I|}{2}$ times, respectively.

EXAMPLE 14.28 *To find the pattern P_1 in Fig. 14-7(b), we can use Theorem 14.27 and the atomic patterns presented in Fig. 14-14(b) as follows:*

$$\begin{aligned}
\mathfrak{I}_3(\{1,2,4,5\}) &= \mathfrak{I}_3(\{1,2\}) \otimes \mathfrak{I}_3(\{1,4\}) \otimes \mathfrak{I}_3(\{1,5\}) \\
&\quad \otimes \mathfrak{I}_3(\{2,4\}) \otimes \mathfrak{I}_3(\{2,5\}) \otimes \mathfrak{I}_3(\{4,5\}) \\
&= \{\{3,5\}\} \otimes \{\{1,4\},\{3,5\}\} \otimes \{\{1,2\},\{3,5\}\} \\
&\quad \otimes \{\{1,2,3,5\},\{4,5\}\} \otimes \{\{3,4,5\}\} \otimes \{\{3,4,5\}\} \\
&= \{\{3,5\}\}.
\end{aligned}$$

Enhancement by dynamic programming. We can reduce the number of the $\otimes$ operations required to evaluate the equations in Theorem 14.27 by storing and re-using intermediate results. This idea is similar to the concept of dynamic programming. In the equations in Theorem 14.27, we can see that the *optimal substructure* [11] appears, which is a hallmark of the applicability of dynamic programming.

For example, the process of realizing $\mathfrak{I}_3$ can be compared to that of decomposing a complete graph into its cliques. We start our explanation with revisiting Example 14.28. Let K_k denote the complete graph with k graph vertices. Suppose that we have the graph K_4, in which the vertices represent the elements of $I = \{1, 2, 4, 5\}$ as shown in Fig. 14-15(a). In this figure, we can decompose the graph K_4 into $\binom{4}{2} = 6$ different K_2. This decomposition corresponds to evaluating Equation (14.9). We thus evaluated $\mathfrak{I}_3(\{1, 2, 4, 5\})$ using $\mathfrak{I}_3(\{1, 2\}), \mathfrak{I}_3(\{1, 4\}), \ldots, \mathfrak{I}_3(\{4, 5\})$ in Example 14.28.

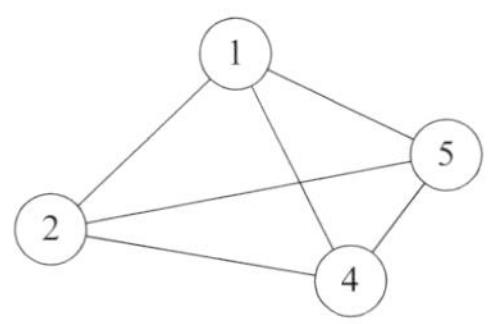

(a) Example 14.28: Theorem 14.27

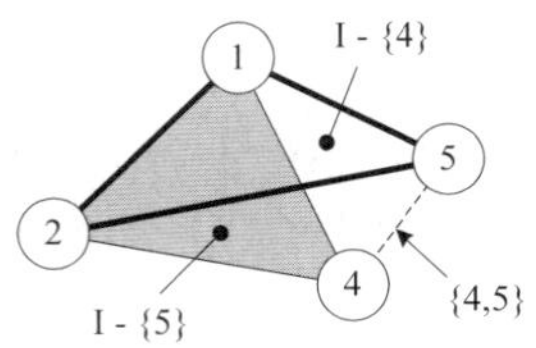

(b) Example 14.30: Corollary 14.29

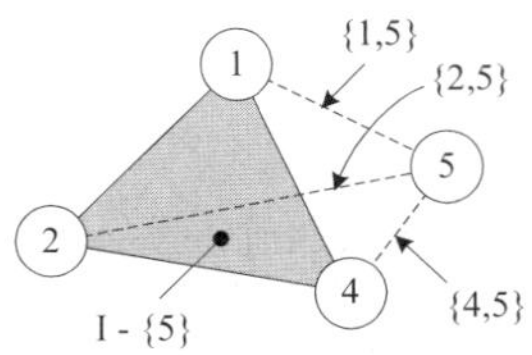

(c) Example 14.32: Corollary 14.31

Figure 14-15. Decomposition of the complete graph K_4.

Alternatively, we can decompose K_4 into two different K_3 and one K_2 as shown in Fig. 14-15(b). The shaded triangle represents the set $I - \{5\} = \{1, 2, 4\}$ and the triangle indicated by bold lines represents $I - \{4\} = \{1, 2, 5\}$. This suggests a different way of evaluating $\mathfrak{I}_3(\{1, 2, 4, 5\})$, namely, the evaluation using $\mathfrak{I}_3(\{1, 2, 4\}), \mathfrak{I}_3(\{1, 2, 5\})$, and $\mathfrak{I}_3(\{4, 5\})$.

The alternative decomposition of $\mathfrak{I}_1$ and $\mathfrak{I}_2$ corresponding to Fig. 14-15(b) is simpler. Since $I = (I - \{4\}) \cup (I - \{5\})$, $\mathfrak{I}_t(I)$ is merely $\mathfrak{I}_t(I - \{4\}) \otimes \mathfrak{I}_t(I - \{5\})$, for each $t \in \{1, 2\}$.

COROLLARY 14.29 *Given input data $A = (R, C)$, let set $I \in 2^R$ and suppose that $i, k \in I$ and $i \neq k$. Then, the image $\mathfrak{I}_t(I)$ for each type $t \in \{1, 2, 3\}$ can be represented as follows:*

$$\mathfrak{I}_1(I) = \mathfrak{I}_1(I - \{i\}) \otimes \mathfrak{I}_1(I - \{k\}) \qquad (14.10)$$

$$\mathfrak{I}_2(I) = \mathfrak{I}_2(I - \{i\}) \otimes \mathfrak{I}_2(I - \{k\}) \qquad (14.11)$$

$$\mathfrak{I}_3(I) = \mathfrak{I}_3(I - \{i\}) \otimes \mathfrak{I}_3(I - \{k\}) \otimes \mathfrak{I}_3(\{i, k\}) \qquad (14.12)$$

When applying Corollary 14.29, we need to call the operator $\otimes$ only once (Types 1 and 2) or twice (Type 3), as long as the intermediate results $\mathfrak{I}(I - \{i\})$

and $\mathfrak{I}(I - \{k\})$ are available. In Section 5.3.0, we explain how to store and re-use intermediate results efficiently using a breadth-first search algorithm.

EXAMPLE 14.30 *We can apply Corollary 14.29 to the previous example as follows:*

$$\begin{aligned}
\mathfrak{I}_3(\{1,2,4,5\}) &= \mathfrak{I}_3(\{1,2,4\}) \otimes \mathfrak{I}_3(\{1,2,5\}) \otimes \mathfrak{I}_3(\{4,5\}) \\
&= \{\{3,5\}\} \otimes \{\{3,5\}\} \otimes \{\{3,4,5\}\} \\
&= \{\{3,5\}\}.
\end{aligned}$$

Figure 14-15(c) shows another method to decompose the graph K_4 for Type 3 patterns. Here K_4 is decomposed into one K_3 and three different K_2. The shaded triangle represents the set $I - \{5\} = \{1,2,4\}$ and the dotted lines the sets $\{1,5\}, \{2,5\}$, and $\{4,5\}$. This suggests a different way of evaluating $\mathfrak{I}_3(\{1,2,4,5\})$ using $\mathfrak{I}_3(\{1,2,4\})$, $\mathfrak{I}_3(\{1,5\})$, $\mathfrak{I}_3(\{2,5\})$, and $\mathfrak{I}_3(\{4,5\})$. The decomposition of $\mathfrak{I}_1$ and $\mathfrak{I}_2$ corresponding to Fig. 14-15(c) remains in essence the same as the previous case.

COROLLARY 14.31 *Given input data $A = (R, C)$, let set $I \in 2^R$ and suppose that $k, l \in I$ and $k \neq l$. Then, the image $\mathfrak{I}_t(I)$ for each type $t \in \{1,2,3\}$ can be represented as follows:*

$$\mathfrak{I}_1(I) = \mathfrak{I}_1(I - \{k\}) \otimes \mathfrak{I}_1(\{k\}) \tag{14.13}$$

$$\mathfrak{I}_2(I) = \mathfrak{I}_2(I - \{k\}) \otimes \mathfrak{I}_2(\{k,l\}) \tag{14.14}$$

$$\mathfrak{I}_3(I) = \mathfrak{I}_3(I - \{k\}) \otimes \left\{ \bigotimes_{\forall i \in I, i \neq k} \mathfrak{I}_3(\{i,k\}) \right\} \tag{14.15}$$

In order to apply Corollary 14.31, we need to execute the operator $\otimes$ twice (Types 1 and 2) or at most $(|I| - 1)$ times (Type 3), as long as the result of $\mathfrak{I}(I - \{k\})$ is available. The number of $\otimes$ operations involved in the computation of $\mathfrak{I}_3$ in Corollary 14.31 is thus more than that in Corollary 14.29. However, it is easier to manage the partial results in Corollary 14.31, thus compensating the larger number of $\otimes$ operations required. Section 5.3.0 presents a depth-first search algorithm, which exploits Corollary 14.31 to evaluate $\mathfrak{I}$ efficiently.

EXAMPLE 14.32 *We can apply Corollary 14.31 to Example 14.28 as follows:*

$$\begin{aligned}
\mathfrak{I}_3(\{1,2,4,5\}) &= \mathfrak{I}_3(\{1,2,4\}) \otimes \mathfrak{I}_3(\{1,5\}) \otimes \mathfrak{I}_3(\{2,5\}) \otimes \mathfrak{I}_3(\{4,5\}) \\
&= \{\{3,5\}\} \otimes \{\{1,2\},\{3,5\}\} \otimes \{\{3,4,5\}\} \otimes \{\{3,4,5\}\} \\
&= \{\{3,5\}\}.
\end{aligned}$$

Efficient implementation of the operator $\otimes$ using ZBDDs. We assume the reader to be familiar with the basic concepts of Boolean functions and with the data structures commonly used for the symbolic manipulation of such functions such as *Binary Decision Diagrams* (BDDs) [5] and in particular *Zero-suppressed Binary Decision Diagrams* (ZBDDs) [26]. Appendix 14.A provides a brief introduction to ZBDDs. More extensive background material on this subject can be found in [12, 27, 26, 25].

In order to use ZBDDs to implement the operator $\otimes$, we first need to represent the operands of $\otimes$ by ZBDDs. A *combination* of m elements is an m-bit vector $\langle b_1, b_2, \ldots, b_m \rangle \in \mathbb{B}^m$, where $\mathbb{B} = \{0, 1\}$. The i-th bit reports whether the i-th element is contained in the combination. Thus, a set of combinations corresponds to a Boolean function $\mathfrak{f} : \mathbb{B}^m \to \mathbb{B}$ and can be represented by ZBDDs. The operand of $\otimes$ is a set of column sets, $\mathfrak{J}(I)$, and each column set $J \in \mathfrak{J}(I)$ can easily be converted to a combination as follows. Given input data $A = (R, C)$, assume $C = \{1, 2, \ldots, m\}$. Then, the set J corresponds to an m-bit vector $\langle b_1, b_2, \ldots, b_m \rangle$, where $b_i = 1$ if $i \in J$, and $b_i = 0$ otherwise. Representing this m-bit vector by ZBDDs is a standard procedure and is thus beyond the scope of this paper. We refer the interested reader to Appendix 14.A and [27, 26, 25] for further details.

EXAMPLE 14.33 *In Fig. 14-14(b),* $\mathfrak{J}_3(\{2,5\}) = \{\{3,4,5\}\}$. *The set* $\{3,4,5\}$ *can be converted to 5-bit vector* (00111) *and represented by the ZBDD in Fig. 14-16(a). In the same example,* $\mathfrak{J}_3(\{4,5\}) = \mathfrak{J}_3(\{2,5\})$. *Thus,* $\mathfrak{J}_3(\{4,5\})$ *can be represented by the identical ZBDD for* $\mathfrak{J}_3(\{2,5\})$ *without creating a new one.*

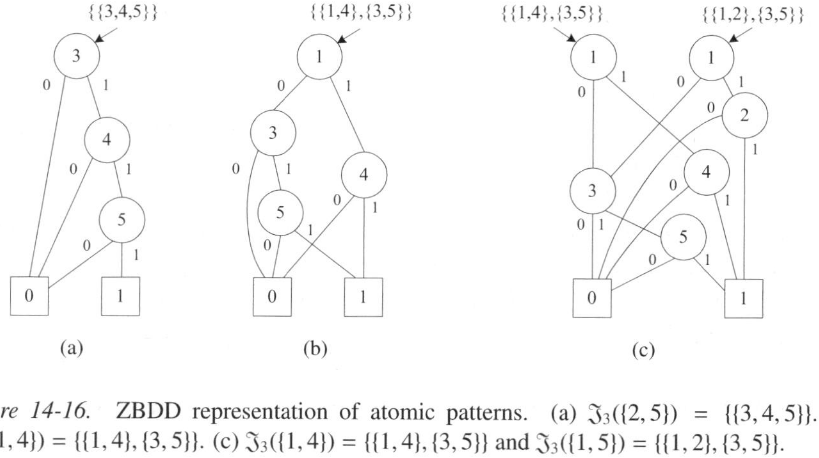

Figure 14-16. ZBDD representation of atomic patterns. (a) $\mathfrak{J}_3(\{2,5\}) = \{\{3,4,5\}\}$. (b) $\mathfrak{J}_3(\{1,4\}) = \{\{1,4\},\{3,5\}\}$. (c) $\mathfrak{J}_3(\{1,4\}) = \{\{1,4\},\{3,5\}\}$ and $\mathfrak{J}_3(\{1,5\}) = \{\{1,2\},\{3,5\}\}$.

EXAMPLE 14.34 *In Fig. 14-14(b),* $\mathfrak{J}_3(\{1,4\}) = \{\{1,4\},\{3,5\}\}$. *This corresponds to the set of combinations* $\{10010, 00101\}$ *and can be represented by the ZBDD in Fig. 14-16(b). Also,* $\mathfrak{I}_3(\{1,5\}) = \{\{1,2\},\{3,5\}\}$ *can share the part of the ZBDD for* $\mathfrak{I}_3(\{1,4\})$*, as shown in Fig. 14-16(c).*

Next, we implement the operator $\otimes$ by directly manipulating the ZBDDs representing the operands. This allows us to avoid explicit enumeration of the intermediate results, thus providing a large speed-up over the conventional methods to represent and manipulate sets [27, 26]. As is often the case with the operators defined on ZBDDs, we define the operator $\otimes$ recursively. We first partition a set of combinations into two smaller sets of combinations. Let $\mathcal{T}$ be a set of combinations. We partition $\mathcal{T}$ into $\mathcal{T}_1$ and $\mathcal{T}_0$ with respect to the i-th element b_i in such a way that $\mathcal{T}_1$ have all the combinations where $b_i = 1$, and $\mathcal{T}_0$ includes all the other combinations where $b_i = 0$. This partition can easily be done in a ZBDD by simply recognizing two subgraphs with respect to the topmost vertex. The subgraphs connected by the 1-edge and 0-edge correspond to $\mathcal{T}_1$ and $\mathcal{T}_0$, respectively. Based upon this partitioning, it follows that $\mathcal{T} \otimes \mathcal{U} = (\mathcal{T}_0 \otimes \mathcal{U}_0) \cup (\mathcal{T}_1 \otimes \mathcal{U}_0) \cup (\mathcal{T}_0 \otimes \mathcal{U}_1) \cup (\mathcal{T}_1 \otimes \mathcal{U}_1)$. Further implementation details can be found in [27, 26, 5, 4].

5.3 Finding Homogeneous Patterns

We present two methods to find homogeneous patterns. Both methods utilize the function $\mathfrak{I}$ previously developed. Before providing the details of these methods in Sections 5.3.0 and 5.3.0, we present an example to explain the fundamental ideas common in both methods.

An example of finding Type 3 patterns. Figure 14-17 shows the process to find the patterns in Fig. 14-7(b) from the data matrix in Fig. 14-7(a), in which the set of rows $R = \{1,2,3,4,5\}$. In the graphs shown in the figure, each vertex v has two associated fields, namely $v.I$ and $v.\mathcal{J}$. The field $v.I$ is to save a set of rows, and the field $v.\mathcal{J}$ is to store the image $\mathfrak{I}(v.I)$. The *level* of the vertex v is defined as the cardinality of $v.I$. Also, we connect vertex v_1 at level l and vertex v_2 at level $l+1$ by an edge if $v_1.I \subset v_2.I$.

Figure 14-17(a) presents a graph in which each vertex represents an elements in 2^R and a vertex is connected to others by the above rule. For example, $v.I = \{1,2\}$ for the vertex v indicated by "12". This vertex is connected to the vertices indicated by "123", "124", and "125".

We can make two key observations in the graph constructed as above. First, not all vertices need to be examined. Thus, we can avoid exhaustive enumeration of $I \in 2^R$. Second, the intermediate results required to apply Corollaries 14.29 and 14.31 are available from the vertices at the previous level.

The first observation is based upon the following fact: If $\mathfrak{I}(I) = \emptyset$, then $\mathfrak{I}(I') = \emptyset$ for all $I' \supseteq I$. This is because if the pair (I, J) does not represent a homogeneous pattern, then the pair (I', J) with $I' \supseteq I$ cannot be a homogeneous

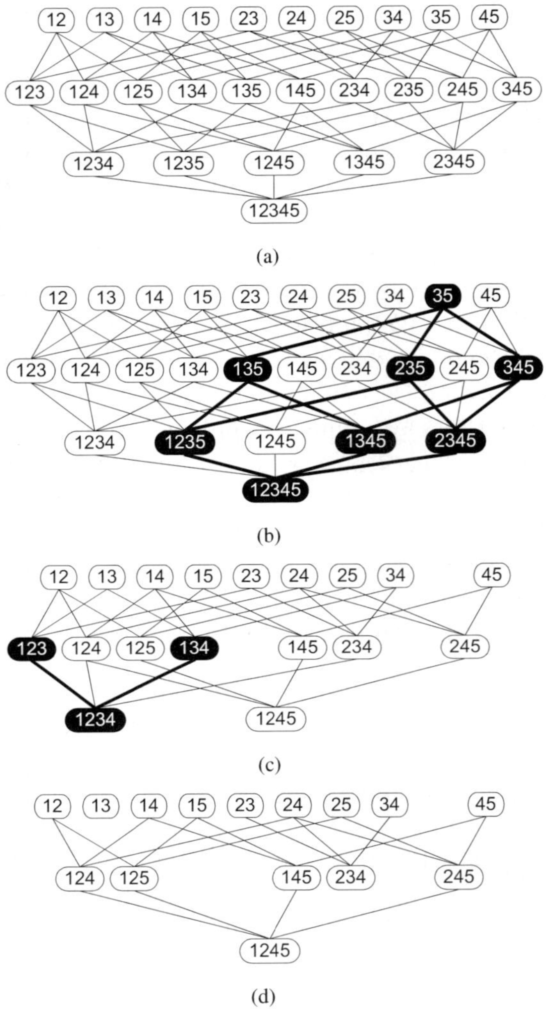

Figure 14-17. The process to find the patterns presented in Fig. 14-7(b). This is only for explanation of the idea, and in practice, we do not need the graph in its entirety all the time. Refer to Algorithms 4 and 5 for more details.

pattern, either. For example, in Fig. 14-14(b) we know that $\mathfrak{J}_3(\{3,5\}) = \emptyset$. Thus, it is possible to conclude that $\mathfrak{J}_3(I) = \emptyset$ for all $I \supseteq \{3,5\}$. Consequently, we can eliminate any vertex v such that $v.I \supseteq \{3,5\}$. In the graph in Fig. 14-17(b), the vertices to be deleted are indicated.

The example in Fig. 14-17(c) shows that another vertex elimination process is possible, starting from the vertices at level 3, namely, "123" and "134". The vertex "123" should be deleted because $\mathfrak{J}_3(\{1,2,3\}) = \mathfrak{J}_3(\{1,2\}) \otimes \mathfrak{J}_3(\{1,3\}) \otimes \mathfrak{J}_3(\{2,3\}) = \emptyset$. We can remove the vertex "134" similarly. Thus, any vertex

v such that $v.I \supseteq \{1, 2, 3\}$ or $v.I \supseteq \{1, 3, 4\}$ can be deleted. Finally, the graph in Fig. 14-17(d) shows the vertices that remain undeleted. It is these vertices to which we apply the function $\mathfrak{J}$ to find homogeneous patterns. Since the remaining vertices correspond to all $I \in 2^R$ that can potentially be the row set of a homogeneous pattern, applying the function $\mathfrak{J}$ to these vertices enables us to find all the homogeneous patterns that satisfy the input parameters specified.

To exploit the intermediate results stored in the vertices at the previous level, the breadth-first algorithm in Section 5.3.0 starts with the vertices at level 2 and proceed to level $l + 1$ from level l only after no vertex at level l is left. This is compatible with the decomposition of $\mathfrak{J}$ in Corollary 14.29. In contrast, the depth-first algorithm in Section 5.3.0 starts with vertex v at level 2 and proceeds until the algorithm examines all the vertices whose I set contains $v.I$. Then the algorithm starts with another vertex at level 2. This algorithm fits with the decomposition of $\mathfrak{J}$ in Corollary 14.31. Both algorithms find the same homogeneous patterns, although one can be faster than the other, depending upon the specific input data matrix and parameters used.

One important comment is in order. Obviously, it is not realistic to construct the graph like the one in Fig. 14-17(a) in its entirety, especially when the set R has many elements. The examples in Fig. 14-17 are only for explanation. As will be described in Algorithms 4 and 5, the breadth-first and the depth-first algorithms do not need to examine all the vertices simultaneously.

Breadth-first algorithm. Algorithm 4 details our breadth-first approach to find homogeneous patterns. The input is a data matrix, pattern type, and parameters for atomic pattern generation. The output are homogeneous patterns found from the input data matrix.

In **Line 1**, atomic patterns are generated by the algorithms explained in Section 4 with the input parameters.

In **Lines 2–11**, the base vertices at level 2 are generated. Each vertex v has three associated data fields. The fields $v.I$ and $v.\mathcal{J}$ are the same as explained in the previous section. The field $v.level$ is to store the level of the vertex v. For Type 2 or Type 3 patterns, a new vertex is created for each pair of rows, unless no atomic pattern exists for the pair. The base vertices for Type 1 patterns also start at level 2 by merging two atomic patterns.

In **Lines 13–34**, the algorithm iterates for each level and performs the following for each vertex at level l. **In Lines 16–17**, the algorithm reports any candidate patterns obtained from the previous iteration. In **Lines 18–34**, new vertices appearing at level $l + 1$ are generated. To this end, the algorithm examines two vertices v_i and v_j at level l. **Lines 21–22** are to test if the two vertices are qualified to create a new vertex at level $l + 1$. As long as the sets $v_i.I$ and $v_j.I$ have the same elements but one, the vertices v_i and v_j can create a new vertex at the next level. Since a vertex at level $l + 1$ should have only

Algorithm 4: Breadth-first pattern mining algorithm

```
input  : A = (R, C), a data matrix
input  : type ∈ {1, 2, 3}, pattern type
input  : parameters for atomic pattern generation
output: homogeneous patterns
1  Generate atomic patterns (see Section 4);
2  foreach {q, r} ⊆ R do
3      if type = 1 then
4          𝒥 := ℑ1({q}) ⊗ ℑ1({r});
5      else
6          𝒥 := ℑtype({q, r});
7      if 𝒥 ≠ ∅ then
8          Create vertex v;
9          v.I := {q, r};
10         v.𝒥 := 𝒥;
11         v.level := 2;
12 maxLevel := |R|;
13 for l = 2 to maxLevel do
14     for i = 1 to numVertices in level l do
15         v_i := i-th vertex in level l;
16         foreach J ∈ v_i.𝒥 do
17             Collect pattern (v_i.I, J);
18         if l < maxLevel then
19             for j = i + 1 to numVertices in level l do
20                 v_j := j-th vertex in level l;
21                 I := v_i.I ∪ v_j.I;
22                 if |I| ≠ l + 1 then next;
23                 if a vertex for I exists then next;
24                 if type = 3 then
25                     e_i := the element in I − v_i.I;
26                     e_j := the element in I − v_j.I;
27                     𝒥 := v_i.𝒥 ⊗ v_j.𝒥 ⊗ ℑ3({e_i, e_j});
28                 else
29                     𝒥 := v_i.𝒥 ⊗ v_j.𝒥;
30                 if 𝒥 ≠ ∅ then
31                     Create vertex v;
32                     v.I := I;
33                     v.𝒥 := 𝒥;
34                     v.level := l + 1;
35         Remove v_i;
36 Remove redundancy and return remaining patterns;
```

Figure 14-18. Algorithm 4.

one more rows than a vertex at level l, if the union of $v_i.I$ and $v_j.I$ has more than $l + 1$ elements, the two vertices v_i and v_j cannot spawn a new vertex in the next level. For example, if $v_i.I = \{1, 2, 3\}$ and $v_j.I = \{1, 2, 4\}$ then these two vertex can create a new vertex, v, at level 4 with $v.I = \{1, 2, 3, 4\}$. In contrast, if $v_i.I = \{1, 2, 3\}$ and $v_j.I = \{1, 4, 5\}$, then they cannot generate a new vertex at level 4, because the row sets differ by two elements. This way of creating new vertices is to avoid exhaustive enumeration. In **Line 23**, if the two vertices v_i and v_j are eligible for creating a new vertex, the algorithm sees whether the corresponding vertex already exists or not. If not, the algorithm computes the set $\mathcal{J}$ for this new vertex by Corollary 14.29 in **Lines 24–29**. In **Lines 30–34**, the new vertex v is actually created and stored for further reference in the next iteration, if the set $\mathcal{J}$ is not empty. Otherwise, no new vertex is created. This corresponds to removing all the downstream of the vertex v in which $v.\mathcal{J} = \emptyset$ (e.g., Fig. 14-17(b) and 14-17(c)).

In **Line 35**, a vertex is deleted as soon as it becomes of no use. Thus, the algorithm can keep at most two levels of the vertices at a time, rather than the entire graph.

In **Line 36**, any redundant patterns are removed and the remaining patterns are returned.

Depth-first algorithm. In Section 5.3.0, we explained our breadth-first approach. In this section, we introduce an alternative pattern mining algorithm using a depth-first approach. We start the description with the examples in Fig. 14-17 and 14-19. In order to visit the vertices in the depth-first sense, we need to restructure the graph. In particular, we remove some edges from the graph in Fig. 14-17(a) so that the graph becomes the trie in Fig. 14-19(a). A *trie* [1] is a special structure for representing sets of words. Here we regard the set $I \in 2^R$ as a word assuming a total order among the elements in R. For instance, we can assume the total order $1 \prec 2 \prec 3 \prec 4 \prec 5$ for the set $R = \{1, 2, 3, 4, 5\}$. Hence, the set $I = \{1, 2, 3\}$ corresponds to the word "123". This word is inserted in to the trie as the descendant of the word "12" and as the parent of the words "1234" and "1235", as shown in Fig. 14-19(a).

Algorithm 4 provides the details of our depth-first approach. In **Lines 2–6**, the algorithm traverse the trie in preorder. More precisely, our algorithm *constructs* the trie in preorder rather than *traverses* it. In other words, the algorithm creates vertices whenever necessary and deletes them afterwards rather than keeping the trie in its entirety all the time. In **Line 7**, the algorithm reports the homogeneous patterns produced after removing redundant patterns, if any.

For each vertex v encountered in this *preorder* construction of the trie, the algorithm performs the following (**Lines 8–36**). The algorithm computes $v.\mathcal{J}$ by Corollary 14.31 in **Line 18–24**. If the set $v.\mathcal{J}$ is empty, the algorithm does not proceed to examining the descendant vertices and returns to the parent vertex (**Line 25**). This is equivalent to deleting all the descendant vertices in Fig. 14-19(b). If the set $v.\mathcal{J}$ is not empty, the algorithm produces homogeneous patterns $(v.I, J)$ for all $J \in \mathfrak{J}(v.I)$ in **Lines 26–27**. Then the algorithm creates a list of the descendant vertices in **Lines 28–29**. This step is necessary because the algorithm does not keep the entire trie all the time, and thus the vertex v is not already connected to its children. The "largest" element in Line 28 means the "largest" element in the total order we are assuming among the elements of R. For example, the largest element of the set $\{1, 2, 4\}$ is the element "4",

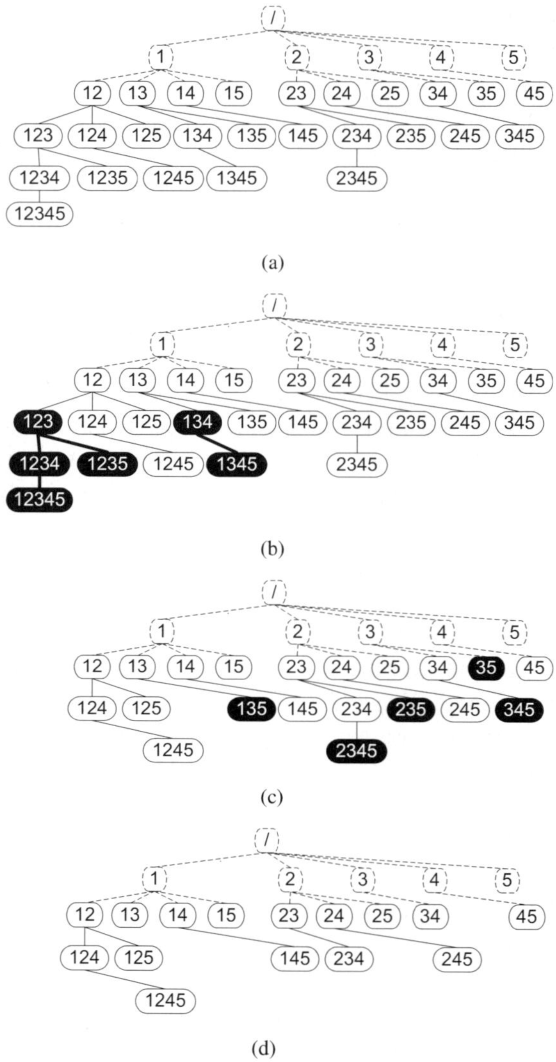

Figure 14-19. An example to explain the depth-first pattern mining algorithm.

assuming 1 < 2 < 4. In **Lines 30–35**, the algorithm creates the descendant vertices and visit them to repeat the steps performed in Lines 8–36.

5.4 Remarks

The pattern mining problem addressed in this paper is related to the problem of finding the maximum edge biclique in a bipartite graph, a problem known to be NP-complete [23, 30]. Although the worst-case complexity of Algorithms 4 and 5 is exponential in the number of rows in the input data matrix, the

```
Algorithm 5: Depth-first Pattern Mining Algorithm
  input  : A = (R, C), a data matrix
  input  : type ∈ {1, 2, 3}, pattern type
  input  : parameters for atomic pattern generation
  output: homogeneous patterns

Generate atomic patterns (see Section4);
foreach {q, r} ⊆ R do
    Create vertex v;
    v.I := {q, r};
    ConstructTrieInPreorder(v);
    delete v;
Remove redundancy and return remaining patterns;

procedure ConstructTrieInPreorder (vertex v)
begin
    if |v.I| = 2 then
        if type = 1 then
            v.𝒥 := ℑ_1({q}) ⊗ ℑ_1({r});
        else
            v.𝒥 := ℑ_type(v.I);
    else
        vertex p := v.parent;
        k := the element in v.I − p.I;
        if type = 1 then
            v.𝒥 := p.𝒥 ⊗ ℑ_1({k});
        else if type = 2 then
            k' := any element in p.I;
            v.𝒥 := p.𝒥 ⊗ ℑ_2({k, k'});
        else
            v.𝒥 := p.𝒥 ⊗ (⊗_{∀i∈p.I} ℑ_3({i, k}));
    if v.𝒥 = ∅ then return;
    foreach J in v.𝒥 do
        Collect pattern (v.I, J);
    l := the "largest" element in v.I wrt a total order ≺;
    I := {i | i ∈ R and l ≺ i};
    foreach i ∈ I do
        create vertex w;
        w.I := v.I ∪ {i};
        w.parent := v;
        ConstructTrieInPreorder(w);
        delete w;
end
```

Figure 14-20. Algorithm 5.

execution time on typical benchmarks is practical, as will be shown in Section 6. This is due to the efficient techniques such as the ZBDD-based symbolic manipulations and the dynamic programming approach, which enable us to avoid the exhaustive and explicit enumeration of the intermediate results. In particular, the role of the ZBDDs is crucial in this study. Without using the ZBDDs, it would not be possible to achieve the efficiency that the current implementation of our algorithm shows.

In fact, ROBDDs and variants such as ZBDDs have been widely used to solve many practical instances of intractable problems [12, 27, 25]. Some data analysis methods recently proposed [45, 28] rely on this idea of managing massive data through the symbolic representation of Boolean functions. In particular, the method proposed in this study is a generalization and extension of our earlier work [45], which focused only on finding Type 3 patterns.

Our pattern mining algorithms discussed so far are exact in the sense that they can find all the patterns that satisfy specific input parameters. If desired, it is possible to employ a heuristic algorithm that runs quickly but can find only a subset of the possible patterns. For example, we can implement "greedy" $\otimes$ operator that reports only k largest (in terms of cardinality) sets, which will make the cardinality of $\mathcal{J}$ decrease. We can also utilize a measure of overlap such as Jaccard's coefficient [24] to avoid generating "similar-looking" atomic patterns, thus reducing the number of atomic patterns considered in later steps.

6. EXPERIMENTAL RESULTS

We implemented our method in C++ on a 3.06 GHz Linux machine with 4 GB RAM. We used the libraries provided by the BDD packages CUDD (`http://vlsi.colorado.edu/~fabio/CUDD/`) and EXTRA (`http://www.ee.pdx.edu/~alanmi/research/extra.htm`) for the implementation of the operator $\otimes$ defined in Section 5.2. For comparison, we also developed an implementation of our method without using the ZBDDs. Table 14-5 shows the algorithms used in our experiments. Table 14-6 lists the parameters used for each experiment presented in this section.

Table 14-5. The pattern mining methods tested in the experiments.

ID	*Name/description*	*Algorithm employed*	*Pattern types supported*
METHOD 1	δ-biclustering [9]	Greedy iterative search	Type 3
METHOD 2	δ-pClustering [42]	Exhaustive enumeration	Type 3
METHOD 3	GEMS [43]	Gibbs sampling	Types 1, 3
METHOD 4	Our method[a]	Algorithms 4, 5	Types 1, 2, 3
OUR METHOD	Our method[b]	Algorithms 4, 5	Types 1, 2, 3

[a]Implemented without using ZBDDs.
[b]Fully implemented.

6.1 Synthetic Data Sets

To verify the correctness of our method, we tested it with synthetic data sets that have pre-defined embedded patterns. The synthetic data were prepared as follows. We first created null matrices of 100 rows and 5 different numbers of columns (1K, 3K, 6K, 9K and 12K). We then replaced the elements of each matrix with random numbers ranged from 0 to 500. For the matrix of $n = 100$ rows and $m \in \{1K, 3K, 6K, 9K, 12K\}$ columns, we embedded $0.05m$ pre-defined patterns that have at least $0.1n$ rows and at least $0.01m$ columns.

Table 14-6. The algorithm parameters used for the experiments.

Experiments		*Algorithms and parameters*							
Figure	*Data*	OURS	METHOD 1		METHOD 2			METHOD 3	
		τ	α	δ	δ	nr	nc	a	w
14-21	Synthetic	0	1.2	1000	0	10	10–120	0.01	250
14-22	Kidney[18]	20–25	1.2	150	20–25	10–14	18–20	0.01–0.1	25–75
14-23(a)	Yeast[10, 39]	75–80	1.2	300	10–15	80	10–12	0.25	15–30
14-23(b)	Kidney[18]	20–25	1.2	150	20–25	10–14	18–20	0.01–0.1	25–75
14-24(a)	Yeast[10, 39]	75–80	1.2	300	10–15	80	10–12	0.25	15–30
14-24(b)	Kidney[18]	20–25	1.2	150	20–25	10–14	18–20	0.01–0.1	25–75
14-26(a)	Yeast[10, 39]	75–80	1.2	300	10–15	80	10–12	0.25	15–30
14-26(b)	Kidney[18]	20–25	1.2	150	20–25	10–14	18–20	0.01–0.1	25–75

Each pre-defined pattern was created in such a way that the values in every row or column fluctuate in harmony[3] and that all the methods involved in the experiment can detect it.

We invoked the methods listed in Table 14-5 with the parameters specified in Table 14-6. The results are depicted in Fig. 14-21. Fig. 14-21(a) shows the response time spent by each method in order to find all the embedded patterns. Fig. 14-21(b) shows the plot of the total number of patterns discovered by each method given the same time as spent by our method implemented with ZBDDs. We can see in the experiments that it takes less time for our method to find all the embedded patterns and that our method can find more patterns given the same time, compared with the other methods tested. Especially, we observed that the use of ZBDDs indeed provides a substantial speed-up over the alternative implementation without ZBDDs.

6.2 Biological Data Sets

We tested our methods and the alternatives with a couple of large-scale data set obtained from actual biological experiments. Specifically, we used the gene expression data sets produced by Affymetrix gene chips and cDNA microarrays, since this type of data is one of the largest and most widely available. As previously emphasized, our method is applicable to other types of data as well, as long as they can be represented by a matrix of real numbers. For more information on gene expression data, we refer the interested to [20].

Data preparation. We used two different data sets. The first was the yeast *Saccharomyces cerevisiae* cell cycle expression data [10, 39] produced by Affymetrix gene chip experiments. This data set contains the expression information of 2,884 genes under 17 experimental conditions. The second was the cDNA microarray data for renal cell carcinoma [18], which represents

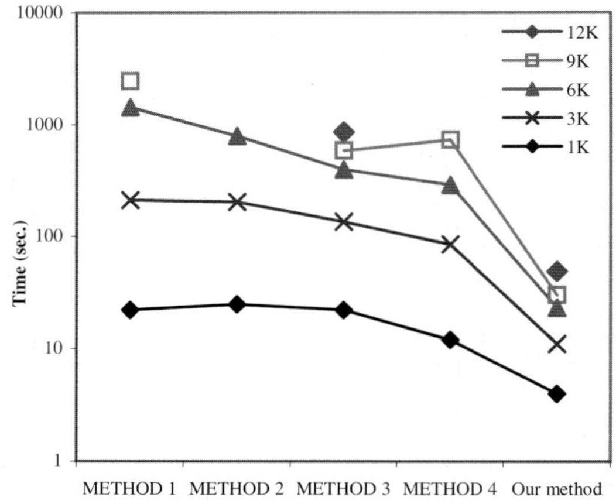

(a)

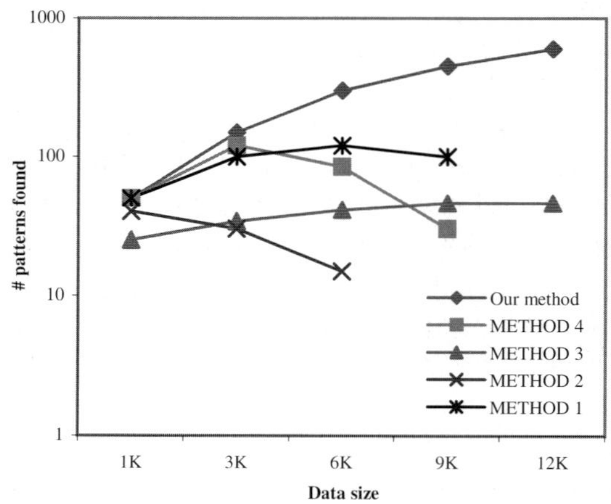

(b)

Figure 14-21. Performance comparison using synthetic data sets. The missing points on the plots mean that the corresponding experiment could not be finished in reasonable time. (a) The response time spent by each method in order to find all the embedded patterns from the synthetic data sets of various sizes. (b) The number of patterns found by each method within the same time spent as our method.

the expression levels of 1,876 genes under 27 different experimental conditions. Usually, gene expression data is arranged in a data matrix, in which each row corresponds to one gene and each column to one experimental condition. Fig. 14-22 shows the heat map of this data set and some patterns found by our method.

Running time comparison. We ran the methods listed in Table 14-5 with the parameters specified in Table 14-6. In the plots in Fig. 14-23(a) and 14-

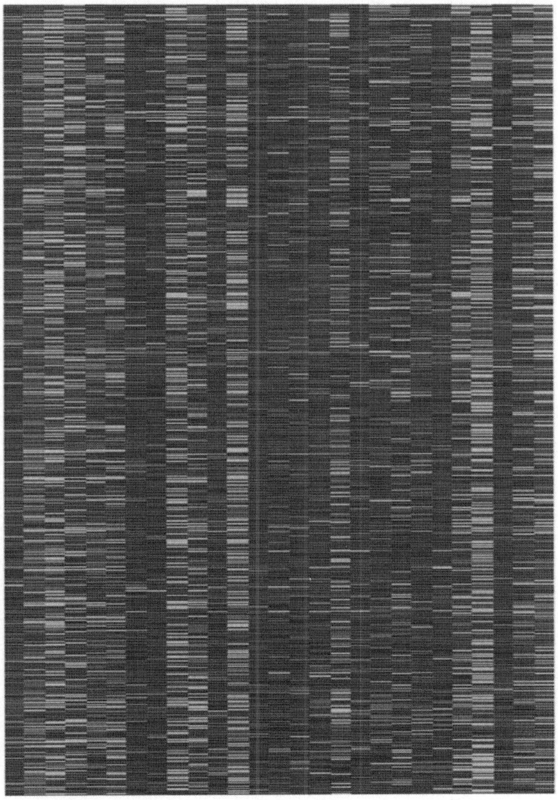

Figure 14-22. The heat map of the renal cell carcinoma data [18] and some patterns found by our method. The legend for the heat map is also presented in the upper right corner. The red color indicates up-regulation whereas the green color represents down-regulation. The black color means no change in regulation level. The entire data matrix has 1,876 rows (genes) and 27 columns (experimental conditions).

23(b) we compared the time to find the first *k* patterns from the yeast cell cycle data and the renal cell carcinoma data, respectively. The *x*-axis is the number of patterns produced and the *y*-axis is the response time to find these patterns. Our methods as well as METHOD 2 and METHOD 4 (see Table 14-5) do not take as input the exact number of patterns to find. Thus, we ran these algorithms multiple times with different parameter values to find approximately *k* patterns. For METHOD 1 and METHOD 3, the exact number of patterns to find was specified as input parameters.

Pattern quality evaluation. The experiments presented so far have demonstrated that our method outperforms the alternatives in terms of efficiency and the number of patterns. Here we present more experimental results to show that our method can produce statistically more significant and

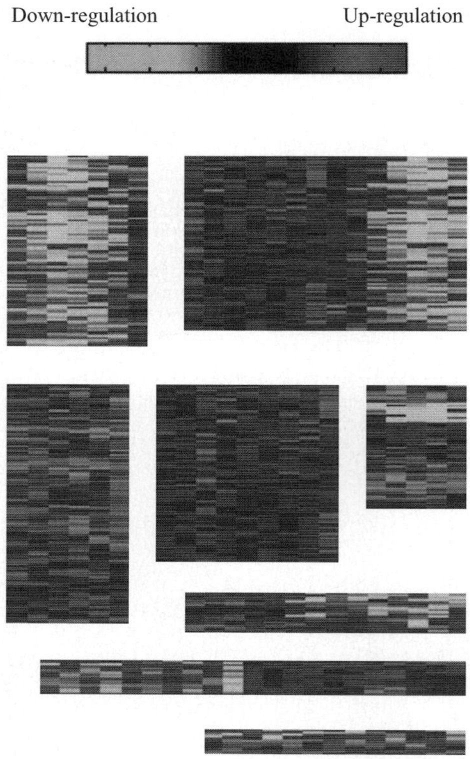

Figure 14-22 (continued). Some patterns (submatrices) discovered by our method.

biologically more meaningful patterns, thus suggesting that our method can be helpful to the researchers in biomedicine as well. To this end, we utilize the concept of *correspondence plots* [38] and the *Mean Squared Residue* (MSR) scores [9].

Correspondence plot. To assess the statistical significance and biological meaning of discovered patterns, we employed a technique [38] that enables us to compute the p-value of each pattern with respect to known (putatively correct) biological knowledge. Suppose prior knowledge classifies N genes into M classes, $H_1, H_2, \ldots H_M$. Let P be a pattern with g genes and assume that out of those g genes, g_j genes belong to class H_j. Assuming the most abundant class for the genes in P is H_i, the hypergeometric distribution is used

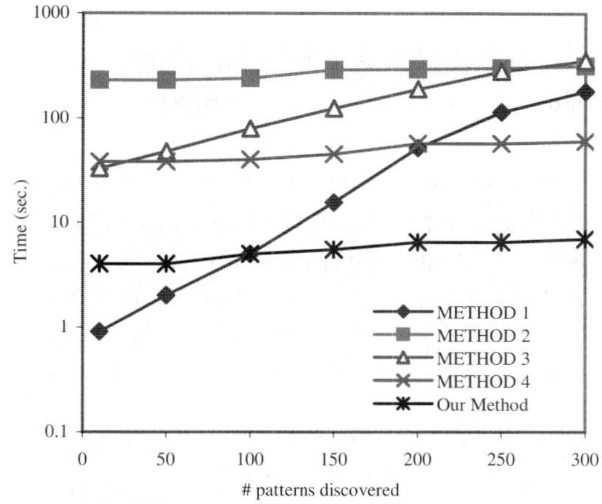

(a) Yeast cell cycle data [10, 39]

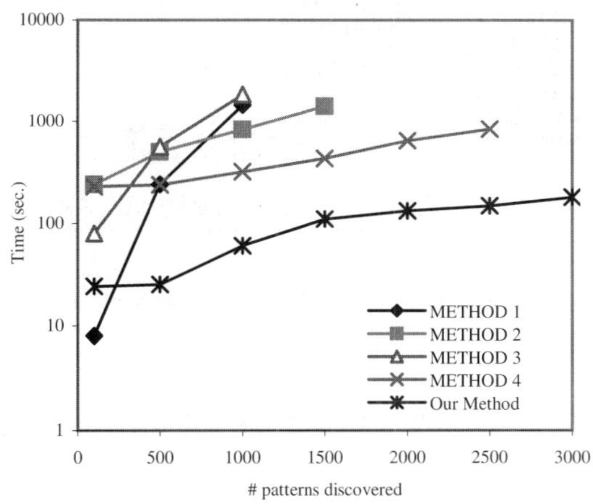

(b) Renal cell carcinoma data [18]

Figure 14-23. Performance comparison using biological data sets. The missing points on the plots mean that the corresponding experiment could not be finished in reasonable time.

to calculate p, the p-value of the pattern P:

$$p = \sum_{k=g_i}^{g} \frac{\binom{|H_i|}{k}\binom{N-|H_i|}{g-k}}{\binom{N}{g}}. \tag{14.16}$$

That is, the p-value corresponds to the probability of obtaining at least g_i elements of the class H_i in a random set of size g. As the known biological knowledge, the categories of yeast genes proposed by Tavazoie et al. [39] and the human genes classes reported by Higgins et al. [18] were used for our experiments.

In the correspondence plot, early departure of a curve from the x-axis indicates the existence of patterns with low p-values. Consequently, the area under a curve approximately shows the degree of statistical significance of the patterns used to draw the curve.

Figure 14-24 presents the correspondence plots for the patterns generated by several different methods on the yeast data and the renal cell carcinoma data. The plots also include randomly generated patterns. Both plots indicate that the patterns shown are all far from the random noise. It is also demonstrated that the patterns found by our algorithm tend to be more statistically significant than the others, meaning that our patterns conform to the known biological classification more accurately.

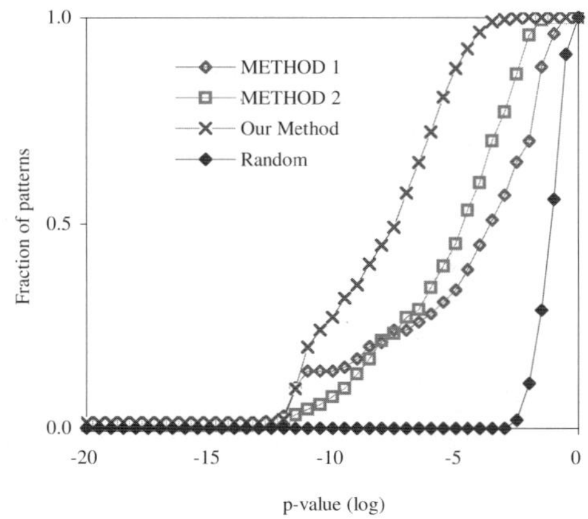

(a) Yeast cell cycle [39]

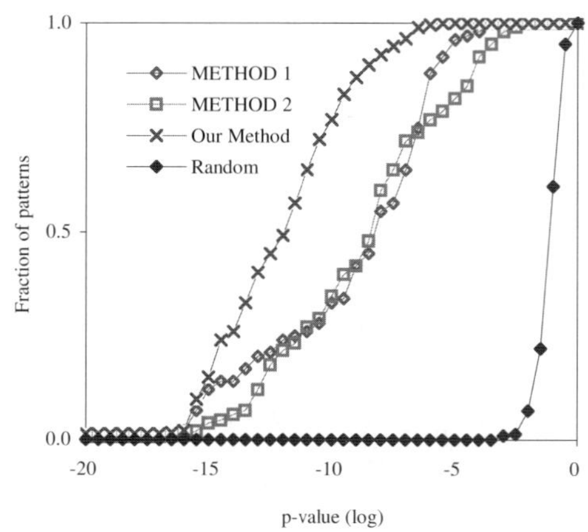

(b) Renal cell carcinoma [18]

Figure 14-24. Correspondence plots that show the distribution of p-values of the produced patterns with respect to prior biological knowledge. A point (x, y) on the plot presents the fraction (y) of patterns whose p-value is at most x.

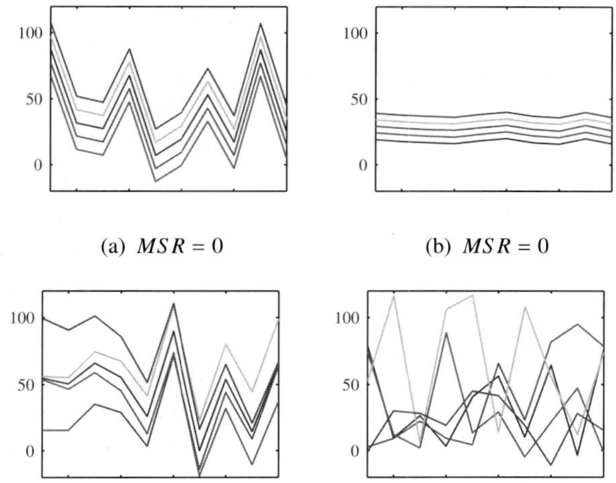

(a) $MSR = 0$ (b) $MSR = 0$

(c) $MSR = 103$ (d) $MSR = 946$

Figure 14-25. MSR scores as a measure of pattern quality. A low MSR value typically means a high level of coherence, and vice versa [9].

MSR scores. The *Mean Squared Residue* (MSR) sores can measure the degree of coherence exhibited by the elements in a matrix [9]. In the *analysis of variance* (ANOVA), a *residue* is defined for each element in a matrix as the difference between the element and the mean of all elements of the matrix [32]. The residue of element a_{ij} of a matrix denoted by pair (I, J) is $r_{ij} = a_{ij} - \overline{a}_{i\bullet} - \overline{a}_{\bullet j} + \overline{a}_{\bullet\bullet}$, where $\overline{a}_{i\bullet}$ is the mean of the ith row, $\overline{a}_{\bullet j}$ the mean of the jth column, and $\overline{a}_{\bullet\bullet}$ is the mean of all elements in A. The MSR of the matrix is then defined as

$$MSR(I, J) = \frac{1}{|I||J|} \sum_{i \in I, j \in J} r_{ij}^2. \tag{14.17}$$

Thus, a low value of residue typically means a high level of coherence, and vice versa [9]. For example, the MSR score of the patterns depicted in Fig. 14-25(a) and 14-25(b) is zero, since the values fluctuate in harmony. In contrast, the pattern shown in Fig. 14-25(d) is very noisy and thus has a higher MSR score. The pattern in Fig. 14-25(c) has an intermediate MSR score. Consequently, the MSR scores can be useful to evaluate the quality of patterns of all types defined in this study.

Figure 14-26(a) and 14-26(b) show box plots to compare the MSR scores of the patterns discovered from the yeast cell cycle and the renal cell carcinoma data, respectively. A *box plot* is a plot that represents graphically several descriptive statistics such as median and percentiles of a data sample [14]. The reader can refer to the caption of Fig. 14-26 to find how to read a box plot.

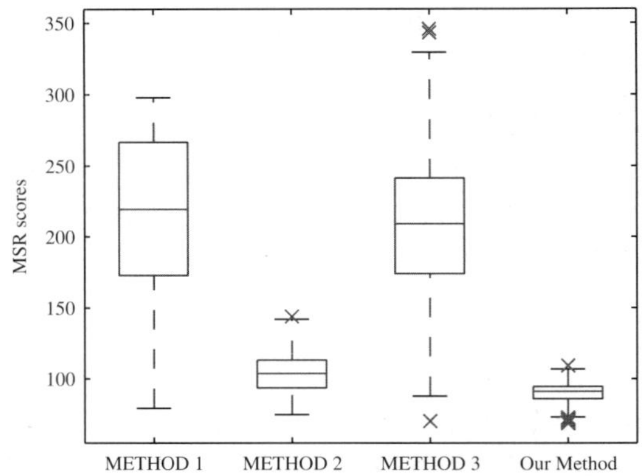

(a) Yeast cell cycle [39]

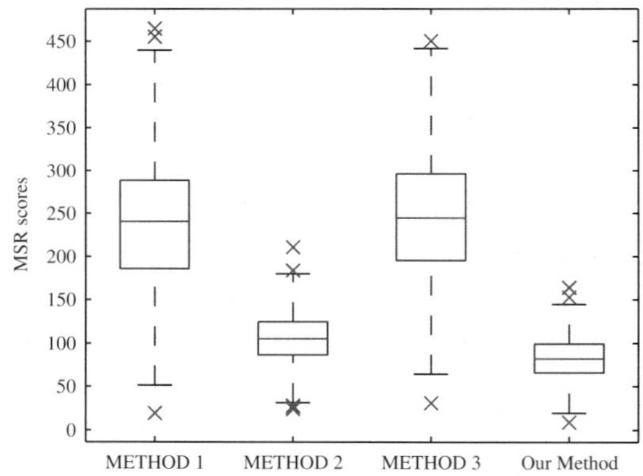

(b) Renal cell carcinoma [18]

Figure 14-26. Box plots for MSR comparison. The line in the middle of a box indicates the position of the median. The upper and lower boundaries of the box represent the location of the 75th and 25th percentiles, respectively. The symbol 'x' outside the ends of the tails corresponds to outliers.

As is evident from the box plots, the patterns found by our method have the lowest median MSR scores in our experiments. To quantitatively establish this observation, we performed the Wilcoxon rank sum test [14] to compare the patterns discovered by our method with the others. We generated approximately 500 patterns per method for each data set and compared a group of patterns found by our method with another group of patterns detected by an alternative method. In all the cases we tested, the difference in the median was statistically significant at 0.01% level ($P < 0.0001$). This result shows that our method tends to find better patterns with respect to the MSR scores.

7. CONCLUSIONS

Compared with conventional biological data acquisition techniques, better productivity, reliability and speed are possible through the miniaturization and integration realized in microfluidics-based biochips. Given that the throughput of these fascinating technologies is growing fast, it is crucial to have efficient computational tools to analyze the large-scale biological data obtained. In this paper, we proposed an effective pattern mining method that can be useful for a variety of biochip applications. Given a data matrix, the proposed method can find patterns appearing as a submatrix of the data matrix. In particular, we introduced the notion of homogeneous patterns and formulated the problem of finding three types of homogeneous patterns frequently encountered in the literature. We also mathematically characterized the problem and developed a novel method applicable to large-scale biological data. The proposed method employed dynamic programming as well as efficient data structures such as zero-suppressed decision diagrams (ZBDDs), which were particularly useful to extend the scalability of our method. Consequently, given a data matrix of practical scale, our approach can find with great efficiency all the homogeneous patterns that satisfy specific input parameters. We tested our method with the biochip data produced by Affymetrix gene chips and cDNA microarrays and confirmed the effectiveness of our approach. Therefore, we conclude that our method can provide the designers of high-throughput biochips with the necessary feedback for the next design iteration in timely manner.

Appendix: Zero-suppressed Binary Decision Diagrams

In most combinatorial applications, sets of combinations (see Section 5.2.0) are *sparse*, which is defined as follows [25]:

- The sets contain only a small fraction of the 2^n possible bit vectors;
- Each bit vector in the sets has many zeroes.

The *Zero-suppressed Binary Decision Diagram* (ZBDD) [26, 27] is an efficient data structure to represent and manipulate a set of combinations. Minato [26, 27] proposed two reduction rules to reduce ordinary BDDs to ZBDDs: (1) merge equivalent sub-graphs, and (2) if the 1-edge of a node v points to the 0-terminal vertex, then eliminate v and redirect all incoming edges of v to the 0-successor of v. Consequently, ZBDDs can exploit both types of sparsity defined above and provide an efficient representation for manipulating large-scale sets of combinations [25].

For instance, the ROBDD in Fig. 14.A-1(a) represents a set of combinations {1000, 0100} for four input variables (*abcd*). Each path from the root vertex to the 1-leaf corresponds to a combination. By applying the ZBDD reduction rules, we can reduce the BDD in Fig. 14.A-1(a) to the ZBDD in Fig. 14.A-1(b), which is more compact in terms of the number of vertices. As shown in Fig. 14.A-1(c), Minato [27] compared the size of a ZBDD with that of an ROBDD for a large set of combinations and showed that ZBDDs provide a much more compact representation of sets of combinations in most cases.

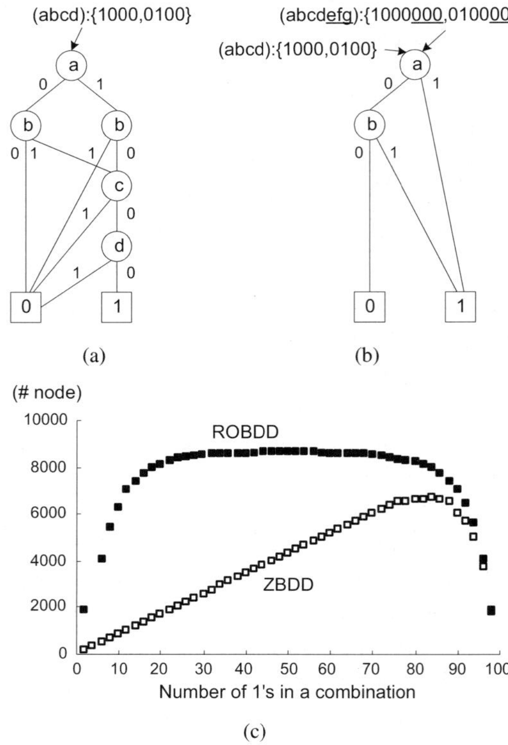

Figure 14.A-1. Representation of a set of combinations. (a) ROBDD representation. (b) ZBDD representation. (c) Comparison of ROBDD and ZBDD [27].

ZBDD representations are independent of the number of input variables as long as the combination remains the same, which is due to the "zero-suppression" effect. Consequently, we do not need to fix the number of input variables before generating graphs, and ZBDDs automatically suppress the variables that never appear in any combination [27]. For example, a set of combinations {1000000, 0100000} for seven variables ($abcdefg$) is represented by the same ZBDD in Fig. 14.A-1(b). This property does not hold for other types of BDDs.

Appendix: Proof of Theorem 14.23

We can prove the theorem by showing that (1) $(\mathcal{T} \otimes \mathcal{U}) \otimes \mathcal{W} \supseteq \mathcal{T} \otimes (\mathcal{U} \otimes \mathcal{W})$ and (2) $(\mathcal{T} \otimes \mathcal{U}) \otimes \mathcal{W} \subseteq \mathcal{T} \otimes (\mathcal{U} \otimes \mathcal{W})$.

PROOF 1 *We first prove (1). For the sake of contradiction, assume that* $(\mathcal{T} \otimes \mathcal{U}) \otimes \mathcal{W} \subset \mathcal{T} \otimes (\mathcal{U} \otimes \mathcal{W})$. *This means that there exists a set S such that* $S \in \mathcal{T} \otimes (\mathcal{U} \otimes \mathcal{W})$ *and* $S \notin (\mathcal{T} \otimes \mathcal{U}) \otimes \mathcal{W}$. *Assume that* $S = T \cap (U \cap W)$, *where* $T \in \mathcal{T}$, $U \in \mathcal{U}$, *and* $W \in \mathcal{W}$. *Since* $S \notin (\mathcal{T} \otimes \mathcal{U}) \otimes \mathcal{W}$, *there must exist* $W' \in \mathcal{W}$ *such that* $(T \cap U) \cap W \subset (T \cap U) \cap W'$. *By the associative law for basic set intersection,* $T \cap (U \cap W) = (T \cap U) \cap W \subset (T \cap U) \cap W' = T \cap (U \cap W')$. *In other words, if* $S \notin (\mathcal{T} \otimes \mathcal{U}) \otimes \mathcal{W}$, *then there must exist* $W' \in \mathcal{W}$ *such that* $(U \cap W) \subset (U \cap W')$. *However, since*

$S \in \mathcal{T} \otimes (\mathcal{U} \otimes \mathcal{W})$, there cannot exist $W' \in \mathcal{W}$ such that $(U \cap W) \subset (U \cap W')$. We have reached a contradiction and thus our original assumption that $(\mathcal{T} \otimes \mathcal{U}) \otimes \mathcal{W} \subset \mathcal{T} \otimes (\mathcal{U} \otimes \mathcal{W})$ must be false. Therefore, $(\mathcal{T} \otimes \mathcal{U}) \otimes \mathcal{W} \supseteq \mathcal{T} \otimes (\mathcal{U} \otimes \mathcal{W})$. By symmetry, we can prove $(\mathcal{T} \otimes \mathcal{U}) \otimes \mathcal{W} \subseteq \mathcal{T} \otimes (\mathcal{U} \otimes \mathcal{W})$ in a similar way.

We have shown that $(\mathcal{T} \otimes \mathcal{U}) \otimes \mathcal{W} \supseteq \mathcal{T} \otimes (\mathcal{U} \otimes \mathcal{W})$ and $(\mathcal{T} \otimes \mathcal{U}) \otimes \mathcal{W} \subseteq \mathcal{T} \otimes (\mathcal{U} \otimes \mathcal{W})$, which completes the proof. ■

Appendix: Proof of Theorem 14.27

The derivation of Equations (14.4), (14.5), and (14.6) is straightforward from the definition of atomic patterns. If the set I has only one row (Type 1) or two (Types 2 and 3), the image $\Im(I)$ simply consists of the column set of atomic patterns for the row(s) in I. Equations (14.7) and (14.8) can be derived from the generalization of Properties 1 and 2, respectively, by replacing the operator $\cap$ with the operator $\otimes$ defined in Section 5.2.

Here we focus on the derivation of Equation (14.9). To this end, we first propose the following lemma.

LEMMA C.1 *Let (I, J) be a homogeneous pattern. If $\{i, k\} \subseteq I$, then there exists at least one set $J' \in \Im_3(\{i, k\})$ such that $J \subseteq J'$.*

PROOF 2 *Assume $J \supset J'$ for all $J' \in \Im_3(\{i, k\})$. Since (I, J) is a homogeneous pattern and $I \supseteq \{i, k\}$, its sub-pattern $(\{i, k\}, J)$ is also a homogeneous pattern under the same definition. By definition, if $J' \in \Im_3(\{i, k\})$, then there exists no $J'' \supset J'$ such that $(\{i, k\}, J'')$ is yet another homogeneous pattern under the same definition. We have reached a contradiction and thus our original assumption that $J \supset J'$ for all $J' \in \Im_3(\{i, k\})$ must be false. Therefore, there must be at least one instance of $J' \in \Im_3(\{i, k\})$ such that $J \subseteq J'$.*

Now we derive Equation (14.9). Let $P = (I, J)$ be a maximal homogeneous pattern. Then, by Lemma C.1, for each $\{i, k\} \subseteq I$, there exists *at least one* set $J_{\{i,k\}} \in \Im_3(\{i, k\})$ such that $J \subseteq J_{\{i,k\}}$. For the sake of explanation, assume for now that *only one* such $J_{\{i,k\}}$ is contained in each $\Im_3(\{i, k\})$. Then, it follows that

$$J \subseteq \bigcap_{\forall \{i,k\} \subseteq I} J_{\{i,k\}}. \tag{14.C.1}$$

Moreover, since the pattern P is maximal, there is no J' such that $J' \supset J$ and $J' \subseteq \bigcap_{\forall \{i,k\} \subseteq I} J_{\{i,k\}}$. Thus, the following equation holds for J:

$$J = \bigcap_{\forall \{i,k\} \subseteq I} J_{\{i,k\}}. \tag{14.C.2}$$

In general, each $\Im_3(\{i, k\})$ can have multiple instances of $J_{\{i,k\}}$, not only one as previously assumed. Thus, we can have multiple instances of Equation (14.C.2), which can be compactly represented using the operator $\otimes$ defined in Section 5.2:

$$J \in \bigotimes_{\forall \{i,k\} \subseteq I} \left\{ J_{\{i,k\}} | J_{\{i,k\}} \in \Im_3(\{i, k\}), J \subseteq J_{\{i,k\}} \right\}. \tag{14.C.3}$$

Finally, suppose that we replace the operands of $\otimes$ in Relation 14.C.3 with $\{J_{\{i,k\}} | J_{\{i,k\}} \in \Im_3(\{i, k\})\} = \Im_3(\{i, k\})$, removing the constraint on J. Then, we can find not only the set J but also the other column sets that can form a homogeneous pattern with the row set I:

$$\{\text{all column sets that can form a Type 3 pattern with } I\} = \bigotimes_{\forall \{i,k\} \subseteq I} \Im_3(\{i, k\}). \tag{14.C.4}$$

By definition, the operator ⊗ gives only maximal sets. Therefore, Equation (14.C.4) is equivalent to Equation (14.9). We have derived Equation (14.9), and this completes the proof of Theorem 14.27. ■

Notes

1. δ-biclusters are not homogeneous patterns, since a subcluster of a δ-bicluster is not necessarily a δ-bicluster [9, 42]. However, δ-biclusters are included here because they also aim at modeling coherent behavior of matrix elements, and it has been reported that δ-biclusters are closely related to δ-pClusters in many aspects [42, 46].

2. Formally, a pattern $P = (I, J)$ is called *maximal* if there is no pattern $P' = (I', J')$ such that $I \subseteq I'$ and $J \subseteq J'$ under the identical input conditions.

3. Every row or column is a shifted version of each other; examples are shown in Fig. 14-25(a) and 14-25(b).

REFERENCES

[1] A. V. Aho, J. E. Hopcroft, and J. D. Ullman. *Data Structures and Algorithms*. Addison-Wesley, Reading, Massachusetts, 1983.

[2] R. B. Altman and S. Raychaudhuri. Whole-genome expression analysis: challenges beyond clustering. *Current Opinion in Structural Biology*, 11:340–347, 2001.

[3] A. Ben-Dor, B. Chor, R. Karp, and Z. Yakhini. Discovering local structure in gene expression data: The order-preserving submatrix problem. *Journal of Computational Biology*, 10(3-4):373–384, 2003.

[4] K. S. Brace, R. L. Rudell, and R. E. Bryant. Efficient implementation of a BDD package. In *Proceedings of the 27th Design Automation Conference*, pages 40–45, 1990.

[5] R. E. Bryant. Graph-based algorithms for Boolean function manipulation. *IEEE Trans. Comput.*, C-35(8):677–691, Aug. 1986.

[6] R. E. Bryant. Binary decision diagrams and beyond: Enabling technologies for formal verification. In *IEEE/ACM International Conference on Computer Aided Design, ICCAD, San Jose/CA*, pages 236–243. IEEE CS Press, Los Alamitos, 1995.

[7] A. Califano, G. Stolovitzky, and Y. Tu. Analysis of gene expression microarrays for phenotype classification. In *Proc. Int Conf Intell Syst Mol Biol*, pages 75–85, 2000.

[8] W. Chen, M. Reiss, and D. J. Foran. A prototype for unsupervised analysis of tissue microarrays for cancer research and diagnostics. *IEEE Trans Inf Technol Biomed*, 8(2):89–96, 2004.

[9] Y. Cheng and G. M. Church. Biclustering of expression data. In *Proceedings of ISMB*, pages 93–103, 2000.

[10] R. J. Cho, M. J. Campbell, E. A. Winzeler, L. Steinmetz, A. Conway, L. Wodicka, T. G. Wolfsberg, A. E. Gabrielian, D. Landsman, D. J. Lockhart, and R. W. Davis. A genome-wide transcriptional analysis of the mitotic cell cycle. *Mol. Cell*, 2(1):65–73, July 1998.

[11] T. H. Cormen, C. E. Leiserson, R. L. Rivest, and C. Stein. *Introduction to Algorithms*. The MIT Press, Cambridge, Massachusetts, 2001.

[12] G. De Micheli. *Synthesis and Optimization of Digital Circuits*. McGraw-Hill, New York, 1994.

[13] J. DeRisi, L. Penland, P. O. Brown, M. L. Bittner, P. S. Meltzer, M. Ray, Y. Chen, Y. A. Su, and J. M. Trent. Use of a cDNA microarray to analyse gene expression patterns in human cancer. *Nature Genetics*, 14(4):457–460, 1996.

[14] S. Drăghici. *Data Analysis Tools for DNA Microarrays*. Chapman & Hall/CRC, Florida, 2003.

[15] A. C. R. Grayson, R. S. Shawgo, A. M. Johnson, N. T. Flynn, Y. Li, M. J. Cima, and R. Langer. A BioMEMS review: MEMS technology for physiologically integrated devices. *Proceedings of the IEEE*, 92(1):6–21, 2004.

[16] D. Gusfield. *Algorithms on String, Trees and Sequences: Computer Science and Computational Biology*. Cambridge, New York, 1997.

[17] A. Hassibi and T. H. Lee. A programmable electrochemical biosensor array in $0.18\mu m$ standard CMOS. In *Proceedings of IEEE International Solid-State Circuits Conference*, February 2005.

[18] J. P. T. Higgins, R. Shinghal, H. Gill, J. H. Reese, M. Terris, R. J. Cohen, M. Fero, J. R. Pollack, M. van de Rijn, and J. D. Brooks. Gene expression patterns in renal cell carcinoma assessed by complementary DNA microarray. *American Journal of Pathology*, 162(3):925–932, March 2003.

[19] S. Kiyonaka, K. Sada, I. Yoshimura, S. Shinkai, N. Kato, and I. Hamachi. Semi-wet peptide/protein array using supramolecular hydrogel. *Nature Mater.*, 3(1):58–64, 2004.

[20] I. S. Kohane, A. T. Kho, and A. J. Butte. *Microarrays for an Integrative Genomics*. The MIT Press, Cambridge, Massachusetts, 2003.

[21] J. Liu, J. Yang, and W. Wang. Biclustering in gene expression data by tendency. In *Proceedings of CSB*, pages 182–193, 2004.

[22] D. Lockhart, H. Dong, M. Byrne, M. Follettie, M. Gallo, M. Chee, M. Mittmann, C. Wang, M. Kobayashi, H. Horton, and E. L. Brown. Expression monitoring by hybridization to high-density oligonucleotide arrays. *Nature Biotechnology*, 14(13):1675–1680, 1996.

[23] S. C. Madeira and A. L. Oliveira. Biclustering algorithms for biological data analysis: A survey. *IEEE Transactions on Computational Biology and Bioinformatics*, 1(1):24–45, 2004.

[24] C. D. Manning and H. Schütze. *Foundations of Statistical Natural Language Processing*. MIT Press, Cambridge, Massachusetts, 1999.

[25] C. Meinel and T. Theobald. *Algorithms and Data Structures in VLSI Design*. Springer, Berlin, 1998.

[26] S. Minato. Zero-suppressed BDDs for set manipulation in combinatorial problems. In *IEEE/ACM Design Automation Conference, DAC, Dallas/TX*, pages 272–277. ACM Press, New York, 1993.

[27] S. Minato. *Binary Decision Diagrams and Applications for VLSI CAD*. Kluwer, 1996.

[28] S. Minato and H. Arimura. Combinatorial item set analysis based on zero-supressed BDDs. *Hokkaido University Technical Report*, December 2004.

[29] T. M. Murali and S. Kasif. Extracting conserved gene expression motifs from gene expression data. In *Proceedings of Pacific Symposium on Biocomputing*, pages 77–88, 2003.

[30] R. Peeters. The maximum edge biclique problem is NP-complete. *Discrete Applied Math.*, 131(3):651–654, 2003.

[31] S. Raychaundhuri, P. D. Sutphin, J. T. Chang, and R. B. Altman. Basic microarray analysis: grouping and feature reduction. *Trends in Biotechnology*, 19(5):189–193, May 2001.

[32] J. A. Rice. *Mathematical Statistics and Data Analysis*. Duxbury Press, 1994.

[33] A. Y. Rubina, E. I. Dementieva, A. A. Stomakhin, E. L. Darii, S. V. Pan'kov, V. E. Barsky, S. M. Ivanov, E. V. Konovalova, and A. D. Mirzabekov. Hydrogel-based protein microchips: manufacturing, properties, and applications. *Biotechniques*, 34(5):1008–1014, 2003.

[34] T. Sasao and M. Fujita. *Representations of Discrete Functions*. Kluwer, Massachusetts, 1996.

[35] M. Schienle, C. Paulus, A. Frey, F. Hofmann, B. Holzapfl, P. Schindler-Bauer, and R. Thewes. A fully electronic DNA sensor with 128 positions and in-pixel A/D conversion. *IEEE Journal of Solid-State Circuits*, 39(12):2438–2445, 2004.

[36] E. Scrivener, R. Barry, A. Platt, R. Calvert, G. Masih, P. Hextall, M. Soloviev, and J. Terrett. Peptidomics: A new approach to affinity protein microarrays. *Proteomics*, 3(2):122–128, 2003.

[37] I. S. Shergill, N. K. Shergill, M. Arya, and H. R. Patel. Tissue microarrays: a current medical research tool. *Curr Med Res Opin.*, 20(5):707–712, 2004.

[38] A. Tanay, R. Sharan, and R. Shamir. Discovering statistically significant biclusters in gene expression data. *Bioinformatics*, 18:S136–S144, 2002.

[39] S. Tavazoie, J. D. Hughes, M. J. Campbell, R. J. Cho, and G. M. Church. Systematic determination of genetic network architecture. *Nature Genetics*, 22:281–285, 1999.

[40] E. Verpoorte and N. F. de Rooij. Microfluidics meets MEMS. *Proceedings of the IEEE*, 91(6):930–953, 2003.

[41] C. C. Wang, R. P. Huang, M. Sommer, H. Lisoukov, R. Huang, Y. Lin, T. Miller, and J. Burke. Array-based multiplexed screening and quantitation of human cytokines and chemokines. *J Proteome Res.*, 1(4):337–343, 2002.

[42] H. Wang, W. Wang, J. Yang, and P. S. Yu. Clustering by pattern similarity in large data sets. In *Proceedings of ACM SIGMOD*, pages 394–405, 2002.

[43] C.-J. Wu, Y. Fu, T. M. Murali, and S. Kasif. Gene expression module discovery using gibbs sampling. *Genome Informatics*, 15(1):239–248, 2004.

[44] J. Yang, H. Wang, W. Wang, and P. Yu. Enhanced biclustering on expression data. In *Proc. IEEE 3rd Symposium on Bioinformatics and Bioengineering*, pages 321–327, 2003.

[45] S. Yoon, C. Nardini, L. Benini, and G. De Micheli. An application of zero-suppressed binary decision diagrams to clustering analysis of dna microarray data. In *Proceedings of the 26th Annual International Conference of the IEEE EMBS*, pages 2925–2928, September 2004.

[46] S. Yoon, C. Nardini, L. Benini, and G. De Micheli. Enhanced pClustering and its applications to gene expression data. In *Proc. IEEE 4th Symposium on Bioinformatics and Bioengineering*, pages 275–282, May 2004.

INDEX